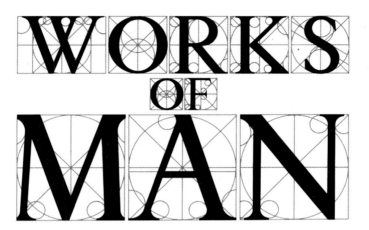

WORKS OF MAN

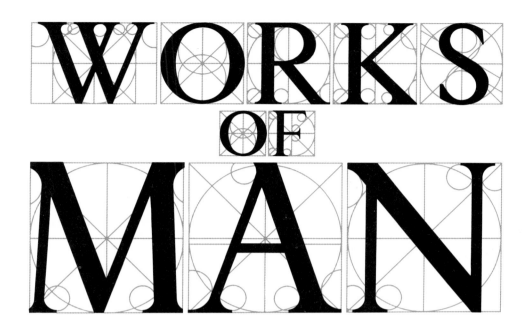

WORKS OF MAN

RONALD W CLARK

CENTURY PUBLISHING
LONDON

Copyright © Ronald W. Clark 1985
All rights reserved

First published in Great Britain in 1985
by Century Publishing Co. Ltd,
Portland House, 12-13 Greek Street,
London W1V 5LE

British Library Cataloguing in Publication Data

Clark, Ronald W.
    Works of man: discoveries, inventions and
    technological achievements that have changed
    the course of history.
    1. Inventions - History
    I. Title
    609   T15

ISBN 0-7126-0767-6

Edited, designed and produced by
Eddison/Sadd Editions Limited
2 Kendall Place, London W1H 3AH

Phototypesetting by Bookworm Typesetting,
Manchester, England
Origination by Columbia Offset, Singapore
Printed and bound in Yugoslavia by
Mladinska Knjiga, Ljubljana

CONTENTS

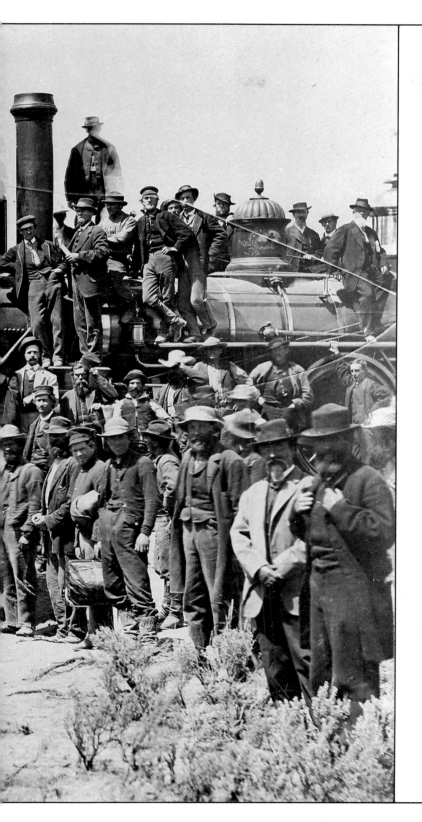

'Nations can be made or undone at the desk of an engineer'

William Swinton in *The Twelve Decisive Battles of the War* [the American Civil War]: *a History of the Eastern and Western Campaigns, in relation to the actions that decided their issue*, 1867

Celebrations at Promontory Point, north of the Great Salt Lake, Utah, after the Central Pacific and the Union Pacific Railways had in 1869 linked up to join the east and west coasts of the United States of America.

INTRODUCTION

Many factors have contributed through the centuries to the stumble forward of progress which has enabled men to exploit the laws of nature by using ever more complex buildings and machines. Not the least has been chance, that arbiter which, as Winston Churchill once said, can govern which side of a tree an officer steps and thus settle whether he ends up a mangled wreck or survives as a successful commander.

Yet, if there is one component on which most if not all technological progress rests, it is probably that described by the late A.P.Rowe, director of Britain's Telecommunications Research Establishment whose ability to keep its wartime staff of radio geniuses and prima donnas working as a team played a significant part in winning the war. The birth of radar, he once wrote, 'awaited contact between those with a need and those with a contact which would meet that need' – in this case contact between H.E.Wimperis, director of scientific research at the Air Ministry, and Robert Watson-Watt of the Radio Research Station, working at the time on a radio technique without thought of military application. At least as far back as 1921, Marconi had noted that the position of ships could be pinpointed by their reflection of radio waves. Edward Appleton, probing the layers of the ionosphere a few years later, had reported similar phenomena. Both naval and Army research workers had speculated on the possibilities of radio-location in the later 1920s. Yet it was only after the 1934 Air Exercises had shown that London was wide open to attack from the air, that radar was pulled into existence as a possible way of meeting the danger.

An urgent need of this kind has invariably been required before a potential technological advance has been changed from dream to reality. Thus the steam-driven caterpillar-tracked vehicle which was actually built in 1888 was forgotten until the machine guns of World War I made development of the tank or its equivalent a necessity.

A need leading to its satisfaction – sometimes by circuitous routes – can be seen clearly enough in many of the chapters, from prehistoric times to the space age, which chronicle man's attempts to change his environment, improve his living conditions, and generally exploit for his own purposes the slowly discerned laws of nature. It is nowhere more obvious than in the story of the revolutions produced by the steam engine, a story which runs as an unbroken thread through so much of historic times. It was the imminent exhaustion of Britain's forests, their timber needed for building the 'wooden walls' of the Royal Navy, which provided the initial spur. As wood became in ever shorter supply, even for ship-building, it was replaced as fuel by the coal which lay in abundance below so much of Britain's countryside. But miners were often faced by one crippling problem: seepage of water into the mine-workings. First hand-pumps were used to drain the water, then pumps operated by animals. But as coal was mined at ever greater depths, something more efficient was required. The experiments which had been carried out on the Continent as the philosophers of the Renaissance world encouraged scientific thought, were conscripted and developed by a succession of Englishmen and Scotsmen for utilitarian purposes and the steam engine came into existence to help keep coalmining alive.

Once the steam age had shown that coal pits could apparently be kept open indefinitely, the distribution of coal throughout a country where communications were little better than those of Roman times demanded two developments: the growth of canals and, once it was shown that steam could provide mobile motive power, the great age of the railways. Simultaneously, the Industrial Revolution was made possible by whole families of ingenious steam-driven machines – not only railway engines and engines to power ocean liners, but the varied equipment that during the nineteenth century made Britain 'the workshop of the world'. The example was followed, elsewhere in Europe and in the United States, more quickly than was expected.

This spread of engineering was made possible by scientific investigation of the materials used and, a natural development, the production of more efficient metals. By the end of the eighteenth century science had provided the base on which the builders of the early steam engines had been able to improve the efficiency of the machinery they made. Then, in the nineteenth century, science made it possible to use safely new materials and new methods in structures as remarkable as London's Crystal Palace, the Eiffel Tower of Paris, and the Forth Bridge, that wonder of the late Victorian age. Here, as could be shown from the prehistoric Iron Age onwards, science and technology moved forward together, spurred on either by political or economic need – both in the case of the Forth Bridge – or, as with the Crystal Palace and the Eiffel Tower, by the genius of an imaginative engineer. Little wonder that before the century was out men working at the sharp end of scientific invention would be creating metals, plastic materials and artificial dyes tailor-made to carry out specific tasks.

At the end of the nineteenth century, development of the petrol engine at last solved the power-to-weight problem of men who wished to fly. Thirty years later the production of metals capable of withstanding temperatures of a new ferocity enabled Frank Whittle decisively to change the air maps of the world with his jet engine. Not long afterwards, and in the wake of the electronic revolution which was a direct outcome of World War II, fear of what the Russians might achieve drove the Americans into the technological successes of the space age with its multiple spin-offs for everyday life. It is ironic that a decade earlier fear of what the Germans might achieve had driven the rest of the world into the nuclear revolution whose transformations of life are not yet complete after four decades.

The works of man which have been changing the face of the planet since prehistoric times can thus be seen as responses to needs which have eventually been satisfied by human thought and ingenuity. These responses have themselves sprung from the ambitions and quiddities of men who have spied a need either by good luck or due to an underlying and useful genetic make-up. But they have invariably had to press on with their work in the face of the setbacks that dog all pioneers. Progress, when all is said and done, usually reveals itself as a matter of character as well as of luck.

BIG BUILDINGS OF THE ANCIENT WORLD

The first engineer was the man who looked at the two upright stones before his cave, believed they would support the weight of a horizontal lintel, and contrived to raise into position a third stone which made the structure safe and sound.

This innovator could also be considered an architect, since in ancient times the architect was the engineer in embryo, and for millennia the duties of each tended to overlap. In due course the control of water, the exploitation first of metals, then of steam, produced a family of specialist engineers, later divided and sub-divided into men who dealt with military and civilian applications of the natural laws; into chemical, electrical and aeronautical engineers and then into the experts of modern times, each creating his own variety of man-made wonders.

But before even the first engineering achievements could come into being, man had to make certain elementary advances on which the interconnected disciplines of engineering, science and technology were to depend. The marvels of the pyramids and of the space shuttle, of the Roman Colosseum and the multi-thousand ton concrete foundations of a nuclear power station laid correct to a ten-thousandth of an inch, all follow fundamental discoveries without which engineering would never have existed.

The first crucial step forward – or, more accurately, succession of steps – was control of fire, needed as a protection against cold and wild animals, then used to make raw food more palatable. Eventually, and after how many thousands of years is unknown, early man discovered that fire could perform two other allied functions. One was to produce metals such as iron from the metallic ore lying in the ground. The other was the heating of different metals until they melted and the production, by mixing them, of alloys which were harder than their separate components. From copper and tin it was possible to produce bronze. Moreover by increasing the tin content bronze could be made harder; by decreasing it the alloy was made more malleable. Thus it became possible to produce materials tailor-made for special purposes.

Next came the wheel. Potters' wheels are known to have been in existence at least 6,000 years ago and it is possible that wheeled transport began equally far back. However, the earliest examples were of wood and are unlikely to have survived. The date when wheels revolving on axles began to take over from the long sleds which were for centuries the only means, other than beasts of burden, for moving heavy loads, remains unknown. Certainly by 2500 BC chariot wheels were used in the Sumerian armies of Mesopotamia. The first wheels, introduced about 4000 BC, were solid. Spoked wheels gradually replaced them and were in fairly common use by about 2000 BC. During the transitional period openings in the solid wheels became spokes, the process continuing in both agricultural vehicles and war chariots.

11

A Mesopotamian chariot of c.2500 BC. Wheels, introduced c.4000 BC, were originally solid, gradually becoming spoked.

At some equally unknown but certainly distant date man began to develop the use of iron, the fourth in quantity of the elements found in the crust of the earth, but one which has been chemically understood only in the last three hundred years. Iron exists in nature as various ores whose components can be separated by heat and it is likely that the first 'smelting' of iron occurred by accident in a camp fire which, on going out, was found to contain a spongy mass, a 'bloom', of iron in a mixture of slag and cinders. Frequent reheating and hammering produced what was to become known as wrought iron. Later it was discovered that quenching in cold water made it harder than bronze.

These basic steps had been taken at least by 3000 BC, the date by which primitive man began to evolve from a food gatherer into a food producer. The transformation meant that he started to live in permanent quarters throughout the year, instead of being constantly on the trot as he followed the animals who moved with the seasons. With settlement, and a growing dependence on crops, there came an increased need for river control and irrigation. These were in most cases two sides of the same coin. But it is in the development of irrigation in Mesopotamia and Egypt – as well as in the Indus Valley of India and the Yellow River Valley of China – that primitive engineers made their first, and essential, contributions to the spread of civilization. Without distribution of the waters which allowed men to grow crops each year, populations would have remained small, towns and cities would have failed to develop and life itself would have been altogether more precarious in an environment where water is the first requirement for survival.

The problems of water control which faced the early engineers were different around the settlements of Mesopotamia, strung out along the Tigris and the Euphrates, from those in Egypt where life was dependent on the Nile. The waters of the Tigris and the Euphrates rose at some unpredictable date between April and June, and could bring enough silt to clog irrigation channels. The water had, therefore, to be carefully controlled and then stored for use in the months when it could best be distributed. By comparison, the Nile rose more predictably, less fiercely and at a more convenient time, so that its waters could be led direct into a network of irrigation channels. But its annual inundations were so great that property boundaries had to be regularly redelineated, a requirement that led to the art of surveying.

It is only of the more spectacular irrigation schemes of ancient times that anything is known today. But throughout thousands of square miles in the great river valleys where the earliest civilizations flourished, a wide variety of such schemes were dug during the millennia before Christ. Some supplied surface irrigation by means of a network of furrows criss-crossing the land. Others provided subsoil irrigation and consisted of deep channels dug round the entire plot and kept filled with water which gradually but continually seeped into the surrounding ground.

Some schemes were remarkably ambitious. Nebuchadnezzar in the eleventh century BC drained the Nahmalka, an arm of the Euphrates, and built above the city of Sippara a reservoir 20 fathoms (36 metres) deep from which waters could be let out to irrigate the surrounding plain. Three hundred years later King Assuranzairoak of Assyria cut a mile-long channel through the rock at Negoub as part of a scheme to divert the waters of Zab to Numrud. And in the seventh century BC King Sennacherib built the astonishing stone canal which brought water to Nineveh from Bavia, some 50 miles (80 kilometres) away. Sennacherib banked in the Tigris, built a great temple at Nineveh and carried out a big programme of public works. But his most important achievement was the 50 mile (80 kilometre) canal, lined with more than two million blocks of stone. At one place, where it was 20 yards (18.3 metres) in width, it crossed a 300 yard (274 metre) wide river on an aqueduct supported by five arches, while outside Nineveh a complex system of dams and sluices enabled it to irrigate the surrounding country. The waterproofing of the canal was extremely sophisticated, consisting of a bed of concrete which was floated on bitumen, and on which there was laid a stone pavement, accurately jointed and sloped to a fall of one in 80. The grading of the canal was made with greater accuracy than was necessary for ensuring an unbroken supply of water and appears to have been planned so that as one section was completed the blocks for the next section could be moved down the watercourse on rollers.

Ancient Egypt had comparable irrigation systems as well as a flood-storage scheme 50 miles (80 kilometres) south of Cairo where the marshy Faiyum depression could be filled and its waters fed into the Nile by the Canal of Joseph, often considered the oldest canal in the world. There were also at least two stone and masonry dams south of Cairo, built about 3000 BC and throughout most of Egypt's history the waters of the Nile were controlled with considerable success.

With few exceptions, irrigation demanded the raising of water from one level to another and in the ancient world manpower, animal power, water power were all used, and sometimes an ingenious combination of all three. The most simple device was the shaduf, an apparatus consisting of a bucket at the end of a counterweighted lever; the operator pressed the bucket into the lower water level, then allowed the counterweight to raise it, after which the water was tipped out into the higher level. There was also the Archimedean screw, a hollow helical cylinder which acted as pump if rotated while one end was submerged below water. Of the various animal-powered methods, one of the most used was the saquiyah in which an endless chain of pots was dropped into a well, and pulled from it, by an animal walking constantly in a circle to keep the chain moving. Among the water-powered methods were the tympanum and the noria. The first was a drum with eight compartments turned treadmill fashion on a horizontal axis so that its lower sections dipped into a stream. Each of the eight compartments in turn scooped up water which was enabled to flow out through holes in the drum into a wooden trough. The noria consisted of a series of bottles, attached to the circumference of a horizontally mounted wheel which dipped into a stream; this kept the succession of bottles constantly on the move, and thus deposited their contents into a trough whence the water was led to where it was needed.

The waters of Mesopotamian rivers and of the Nile not only enabled early man to grow crops but also provided him with one of his most important sources of power. This was probably first used for milling grain, and three kinds of water mill are known from ancient times. The first was almost certainly the vertical shaft mill which consisted of a number of wooden blades, inclined at about 30 degrees to the vertical and attached to a hub which was fixed near the bottom of a vertical shaft. It was necessary for the water, directed on to the blades by a wooden trough, to fall a certain minimum height on to the blades so that the 'used' water could flow away easily. This meant that a pit had sometimes to be dug to ensure the most efficient operation of such mills.

Two other varieties of water mill eventually superseded this primitive kind; the undershot mill, often called the 'Vitruvian' because of its detailed description by the Roman author and architect Marcus Vitruvius Pollio, and the overshot mill which was the most efficient of all. The undershot mill was simply a wheel on a horizontal axle, its circumference bearing vanes or paddles and its lower portion dipping into the moving stream which turned the wheel on its axle. The wheel of the overshot mill was also horizontally mounted, but in this case it was mounted clear of the water while its circumference carried a number of buckets or similar containers. The water from a stream was led on to the top of the wheel where it filled a succession of the buckets; their weight turned the wheel as gravity emptied the buckets and the continuing flow of water filled their successors. Such simple constructions, to which were later added toothed wooden gears enabling horizontal motion to be turned into vertical motion, or vice versa, appear from very early times. Later versions remained in use in western Europe until steam power began to take over from water power in the eighteenth century.

Construction of the early irrigation schemes demanded of their planners, usually priests, little more than rudimentary knowledge, attained by a process of trial and error, which taught them how high and wide enclosing banks should be, and how wide and deep water channels should be to do certain specific jobs. The main difference between the early engineering works in Egypt and those to the east in Mesopotamia was the greater availability of stone in Egypt.

The contrast is shown in their public buildings. In both areas the humble houses of ordinary people were built of sunbaked bricks, but religious buildings were of something more substantial. In Mesopotamia the structure was the ziggurat, usually constructed of brick and rising in a series of rectangular terraces often adorned with plants and flowers. The most famous was the Hanging Gardens of Babylon, covering four acres, based on masonry arches, and with its plants and trees irrigated from a reservoir on top of the ziggurat which was kept perpetually filled with water from the Euphrates. Similar buildings were sometimes of considerable sophistication as well as size. Ur-Nammu's ziggurat at Ur was 79 yards (72 metres) by 59 yards (54 metres) and 28.6 yards (26 metres) high. The façades, made of kiln-dried brick set in bitumen, leant inwards to give an appearance of strength while horizontally they were built with a deliberate convexity to correct the illusion of perspective. To counter the danger of uneven settlement, reeds were used as binding material. At places even more care was taken by the engineers and in a ziggurat at Aqar Quf every fifth course of brickwork was interrupted by a layer of reed matting. The whole structure, moreover, was criss-crossed by cables of tough reeds laid in alternate directions.

In Egypt there were the pyramids, its surviving examples among the most impressive of the early engineers' work to be seen. They were the result, as one scholar has said, of 'the desire for eternal life', and each contained a chamber or chambers in which one or more members of the ruling family would be laid after death. Vast efforts were put into these monuments and Jean François Champollion, one of the founders of modern Egyptology, remarked in the nineteenth century: 'No people, either ancient or modern, have had a national architecture at once so sublime in scale, so grand in expression, and so free from littleness, as that of the ancient Egyptians.' The lid of one sarcophagus, a coffin to be placed in a pyramid, needed 3,000 men for its transport.

The pharaohs of ancient Egypt are today usually grouped into thirty-one dynasties, running from 3188 BC to 332 BC and most of the eighty pyramids, or remains of pyramids which have been identified by archaeologists were built in the third or fourth dynasties, grouped together as the Old Kingdom, between 2815 BC and 2294 BC. During these five hundred years design and method naturally changed although most of the pyramids – including those that were built before and after the main centuries of construction – had certain features in common. All were built from huge blocks of stone whose movement and assembly presented great problems in the days before the use of scaffolding or of block and tackle. Exactly how the builders overcame their difficulties is still open to discussion, and although writing was developed, first by the Sumerians in the barren lands of Mesopotamia, as early as 3000 BC, most of what is known about construction of the pyramids comes from archaeological evidence slowly accumulated over the years.

The pyramid builders, it is now known, understood the use of the wedge and the lever. They used copper tools such as chisels to cut limestone and wooden wedges to split granite blocks from the quarries. The balance beam weighed quantities with considerable accuracy and, perhaps more important, rollers and sledges were employed to move heavy loads. This method was used by other early builders. The great Assyrian winged bulls of Khorsabad, weighing more than 19.6 tons, were first rough-hewn in the quarry, moved by water, then completed their journey by sled, and were moved into position with the help of rollers, ropes, levers and pulleys, before being finished on site. A 60 ton alabaster statue of Dhutotpe was moved on rollers from its quarry to the Nile by 172 men, some of whom poured oil or water on the rollers to lessen friction.

Pyramids were built with primitive equipment and
vast numbers of workers. A modern sound and light display illuminates this one.

After it had been decided to build a pyramid, the first task was to choose a suitable site. It had to be close enough to the Nile to make movement of stone from the river to the site not too difficult; but it could not be so close that there was risk of inundation during the flood season. The site had to be accessible either from the country's contemporary capital or at least from one of the king's residences. The substratum of rock had to be capable of bearing very heavy weights and the ground not too difficult to level.

Once the site had been decided, levelling was carried out – and in the case of the Great Pyramid of Gizeh this appears to have been accurate to within half an inch. The base of the pyramid was then surveyed, care being taken to ensure that the four corners pointed to the cardinal points of the compass. That done, the way would be ready for three operations that were frequently performed simultaneously. One was the construction of a causeway from the Nile to the building site up which the materials would eventually be hauled. The second was quarrying the local limestone for the outer casing of the pyramid, much of it found on the east bank of the river at Tura in the Mukattam hills. At the same time, the granite to be used in the building would be quarried near Aswan, some 300 miles (483 kilometres) upstream.

Quite apart from the organization needed to bring thousands of men to the site at the right time, there also had to be planning to ensure that separate operations could be tied in with each other to complete the job.

Much of the quarrying was followed by rough-hewing on the site. For this there was available the hafted hammer, an instrument which gave access to more power than man had previously had under control. Yet however much 'waste' was removed in the quarry the builders still had the enormous problem of moving the massive blocks of stone.

The most famous of the gigantic structures which they built with primitive equipment was the Great Pyramid of Gizeh, a few miles south-west of Cairo, constructed about 2600 BC to provide a funeral monument for King Cheops (also known as King Khufu). Rising to a height of about 463 feet (141 metres) from its square 779 feet (237 metres) base, it contains about 2,300,000 2½-ton blocks of limestone, the inner ones being of rough stone, the outer ones of fine limestone fitted together with great accuracy. Inside the structure, the chambers are lined with granite blocks from Aswan.

As surprising as the builders' ability to move such immense quantities of material is the precision with which the structure as a whole was built. The base lines are only 7 inches (17.78 centimetres) out of true while the four corners are only $\frac{1}{10}$ of a degree off the four cardinal points of the compass on which they are oriented.

The Greek historian Herodotus gave the earliest account of how the blocks were moved from the Nile along a stone causeway after they had been floated downstream on huge rafts from the quarries from which they had been hewn. 'It took ten years' oppression of the people to make the causeway for the conveyance of the stones, a work not much inferior in my judgment to the pyramid itself,' he wrote. 'The causeway is five furlongs long, ten fathoms wide, and in height at the highest part, eight fathoms. It is built of polished stone and covered with carvings of animals.' Once this approach road from the Nile to the building site had been finished, another twenty years is thought to have been needed for building the pyramid itself, and a total of 100,000 men are said to have been required for the work.

The huge blocks were apparently brought from the quarries with their vertical and horizontal joints already prepared and their edges squared. They were then hauled from the river's edge to the site with the help of rollers and sledges and manhandled into position. When one course, or layer, was finished, an earth ramp was built to its upper level and the stone blocks for the next course were hauled up it into position. There was no pulley or tackle but the task of the steadily hauling humans was eased by the use of viscid mortar on which each successive layer was 'floated' into position.

Considerable use was also made of 'rockers', designed somewhat like the homely rocking chair. The rocker was tilted and then held in position by logs while the stone to be moved was slid on to it. Movement of the logs, or of others placed at the opposite end of the apparatus, put the stone where it was wanted. The process of raising first one end of the rocker, and then the other, could lift stones as required. Mortar was rarely used, except to help move the large blocks into position after which dowels of sycamore wood or sometimes of iron, helped hold them in place.

Within the Great Pyramid of Gizeh there are three funeral chambers. The first, below ground level, was hewn out of the solid rock on which the pyramid was to be built and was reached by a descending passage. The second, the Queen's

Chamber, was positioned below the apex of the pyramid and reached by an ascending passageway. Finally, there was the King's Chamber, built roughly at the centre of the structure and connected by a passageway with the Queen's Chamber almost immediately beneath it.

The accurate siting of the sides of the Great Pyramid to north, east, south and west is proof that its builders had at least an elementary knowledge of astronomy, since they had no magnetic compass. However, there is no foundation for any of the mystical theories which have been devised to explain the situation or the size of any of the pyramids. Such theories were sometimes taken to extreme lengths and the famous Egyptologist, Sir William Flinders Petrie, once recalled how he had met a man measuring one side of the Great Pyramid, but armed with a chisel as well as a tape measure. On asking about the need for the chisel, Petrie was told it was to 'adjust' the length of the side that did not conform to the latest theory.

The Egyptian rulers, as well as their engineers, took their pyramids seriously, and in the time of Rameses the Great, during the third millennium BC, an expedition to secure material for one pyramid involved the assembling of nearly 9,000 men. Five thousand were soldiers, 2,000 were temple staff and the balance included 800 foreign auxiliaries, 900 officials, 130 quarrymen and stone-dressers, three master quarrymen and four sculptors.

The facilities which such armies of workmen had at their disposal were non-existent in some spheres, very good in others. Mechanical equipment was of the most elementary sort, and, while thousands of men were available for moving massive quantities and weights, organization must have been stretched to its limits. Yet the Egyptians produced amazing results. Petrie has pointed out that the granite sarcophagus of Senusert II of 3350 BC was 'ground flat on the sides with a matt face like ground glass that only has about a 200th of an inch error of flatness and parallelism of the side'.

The propensity of the Egyptians, and of the Babylonians and Assyrians too, for building massive monuments was not unique. In fact, throughout most of the ancient world, a high percentage of the available engineering-cum-architectural resources was devoted to the construction of buildings raised for semi-religious reasons or as genuine temples. This concentration of activity was not surprising since many of the works automatically connected with engineering in contemporary life had not yet come into existence. Machinery, in the modern meaning of the word, did not yet exist. The chariot was the most complicated piece of transport mechanism that had yet been devised while the ship, whether pulled through the water by human rowers or using the wind on its sails, was comparatively simple. The equivalent of engineering expertise went into the construction of massive stone buildings and this was true not only of the Middle East and the Far East, but also of Europe where the builders of the Bronze Age erected a number of monuments that are still, thousands of years later, among the most impressive works that man has created. Among them are the thousand shaped granite monoliths arranged in complex stone alignments at Carnac, in south-west France, and the unique group of monuments still standing on Salisbury Plain in the south of England.

The most famous of the British monuments is Stonehenge, 8 miles (12.5 kilometres) from the city of Salisbury and the greatest prehistoric temple of its kind in the world. A few miles away there is Avebury, described by John Aubrey in the seventeenth century as surpassing Stonehenge as much as a cathedral surpasses a village church. Between these two centrepieces is a countryside almost littered with the long barrows in which prehistoric people buried their dead, remains of sanctuaries and of other relics from an age when this chalk upland was the main 'metropolis' of England. A geological map shows how the comparatively high chalk ridge running from East Anglia to the south coast in Dorset meets those of the North Downs and the South Downs in the area of Salisbury Plain. These chalk ridges, together with other high ground, provided the highroads of prehistoric times, useable in bad weather as well as good, and it is inevitable that prehistoric man should have left some of his most impressive monuments along them.

The first sight of Stonehenge genuinely merits the term awe-inspiring, – concentric circles of standing stones, gigantic in size and obviously enormous in weight, many of them approaching 50 tons, and some of them capped by horizontal lintels and thus forming what are known as trilithons.

The history of Stonehenge is a complicated one. Surrounding the circles are the remains of an earlier bank and ditch, built about 2800 BC – some time before the Great Pyramid. Inside the bank, small circles of bare chalk mark the position of a ring of fifty-six pits, known as the Aubrey Holes after their discoverer John Aubrey, and apparently dug when the bank was raised.

The first Stonehenge could boast of only three stones – the Heel Stone standing some way outside the bank, and two smaller stones beside it. Five centuries or so later, eighty bluestones, each weighing up to 4 tons, were brought to Stonehenge from the Prescelly Mountains in South Wales – the only place in Britain where this stone is found – and set up to form a double circle inside the ditch. At the same time an approach in the form of the Avenue was made from the nearby River Avon by an earthwork avenue no longer in existence. About a century later the double circle of bluestones was dismantled and the stones replaced by about eighty sarsens, some of them weighing as much as 50 tons. The sarsens – sandstone boulders lying on the Marlborough Downs 20 miles (32 kilometres) north of Stonehenge – were arranged as a circle of uprights capped by a continuous stone lintel, and as a horseshoe of stones within the circle. Possibly as part of the same plan, about twenty of the discarded bluestones were later erected as an oval. This in turn was soon demolished and the bluestones re-erected in a fresh circle. The axis of the final rearrangement points in one direction to the midsummer sunrise and in the other to the midwinter sunset. It follows the line of its predecessor and may have originated even earlier.

Impressive as is Stonehenge today, the sarsens and bluestones still standing are only the remnants of the five successive building operations that went on between 2800 BC and 1550 BC. There is no building stone in the area and for centuries both sarsens and bluestones were broken up for use in farms or to repair farm tracks. Visitors also damaged the monument and for some while it was possible to hire in the nearby small town of Amesbury a hammer suited to chipping off parts of the stones.

Of the engineering riddles raised by Stonehenge, the most difficult to answer was for long the problem of how the bluestones were brought to Salisbury Plain from the Prescelly Mountains, about 240 miles (384 kilometres) away and on the far side of the River Severn and the Bristol Channel. But in 1954 archaeologists conjectured that it would have been possible to move the bluestones by water

Stones of Avebury, Wiltshire, an enclosure
covering almost 30 acres (12 hectares) built between 2600 BC and 2000 BC.

down to Milford Haven on the Bristol Channel, up the Severn to its junction with the River Avon and then up that river to within a few miles of Stonehenge, whence they could be dragged on sledges. The theory was tested by an elaborate experiment. A dummy bluestone, made of concrete but of correct size and weight, was loaded on to three replicas of prehistoric dugout canoes lashed together. The craft was successfully poled down and up the watercourses to a landing point on the Avon, the only implements used being those available in prehistoric times. Finally a sledge on rollers was brought in to help manhandle the load to Stonehenge. Shortly afterwards similar rollers were used to move the equivalent of 50-ton sarsens from their original site on the Marlborough Downs.

Construction of the trilithons at Stonehenge had also presented engineering riddles, but experiments now suggested that it could have been carried out by first moving an upright on rollers until what was to be its bottom end hung over a previously dug pit. The upright was then dragged by ropes until the lower end dropped into the pit, after which it was raised vertically by further ropework. After a second upright had been raised in the same way the horizontal lintel was first rollered into position and then raised, a few feet at a time, by being levered on to a wooden platform that was progressively raised in height until it reached the top of the vertical sarsens. Further ropework then moved it into its final position.

Three features of Stonehenge's trilithons reveal that the engineers who planned their erection had more knowledge and skill than is at first apparent. Thus the uprights have been finished – with only the primitive tools of the day – so that they are wider at the top than at the bottom; the result is that the effects of foreshortening are removed when the uprights are viewed from the ground. In addition, none of the horizontal lintels is quite straight; instead, they are curved lengthwise so that they fit the curve of the completed circle. A greater demonstration of the early builders' skill is given by the hollowed-out mortises at each end of the lintels and the 9 inch (23 centimetre) tenons on the uprights which fit into the mortise holes.

To the north of Stonehenge, near Silbury Hill, the largest artificial mound in Europe, 130 feet (40 metres) high and covering 5½ acres (2.2 hectares), lies the huge embanked enclosure of Avebury, older and larger than Stonehenge. Greatly impressive, despite the depredations of the last two centuries, the bank and ditch enclose almost 30 acres (12 hectares), and from the top of the bank there is visible the remains of the hundred-sarsen circle inside the ditch and of the two smaller circles within it. Leading from the southern entrance to the site lies what is left of the Avenue, a double line of sarsens leading to what was once the Sanctuary, believed to have been a wooden temple. Only modern markers show where the posts of the Sanctuary once stood.

Built in two bursts of activity between 2600 BC and 2000 BC, Avebury, unlike Stonehenge, contains only undressed stones. But in the circles, as in the Avenue, upright stones often alternate with diamond-shaped sarsens – 'male' and 'female' stones according to some theories.

The great achievements of the earliest builders, in the Middle East as well as in Britain, were carried out with little help from what would today be considered real engineering knowledge. Most of this became available only slowly, over the centuries, during the rise and fall of Greece. The Parthenon, built on the Acropolis at Athens, was among the most famous examples of Greek architecture. Phidias was the sculptor in charge and under him worked the architects Actinus and Callicrates. The building was completed between 447 BC and 438 BC. In modern times various complicated systems on which the building is claimed to have been erected have been put forward to explain the visual beauty of the result.

The men who are often first thought of as the shining examples of Greek civilization are the philosophers, mathematicians and others who concentrated on theoretical knowledge. They were certainly to help lay the foundation on which engineering was to be built. Yet there were those, some much less well known, who were genuine engineers in their own right. Among them was Eupalinos, sometimes called the first civil engineer. About 530 BC Eupalinos built a water-supply system for his native town of Nagara, but he is more famous for the construction on the island of Samos of an aqueduct that required a tunnel, three-quarters of a mile long, through an intervening hill. The tunnel, of 8 feet (2.44 metres) square cross-section, was started simultaneously from both ends, and the survey work was so good that the two tunnels met only 2 feet (0.6 metres) off true.

Two and a half centuries later Archimedes worked out in mathematical detail the principle of the lever; founded the science of statics that was to be of importance in the Renaissance revolution of the sixteenth century when his works were translated from Greek into Latin; and invented the hollow helical screw known as the Archimedean screw which was to be used as a water pump.

To this period there belongs, also, the Pharos of Alexandria, the huge lighthouse which was sometimes named as one of the seven wonders of the world. Pharos was the name of the island which lay about a mile off the coast and which was joined to the mainland by a causeway three-quarters of a mile long, awash when the Nile was high. Here, about 270 BC, in the reign of Ptolemy II, called Philadelphos, the architect Sostrastos of Cnidos was commissioned to build a tower topped by a warning light that would be as much a symbol of Alexandria's commercial power as a guidance to seafarers.

The base of the Pharos – a word which as *phare* in French and *faro* in Italian and Spanish, eventually became a generic term for lighthouse – was a 24 foot (7.4 metre) high heavy stone platform. On this three sections were erected, respectively square, octagonal and cylindrical, thereby bringing the height of the building to between 300 and 400 feet (91.4 and 122 metres). Ramps and stairs led to the top platform where a bright fire of wood was kept burning at night. The Pharos was by far the tallest tower in existence and its construction was so good that it lasted until the thirteenth century when it was destroyed by an earthquake.

Size alone sometimes put the Pharos among 'the seven spectacles', first named by Philon of Byzantium at the turn of the third and second centuries, and it is a tribute to the ancient world's respect for size alone that the world's wonders, of which several lists were made, depended so much on this one factor. The works of man which so impressed the ancients were the 'hanging gardens' of Babylon; the pyramids; the statue of Zeus executed at Olympia by the Athenian sculptor Phidias; the massive Colossus of Rhodes, a statue of Helios the Sun god whose legs allegedly bestrode the entrance to the harbour; the Temple of Ephesus; the Mausoleum of Halicarnassos; and either the walls of Babylon constructed by Nebuchadnezzar, or the Pharos of Alexandria.

Following Archimedes there came Ktesibios who in the second century BC helped to found the engineering tradition at Alexandria. His most famous work was a reconstruction of the *clepsydra*, the water clock of the Egyptians. This was a jar of water at whose base there was a hole through which the water dripped into a second container. A pointer floating in the second container indicated the passage of time, and an adjustment could be made to keep the passage of hours correct – the length of the hours varying between summer and winter, each day between sunrise and sunset being divided into twelve equal parts.

Ktesibios realized that as the level, and pressure, of water on the main jar dropped, the outflow would vary, and that later hourly intervals would be longer than earlier ones. He removed the inaccuracy by providing a constant water supply to the main jar and by the use of a float and valve ensuring that the head of water was constant. The water dripping into the second jar then accurately recorded the passage of time. Simple as the method appeared to be, it remained the most accurate time-keeping device until the invention of the pendulum clock in the seventeenth century.

Also important was Hero – sometimes described as Heros or Heron – of Alexandria, the Greek mathematician and writer. Although the general lines along which engineering developed in the ancient world are known through the efforts of archaeologists, there have always been large gaps in detailed knowledge. That is true not only of the dates and places at which newly developed equipment was first used. It has been true, also, of the most influential men whose work occupied that area, difficult to define, which was bounded by science, engineering, architecture and technology. This is illustrated by Hero. Although he wrote three books dealing with mechanics, the properties of air, and the making of automata, the years during which he lived and worked were constantly argued about until the late 1930s. Some experts placed him in the second century BC, others as late as the second century AD and yet others gave him intermediate dates. Only in 1938 was it pointed out that his method of calculating the Great Circle distance between Rome and Alexandria rested on an eclipse that happened only once between 200 BC and AD 300 – on 13 March AD 62. This and other evidence was proof that he lived during the second half of the first century.

Few of Hero's writings have survived intact, and some of those only through Arab translations discovered in modern times. However, certain of his inventions and devices are known, the most famous being a device that is sometimes called the world's first steam engine. This consisted of a hollow sphere from which there projected two bent tubes. When water in the sphere was heated the steam coming from the tubes caused the sphere to revolve. His conversion of steam power into motion – as a result of the law of action and reaction spelt out by Isaac Newton sixteen centuries later – was never exploited by Hero. He did, however, describe a number of machines incorporating levers, pulleys, inclined planes and wedges by which human forces could be directed or multiplied. One that he is believed to have built opened temple doors automatically and apparently without human intervention. This was achieved by using an altar fire to warm air which drove water from a hollow sphere into a hanging cauldron. Weighted down by the water, the cauldron worked a series of pulleys which opened the doors and allowed them to remain open until the fire died down, the water was siphoned from the cauldron and the doors were automatically pulled shut.

Of the knowledge built up by the Greeks over the centuries little was put to practical use until the rise of the Romans. Much of what is known about their contribution to civil engineering comes from what are often called 'the ten books' on 'Architecture' by Marcus Vitruvius Pollio, a contemporary of Julius Caesar. In fact, it is only one smallish book with ten divisions dealing with such subjects as water supply and aqueducts, mechanics, building materials and public buildings. Employed as a military engineer and professional architect, a term which still doubled for that of civil engineer, Vitruvius laid down stiff qualifications for the professional. He should, he wrote, 'possess not only natural gifts, but also keenness to learn, for neither genius without knowledge, nor knowledge without genius suffices for the complete artist.' He must, he went on, ' be ready with a pen, skilled in drawing, trained in geometry, not ignorant of optics, acquainted with arithmetic, learned in history, diligent in listening to philosophers, understand music, have some knowledge of medicine and of law, and must have studied the stars and the courses of the heavenly bodies.'

Throughout the rise and fall of the Roman Empire, the activities of the engineers possessing these demanding qualifications were mainly devoted to the State; to the roads which the Romans successfully engineered from one end of the Empire to the other, to the aqueducts which supplied so many large Roman towns with water, and to great public buildings. From the skies above England it is still possible to see the undeviatingly straight lines of the roads which were built from London to the Roman cities now known as Bath, Chester and York. And in southern France, the treble tier of the 882 foot (269 metre) long arches of the Pont du Gard still stand as they did when serving as an aqueduct to carry water 160 feet (49 metres) above the River Gard.

At the height of its power, the Roman Empire included 1,197 cities in Italy, 360 in Spain, 300 in Africa, 500 in Asia, 1,200 in Gaul and a number in Britain. They were linked by an astonishing network of twenty-nine main roads radiating

The Pont du Gard aqueduct, France, built in 19 BC across the ravine of Bornègre. It is part of a 31-mile (50-kilometre) aqueduct carrying water to Nîmes.

from Rome which were themselves connected by scores of lesser roads. The care and ingenuity exercised in laying out and constructing this network on which depended the swift movement of troops, and, therefore, adequate military control of the Empire, has rarely been equalled in the history of civil engineering.

First, two parallel trenches would be dug on either side of what was to be the line of the road. The top soil between the trenches would then be removed down to the level of the hard ground. Next, a layer of dry earth would be rammed down to form what was called the *pavimentum*. On top of this there came a layer of small stones forming a watertight course called the *statumen*, above which there came the nucleus, a layer of whatever local material was available – chalk, lime, broken tiles or stone. Finally the agger, as the causeway was called, was sometimes topped by a *summum dorsum* of flagstones so laid as to provide a slight camber to the road, and drainage ditches were dug on either side of it.

Such was the standard for main roads. But while minor roads fell below the standard, the stretches of main roads near Rome were usually finished even more carefully, notably the northern portions of the Appian Way, started by Appius Claudius Caecus in 312 BC and running from Rome to Sardinia, the 310 miles nearest Rome being topped by specially dressed blocks of stone. 'The work was so perfect', it has been claimed, 'that it seemed as if nature had performed it rather than man, for the very joints were hardly perceptible.'

With road building there went bridge building, and throughout western Europe there still exist numerous examples of the Romans' art. Invariably consisting of semicircular arches composed of voussoirs – wedge-shaped stones naturally held in place by gravity – these bridges frequently utilized during construction the wooden centring that held up the components of the arch until the arch itself had been completed. Where the river to be crossed lacked any stone or an island which could be used as foundation for a pier, a coffer dam of the sort described by Vitruvius for harbour works would be built. 'A coffer dam with double sides, composed of charred stakes fastened together with ties, should be constructed in the appointed place,' he wrote, 'and clay in wicker baskets made of swamp rushes should be packed in among the props. After this has been well packed down and filled in as closely as possible, set up your water-screws, wheels and drums, and let the space now bounded by the enclosure be emptied and dried. Then dig out the bottom within the enclosure. If it proves to be of earth, it must be cleared out and dried till you come to solid bottom and for a space wider than the wall which is to be built upon it, then filled with masonry consisting of rubble, lime and sand.'

The skill of the Roman engineers in road making and bridge building was matched by their ingenuity in inventing auxiliary equipment. One such device was the odometer or measuring wheel. From the perimeter of a chariot wheel there projected a short spoke; with each revolution of the wheel the spoke struck a vertical shaft. The top of the shaft ended in a box containing a number of balls, and a toothed gearing arrangement allowed one of the balls to fall out of the box into a lower container at every four or five hundred revolutions of the wheel. Since the circumference of the chariot wheel was known, the distance travelled could be calculated from the number of balls in the lower container. The device, as described by Vitruvius, could also be adapted for ships in which case paddles, turned by movement through the water, were used to operate the toothed gears.

Where roads ended at a frontier of the Empire, a defensive line was often built forward of them by the military engineers. Thus in central Europe the *Limes Germanicus* ran for 300 miles (480 kilometres) from Rhenbrohi on the Rhine to Heinheim near Regensburg on the Danube. These defences were linear barriers, their size and strength dependent on the expected attack and their material dependent on local sources of supply. The most spectacular example still standing is Hadrian's Wall in the north of England, 72 miles (116 kilometres) long, winding across the craggy hills of Northumberland, and one of the most impressive sights in the country. After the Romans had retreated from southern Scotland about AD 100, it became necessary for them to build a defensive line across 'the narrows' of Britain between the Solway Firth on the west and the River Tyne on the east. 'The Wall' was the result, built about AD 120 of local stone where this was available and of turf where it was not, roughly 15 feet (4.5 metres) high and about 10 feet (3 metres) thick at its base. Every mile there was a small fort, in which lived members of the defending garrison, and between each pair of forts there were two smaller turrets. To the north of the wall, the side from which attack could be expected, there was dug in places a deep defensive ditch and to the south there ran a strategic roadway along which troop reinforcements, normally living in seventeen forts still further south, could be speedily brought up.

Although Hadrian's Wall is even today an impressive example of military engineering, some of the most spectacular Roman works still to be seen are the aqueducts which brought water to many of her cities from considerable distances and across wide river valleys. Technically, the word aqueduct applied to the horizontal channels built round mountain slopes and to the tunnels which were sometimes bored through them to carry water, sometimes for scores of miles. Today, however, the word is more frequently used to describe the long rows of stone arches, sometimes one tier high, sometimes two, and occasionally three,

Hadrian's Wall, built as a defensive line across the
'narrows' of Britain between the Solway Firth (west) and the River Tyne (east).

which support a water-carrying channel across a river valley.

At one time as many as fourteen aqueducts with tunnels, hill-contouring channels and stone supports, served the city of Rome, supplying it with 300 million gallons (1,364 million litres) of water a day. The first of these was the Aqua Appia, begun in 312 BC by Appius Claudius, which brought water from springs in the hills 10 miles (16 kilometres) from the city. Other sources were soon being tapped, the Aqua Marcia bringing water from 56 miles (90 kilometres) away across many fine bridges, including the 360 foot (110 metre) long Aqua Rossa. In 19 BC Agrippa began the Aqua Virgo to supply the public baths he had built, and nineteen years later the Aqua Alsietina was constructed to keep filled the 1,181 foot (360 metre) by 1,800 foot (540 metre) lake dug in the city for the spectacle of *naumachia* or mock sea fights.

As many as seven hundred men served on the Roman water supply board, including technicians who calibrated the bronze nozzles through which water was brought into private houses. Others ensured that the water settled properly in reservoirs before being led down to private houses and to the three main groups of public users: the fountains, the public baths and the public buildings. Water was carefully controlled and in the fourth century, the Emperor was allocated 17 per cent of the available supply, private houses and industry 38.5 per cent, while the remainder, about 44 per cent, supplied ninety-five official Roman buildings and 591 cisterns and fountains in the city.

As the Empire expanded, the aqueducts followed, the most famous being the Pont du Gard in the south of France and the Segovia aqueduct known to the Spanish as the 'Devil's Bridge'. Bringing water from the River Frio, 10 miles (16 kilometres) away, the latter is a two-tier construction 2,657 feet (810 metres) long, the water-carrying channel being almost 118 feet (36 metres) above the ground.

It was not only in the building of aqueducts that the Roman water engineers demonstrated their abilities. They were equally competent in constructing reservoirs, and the Piscina Mirabilis of Bacoli covered almost 2,392 square yards (2,000 square metres). Lined with waterproof cement, it could be divided by gates into separate sections for cleaning. Behind Nero's villa at Subiaco there was another example of Roman water storage, a great dam with walls 45 feet (14 metres) thick, behind which there stretched a lake covering 2 square kilometres. So well made was it that it did not collapse until 1305.

Although many aqueducts were freestone constructions, the Romans were experts in the use of a particularly effective form of cement which they made from lime, sand and clay. They also used various forms of specialist cements, notably pozzuolana, a volcanic clay or tufa containing gravel and about 35 per cent soluble silica. When sized with burnt lime and water it bonds with the lime and forms a waterproof cement of great strength.

The arch, a prominent feature in so many aqueducts, was a key element in Roman building and engineering. The Egyptians and the Assyrians had occasionally used it, and so had the Etruscans, but all these predecessors of the Romans had used it sparingly and with caution. Only with the spread of Rome did it become a familiar sight. Among its finest examples were, and still are, those in the amphitheatre of Vespasia in Rome, later known simply as the Colosseum because of its enormous size. There had been earlier amphitheatres in Rome, the word describing a structure where the seats formed an unbroken circle or oval as distinct from theatres in which a stage filled one side of the structure. A previous amphitheatre of stone and wood was destroyed by fire in AD 64 and was replaced by another wooden building which fell into disuse. It was the Emperor Vespasian who about AD 75 ordered the building of the huge structure whose remains are still so impressive.

The central oval arena covered an acre (0.4 hectares), and around this there was built tier upon rising tier of seats. Up to 80,000 spectators could be packed on to the seats, or provided with standing room behind them, and the building remained the largest of its kind in the world until the Yale Bowl was erected at New Haven, Connecticut, in 1914. The outer walls of the Colosseum, forming an ellipse 616 feet (188 metres) by 513 feet (156 metres) consisted of three rows of arches, eighty in each row, topped by an elliptical wall which brought the height of the structure to 154 feet (47 metres). Between the seats and the outer wall there were three concentric ellipses of arches at each height; staircases connected the various storeys, and a system of corridors enabled gladiators and wild beasts who were to fight in the arena to be brought from cells or cages in the basement.

The complexity of the construction, which contained nearly a thousand large arches and hundreds of smaller ones, is shown by a paragraph written by an engineer who measured the Colosseum in 1818. 'The rain which fell upon the several seats and the arena,' he says, 'and the water which flowed through urinals and other arrangements made for the convenience of the multitude assembled within the walls, drained into wide and spacious sewers, which were conducted into the Cloaca Maxima. A large drain in the second corridor, 30 inches wide, received the water, brought down by perpendicular pipes worked in the solid masonry, or placed in indents, lined with tile. The drain of the third corridor, 17 inches wide, and 3 feet in depth, is lined carefully with tile and coated with a fine

The Colosseum, Rome, was the largest building of its kind
in the world until the Yale Bowl was built at New Haven, Connecticut, in 1914.

cement.' The report then continues:

'On the other side of the third corridor, a similar drain, with a fall towards the last described, caught all the water brought by the several branches from the arena.'

The quality of Roman engineering and building was to be remarked upon nearly 2,000 years later by the engineer James Nasmyth after he had visited the amphitheatre at Nîmes. 'One thing I was especially struck with at Nîmes,' he wrote in his autobiography, 'was the ease with which some thousands of people might issue, without hindrance, from the amphitheatre. The wedge-shaped passages radiate from the centre, and, widening outwards, would facilitate the egress of an immense crowd ... Another thing is remarkable – the care with which the huge blocks of magnesium limestone have been selected. Some of the stone slabs are eighteen feet long; they roof over the corridors; yet they still retain the marks of the Roman chisel. Every individual chip is as crisp as on the day on which it was made; even the delicate "scribe" marks, by which the mason, some 1,900 years ago lined out his work on the blocks of stone he was about to chip into its required form, are still perfectly distinct.'

However, it was not only in their public buildings that the Romans produced works whose quality was not to be equalled for a thousand years and more, since their homes were of a sophistication and often of a comfort that was not to be seen again in Europe until the nineteenth century. Perhaps most remarkable was their system of heating, comparable to the central heating of modern times. Two methods, basically the same, were used, the channelled hypocaust or the pillared hypocaust, similar systems which produced hot air in a furnace outside the building. In the former, the hot air was brought by a concealed trench to the centre of the rooms to be heated, then led to the sides of the rooms and up flues built into the walls. The pillared hypocaust brought the warm air to a shallow basement built below the floor which was set on rows of short columns and was perpetually heated when the furnace was alight.

Equal ingenuity was used in the construction of the public baths and private bath houses which were part of private houses in many large cities, and of the villas used as 'country houses' by richer Roman citizens throughout the Empire. So, too, with such problems as rising damp, dealt with by cavity walls or vertical damp courses. Glass windows were used in some Roman houses, panes being as large as 21 by 40 inches (53 by 101 centimetres).

Great care was taken in the design of every feature, including the 'cellar' where wine was to be made and stored. Here steps would lead up to the treading floor which had sunken 'lakes' on either side to receive the must.

'Masonry channels or earthenware pipes are to run from these "lakes" around the outer edge of the walls, and to supply the large earthenware pots placed outside and below them with the running juice, as it trickles down the adjacent stretch of channel,' declared the Roman author Palladius. 'If there is a big supply of wine, then the open space in the middle will be given over to barrels. So that these do not hinder our walking around, we can place them on fairly high platforms or above sunken pots fairly far apart, so that, if necessary, the man in charge can pass between them. But if we do assign a place to the barrels, let it be built like the treading floor with low kerbs and a tiled pavement, so that even if a barrel is leaky and spills out (unknown to the caretaker), the wine that has escaped will flow into the "lake" below, and not get wasted.'

Santa Sophia in Istanbul was one of
the first buildings to have a circular dome set satisfactorily on a square structure.

Despite their use of the arch in buildings, bridges and aqueducts, the Romans never found a way of utilizing it to span more than moderate distances. Therefore, their larger buildings tended to be filled with a clutter of small arches or with columns supporting the roof. Only in the fourth century when the centre of the Empire under Constantine had been moved to the Greek city of Byzantium, on the Bosphorus, renamed Constantinople in AD 330, and turned into Istanbul in 1930, did the engineers and architects of the new Byzantine Empire solve the problem with what has been called one of the great buildings of all time. This was the church of Santa Sophia, one of the first buildings which demonstrated that it was possible to place satisfactorily a circular dome on a square structure and the example which shows the fully developed Byzantine style.

'In height it rises to the very heavens and overtops neighbouring buildings like a ship anchored among them, appearing above the rest of the city while it adorns and forms a part of it,' wrote the historian Procopius. Today, fourteen and a half centuries after it was completed, and after restoration following damage by more than one earthquake, the building is still unique.

When the circular dome had been developed in earlier Roman times, no better way of supporting it had been found than by use of a circular wall. An awkwardly shaped interior was thus produced and after destruction by fire of an earlier church in Constantinople in 532, the Emperor Justinian demanded that his architects, Anthemius of Tralled and Isodorus of Miletus, solve the problem of using a dome to cover a square. A hundred master builders and more than 10,000 workmen supervised the extraordinary building that now began to rise. The centre of the church consisted of a square with 120 foot (37 metre) sides. At each corner there rose a 100 foot (30.5 metre) high column. Joining these four columns were four 60 foot (18.6 metre) high arches and resting on the tops of the arches there sat the dome, 120 foot (37 metres) across and 46 foot (14 metres) high. On two sides huge buttresses took the thrust of the walls down to the foundations and on the two other sides half-domes on semicircular walls formed the chancel and nave of the church and, at the same time, the outward thrust of the two arches. Part of the latter thrust was also counteracted by 15 foot (4.6 metre)long metal tie bars, operating much as the members of what was to develop as the truss bridge. At each corner, ingenious pendentives – spherical triangles or triangular segments – were used to fill the spaces between the arches which spring from the tops of the columns and the horizontal base of the dome above them.

Brick was used for most of the building, and pumice bricks for the dome on account of their lightness. The interior was magnificently decorated, the ceiling covered with sheets of gold, and marble, mosaic, ivory and enamel were all lavishly employed. The dome, Procopius claimed, appeared to be suspended from heaven rather than resting on a solid foundation and to be so 'singularly full of light and sunshine that you would declare that the place is not lit by the sun without, but that the rays are produced within itself.'

The church was built without the aid of the scientific knowledge developed only after the Middle Ages, and trouble began soon after its completion. The piers started to spread and it became necessary to reinforce the supporting buttresses. Earthquakes aggravated the problem, and the dome had to be partly rebuilt, but Santa Sophia has survived as one of the most remarkable works of its kind.

The Romans who controlled most of the known world when at the height of their power, won that power by the ability of their military engineers. In practice they were often the civil engineers as well, and as was to be so throughout history one department of the profession was usually connected, and often closely connected, with the other. In Roman times, and indeed until much later, siege warfare was of great importance, and it followed that the men who devised the best ways of making strong buildings should also be adept at the best ways of destroying them.

Slings and catapults had been used since earliest times but were brought to a high stage of perfection by the Romans. Their *onagar*, or mule, so-called because in action it resembled a kicking mule, could throw a 4 pound (1.8 kilo) stone shot as far as 1,000 feet (305 metres) while their catapults could throw stone shot or arrows the same distance.

Those conquered by successive generations of Romans were by no means defenceless, since there was a long history of communal hillfort defences from the Neolithic people in the third millennium BC to the pre-Roman Iron Age. Few had the skill in military engineering that the Romans had so assiduously cultivated, but

Maiden Castle, the prehistoric defensive fort built in southern Dorset
after 350 BC when an earlier defensive system was reinforced by four concentric rings of ramparts.

many had a clever appreciation of the lie of the land and some of their earthwork and timber defences are still unique monuments to engineering ingenuity. This is certainly true of the cleverly defended hilltop forts built in Britain and elsewhere which are today, after two millennia of erosion and decay, still hugely impressive in their size and in the use they make of the earth's natural features.

One of the greatest of these forts is Maiden Castle, 45 acres (18 hectares) of chalkland in southern Dorset, near to the port of Weymouth, enclosed by four concentric rings of ramparts which follow the contours of the hill and which must have looked even more formidable about AD 45 to the soldiers of the Second Legion faced with capturing it.

The site had been occupied as far back as the third millennium BC but had then been abandoned for hundreds of years and it was not until about 350 BC that invading Celts, possibly from Brittany, took possession and began the task of transforming it into one of the biggest forts of its kind in the country.

The ramparts were dug from the native chalk and strengthened by wooden posts. To the east there were two gaps in the defences, closed by timber gates which could be approached only by winding tracks bounded on either side by high stone walls. Near the gates was an iron-working establishment and possibly a smithy. Inside the encircling ramparts were the small circular houses of the inhabitants. The complexity of the defences is believed to have been an attempt to counter the rise of the sling as a weapon; but this was used for defence as well as for attack, and when the fort was excavated an 'armoury' of more than 20,000 slingstones collected from the nearby Chesil Bank which runs along the Dorset coast was discovered.

The concentration on massive buildings, either for religious ceremonies or for defence, continued in some parts long after the rest of the world had moved from prehistoric into historic times. In Central America, for instance, the most spectacular remains of Mayan civilization are those of the temple complex of Chichen Itza, rising dramatically above the encroaching Mexican jungle. The first city was founded here as late as the sixth century, but it was soon abandoned and not until the tenth century did there begin to arise the massive monuments and temples that survive a thousand years later. The most interesting is the 100 foot (30 metre) Castillo, the temple of Kukulcan, 'the feathered Serpent'. The four-sided pyramid, covering an acre (0.4 hectares) of ground, rises in nine receding terraces, each being traversed by stone staircases. From the summit the quarter-mile (400 metre) Via Sacra is seen leading to the Cenote of Sacrifice, one of the deep water holes into which human victims were thrown.

Nearby there is the 27 acre (11 hectare) complex known as the Group of a Thousand Columns; square and round, plain and sculptured, they are all that remain of the temples and halls, courts and terraces of Chichen Itza in its heyday. The Mayas who built them had considerable mathematical knowledge which they applied to astronomy, and among the courts and temples there is a 75 foot (23 metre) astronomical observatory within which were built two circular corridors, each having observation doors at the four cardinal points of the compass.

In other continents the same dedication to gigantic building projects can also be seen. At Machu Picchu, 7,650 feet (2,330 metres) up in the Andes, believed to be the remains of a great Inca city – including 'a great flight of beautifully constructed stone-faced terraces, perhaps a hundred of them, each hundreds of feet long and 10 feet high' – can still be seen. Deep in Cambodia there are the remains of Angkor Wat, a huge complex of towers, terraces and temples, of which one alone has more than 1,500 columns hewn from a site more than 30 miles away. And in Orissa, eastern India, the great temple complex of Puri, with its 120 separate temples and its 192 foot (58 metre) pagoda tower for Jagannath, still amazes as one of the most astonishing buildings in the world.

These early works of man – of very different dates in different continents – had little connection with engineering as it was to be developed when man began to investigate scientifically the strengths of materials, the forces required to carry out certain tasks, and the revolutionary possibilities opened up by the use of steam. They did, however, provide useful lessons of the utmost value. The very size of these early building achievements tended to remove any doubts from the minds of those who in subsequent generations acquired the knowledge of the engineer as well as that of the builder. If men with only the most simple tools and equipment could raise the pyramids or the Colosseum, they could surely tackle the problems of restraining the North Sea, of building Europe's Gothic cathedrals or mobilizing civilian armies to build defences against gunpowder. The organization of manpower itself, so vital to success in prehistoric times and in the centuries that immediately followed, demonstrated what could be achieved when the efforts of many thousands of labourers were planned and co-ordinated – as the builders of canals and of the railway networks, as well as the architects of the industrial revolution, were to appreciate in later centuries.

The early builders were hardly engineers in the dictionary definition of those who practised 'the application of scientific knowledge for the control and use of power'. But they paved the way for the engineer.

Machu Picchu, 7,650 feet (2,330 metres) up in the Peruvian Andes and believed to be the site of a great Inca city. Behind can be seen Mount Huayna.

ENDING
'BY GUESS AND BY GOD'

The thousand years that followed the break-up of the Roman Empire were crucial as far as man's engineering progress was concerned. At the start much work was still done 'by guess and by God'. By the end, engineers had begun to discover how the strength of materials, the weights which different structures would bear in different conditions, could often be worked out theoretically and the theory then tested by experiment. The craftsman who had once decided intuitively what the dimensions of a beam would have to be, gave place to the technician who before starting on the job would be guided by facts and figures which he could consult in one of the books produced by the new wonder of the printing press.

While this change was taking place European engineering, science and architecture were receiving an invigorating injection of new ideas following the Moslem conquest of Alexandria in AD 641-2 and the Arab invasions which then, within a few centuries, brought Arab ideas to southern Spain and parts of southern France. The Arabs translated many of the Greek and Roman texts previously unknown in Europe but now made available. They acted as the entrepreneurs of ideas between China and India, and the West. And by substituting the Hindu numerical system for the Roman they transformed mathematical calculations, since multiplying 34 by 23 was a far easier proposition then multiplying XXXIV by XXIII.

The diffusion of fresh methods into the Western world began with infiltration from the Arab world, and continued as the Crusades drew western armies into contact with the defensive structures of the Middle East. The outcome was two groups of buildings which are among the most spectacular examples of the builder-engineer's craft: the great cathedrals whose splendour is still a magnet at Chartres and Paris, Cologne, Wells, Canterbury, Salisbury and Durham and the battlemented castles at Pierrefonds and Château Gaillord in France, and at Windsor, Harlech and Caernarvon in Britain.

The Gothic cathedrals and their military equivalents, the great castles of the Middle Ages, both illustrate the fact that engineering advances are sometimes the result of events in the non-technical world. In the late nineteenth century the need for reconstruction after the American Civil War spurred on a multitude of engineering efforts in the United States and was partly responsible for the rapid spread of the railway network across the Continent. In Europe three decades later Anglo-German rivalry compounded the rise of the ironclad and of rifled guns in the struggle for dominance on the oceans as the main arbiter of a nation's destiny at sea. And in the mid-twentieth century it was American-Russian rivalry and the sputnik of 1957 which led the huge American engineering effort into the space race that was eventually to take men to the moon.

In much the same way, the end of the Roman Empire acted as its own catalyst

The fortress of Château Gaillord built by Richard Coeur de Lion in 1195-6.

The moated Bodiam Castle
built in the valley of the River Rother by Sir Edward Dalyngrudge in 1386.

for events. A strong central government, exercising power from Rome, had been able to finance major road, building and civil engineering projects which had great practical value. The fading away of the Empire's influence left a vacuum. It was filled by the Church with its demands for non-secular architecture, and by the growth of nation states in which military defences were needed to protect the frontiers of new countries or dynasties carved from an Empire which had once stretched from the Bosphorus to Hadrian's Wall.

The cathedrals which arose from the demands of the Church and the castles built first as fortified residences by individual lords and later by leaders of the new nation states, were the most impressive architectural-cum-engineering works of the centuries that followed the fall of Rome. But during this period the everyday world was changed by a number of other developments, superficially of minor importance though, in fact, slowly altering the entire texture of life. One, introduced by the Arabs, was the use of the horse-collar which together with nailed horseshoes made the horse a more effective source of power than the oxen which had more frequently been used in the West. Even more important was the great spread of water-powered mills and, somewhat later, of windmills. Water-powered mills had been used in Illyria for grinding corn as far back as the second century BC, but it was only hundreds of years later that their use began to spread throughout western Europe. They had been used in Persia during the seventh century and had reached Spain three hundred years later, following the Moorish invasions. In Britain the Domesday Book of 1086 recorded some 5,000 water mills and it is known that many hundreds more were in operation, the majority of them either overshot or undershot mills. The early windmills had a vertical axle which eventually, in western Europe, was developed into the horizontal axle. Further improvements were to include various ways of automatically turning the windmill sails into the wind and of setting the angle of the sails so thay they made the best use of it.

The growing use of water or wind power slowly led to improvements in the primitive wooden machinery that could be set in motion by a power-turned axle. By the end of the twelfth century the rotary motion of a water wheel was being used to operate trip hammers for fulling cloth, for crushing materials and for tanning leather. At the same time geared wheels, ratchets and pulleys were being used for a variety of mechanical purposes.

However, the most dramatic result of the development of water or wind power was the birth of the blast furnace, an innovation which brought Europe one step nearer the industrial age. When the Roman Empire collapsed the method of producing iron was much as it had been since earliest times. A small clay furnace would be fuelled by charcoal and small pieces of iron ore would be dropped into the 'bloomery' as the furnace was called. After some hours' heating during which the fire would be kept hot by manually operated bellows, the iron would become smelted from the ore and would become a number of small lumps of iron which could be hammered into shape for various practical uses.

Water mills gave more power to the bellows and, later, to the hammers, but it was only about 1340, in or around Liège, that a decisive improvement was made. Here a combination of taller furnaces and the stronger blast from water-powered bellows produced temperatures high enough to melt the iron once it had been smelted from the ore. The liquid metal accumulated at the bottom of the furnace and was drawn off, or tapped, every few hours. The molten iron then flowed into a long narrow sand-lined trough from which there stretched out a series of smaller parallel troughs. The channels into which the iron flowed, and in which it remained until it had solidified, were thought to resemble sows with their litters of feeding piglets; the chunks of iron thus came to be called 'pigs', a name that has remained in the word 'pig iron', even though such methods of casting have long been superseded.

At first, a great deal of the cast iron was turned into wrought iron by heating in another furnace called a finery, the double production process being justified since the blast furnace turning out cast iron could do so as a continuous process, and thus more cheaply. However, numerous uses were soon being found for the iron that could be cast in moulds of any required shape.

With two kinds of iron available, iron began to be used not only for hinges, fastenings and other small articles which had for long been hammered out from the wrought iron 'blooms', but also for larger articles. Firebacks, railings and cooking utensils were some of them. The extended use of iron, which had been given a dramatic encouragement as cast iron became available from the blast furnaces, went on simultaneously with the building of the great cathedrals – where iron was essential to the construction of the complicated clocks which often adorned them. That at Wells was one of the most famous.

After the Roman withdrawal from western Europe, churches for some while continued to be built along Roman lines with the familiar semicircular arch, beamed roofs, wooden ceilings and barrel vaults. Later the height of the ceiling began to be raised and a vaulting of thin masonry was added below the wooden roof. The new method created its own problems, notably that of supporting the greater weight of the higher building. The first step towards a solution was taken with the introduction of the pointed arch which slowly replaced the semicircular arch of Roman times. Its value had been known to the Assyrians, it was adopted by the Arabs and by the early eleventh century had reached southern France. It was not until the twelfth century, however, that the pointed arch gradually began to come into regular use throughout western Europe. Its value was that its height was independent of its span, in contrast to the semicircular arch in which the two were dependent on each other. It thus enabled arches of equal heights to span different widths. This gave great opportunities to the builders of the cathedrals who were, as it has been stressed, 'more than "engineers" and more than "architects"; they were "masters" in the best sense of the word. They were craftsmen, designers and artists all in one who may not have commanded a very extensive knowledge, but who were supreme masters of their profession.'

But the pointed arch brought its own problems. All arches tend to push outwards the vertical walls or pillars on which they rest. The thrust, as it is called, is stronger with pointed than with semicircular arches, and it naturally became more important as the heights of Gothic cathedrals increased, as did the weight that had to be supported.

The round arch had frequently been reinforced by flat pilasters on the exterior of a building, but it was of limited use and from the first decades of the twelfth century a variety of methods were used to take the increased thrust. Notable was the use of flying buttresses. These were, in effect, struts, built in pairs on opposite sides of the building, springing from points outside the main walls and rising to the eaves of the roof. Holding the roof as it were in a vice, they left the walls to support only the dead weight.

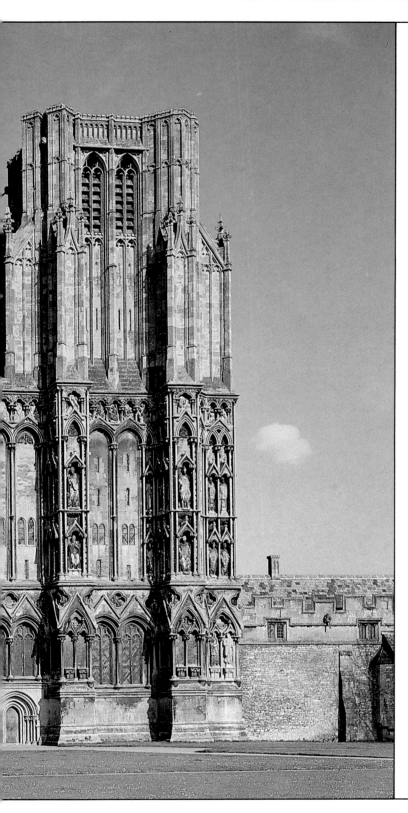

The west front of Wells Cathedral, Somerset, built in 1230-60. The façade was the work of Bishop Joceline, brother of Hugh of Lincoln. The finest west front in England, its statues provide a medieval historical picture book.

Flying buttresses not only performed an engineering task, allowing the architects greater flexibility in the design of their massive stone buildings, but added to the aesthetic appeal of the cathedrals. In addition, ribbed vaulting and buttressing produced regular bays which determined the placing of windows and, therefore, had an effect on the use of the stained glass which was one of the glories of the Middle Ages. The fashion for these larger and more ambitious buildings also increased the need to employ scientific principles. Even the earlier cathedrals required considerable theoretical knowledge as well as practical experience. But the need for more theory was underlined by the disasters of the period. The tower of the great cathedral at Beauvais, finished in 1247, fell once, the roof twice. As one expert on the strength of materials has written, 'It is not surprising that the roofs of the churches continued to fall upon the heads of their congregations with fair regularity throughout the ages of faith.'

The development of Gothic architecture was to have long-term repercussions throughout all Europe. At first the dominating figures behind cathedral architecture had been the priests, not only because of their theological authority and economic power but because they belonged to virtually the only class of men who could read or write, or could understand the constructional problems involved. As Georg Dehio, the German art historian, has described the situation, 'But the bishops and abbots of those times, being mostly of noble extraction, were fully occupied with their twofold duties of spiritual and temporal regency. Even so, they were still able to exert their influence, to a certain degree, on the design of the building ... The dimensions of a church; the materials to be employed; the mustering of labour; the models to be followed, and the innovations to be introduced; all these problems were the responsibility of the taskmaster ... The execution of the work, however, must have been predominantly in the hands of lay-workmen.'

As the pattern changed, as ribbed vault, buttress and flying buttress added complications to the architectural problems, the balance began to alter. Dehio's 'lay-workmen', and in particular the masons, began to exercise more influence. There were three classes of masons: the freemasons who were able to work freestone, the ungrained material such as sandstone and limestone which could be carved in any direction; the rough masons who squared the stone, and the labourers who did the heaving and carrying for their better qualified colleagues. As the masons' importance increased, so did that of the protective guilds into which they organized themselves, tending to speed up the transfer of influence.

The development was echoed in the crafts which produced the increasingly complex stained glass windows adorning the great medieval cathedrals. Windows of glass which had been stained during manufacture were in use at least as early as the ninth century, but it was only with the enterprise and ambitions of the eleventh and twelfth century builders that the technique of the stained glass window began to develop. In the early days the windows contained coloured glass arranged in simple designs; figures came later, then pictorial scenes, all made up from a mosaic of glass pieces which were held in place by H-shaped leaden strips.

First, an outline of the entire window would be drawn on a white board. The shape of the variously coloured glass pieces required would be sketched in, the pieces would be cut and laid in position, then coloured, and finally fired to make the colouring fuse with the glass. When cool, the glass would be inserted in the lead strips which would be soldered together at their meeting points. The

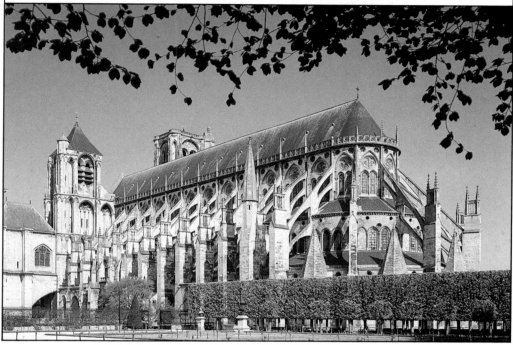

Bourges Cathedral, France, built
at the end of the twelfth century and illustrating the use of flying buttresses.

completed window was then braced with iron crossbars before being mounted in place. The colouring matter was provided by various metallic oxides, of copper or iron for instance, while in the fourteenth century it was found that a derivative of sulphide of silver could be used to give white glass a variety of shades from light yellow to deep orange. The treatment could also be used on coloured glass and thus greatly enlarged the possibilities open to the designer.

Of all the astonishing displays of medieval stained glass that can still be seen, that in Chartres Cathedral is the most remarkable. Here there are nearly 160 individual windows, including two huge rose-windows, respectively 30 feet and 40 feet (9 and 12 metres) across, at the ends of the nave and the transept. Many depict biblical subjects, some incorporate small details indicating that the work was paid for by trade guilds or corporations, such as the basket-makers or the shoemakers. Other windows, paid for by nobles of whom the Count of Chartres and King Louis the Holy, are examples, incorporate the donors' coats of arms.

The dazzling array of colours from stained glass, the long rows of pointed arches leading up to the heights of the nave and, outside, the slim and elegant supports of the flying buttresses are the hallmarks of Gothic, first to be fully developed in the abbey church of St Denis, begun by the Abbé Suger in 1129, the choir being dedicated in 1144. Next came Notre Dame de Paris, and shortly afterwards, the cathedrals of Chartres, Rheims, Amiens, Beauvais and Bourges. In England, the pointed arch was introduced about 1174 by William of Sens who superintended the rebuilding of Canterbury Cathedral. The style spread as quickly in England as it spread in France, although with considerable differences in detail. Germany, Spain and Italy developed their own styles of Gothic.

The North Rose and Lancet windows of the thirteenth-century Chartres Cathedr
building, started in 1194 to replace an earlier cathedral, contains nearly 160 stained gla

Most of these cathedrals continued to be built with only the roughest idea of the strains and stresses involved or of the weights that could be safely borne by separate parts of a building. Leonardo da Vinci, the Florentine artist and scientist, made a lifelong study of natural phenomena. Architectural work on the cathedrals of Milan and Pavia was included in the studies which led him to investigate the strains and stresses of buildings. In the late fifteenth century he calculated the weight that a pillar or cluster of pillars of any particular diameter could properly carry as well as the weight that could be carried by a beam of a particular span.

The contemporary lack of scientific knowledge was also shown by the occasional use of short metal ties to strengthen the masonry. These ties certainly did their job when they were first built into the stonework. But they had no protection against rust and since rust has a greater volume than the iron from which it is formed, tied joints tended to be forced open as soon as wind and weather caused them to rust. But in some buildings the ties were set in lead to counter oxidization.

The greater influence of the Church which followed the collapse of the Roman Empire did not only stimulate the building of cathedrals and increase the importance of engineering to architecture. There was another way in which a stronger Christian attitude was to have its influence on the birth and growth of engineering as the term is understood today. Men and women were gradually to be regarded less and less as tools or serfs and more and more as human beings. A new incentive arose to lessen human labour where this was possible and a climate became established in which any lightening of heavy labour by power-driven machines or mechanical devices was welcomed. But the gradual development of wind and water to carry out a growing number of tasks previously done by hand was, of course, mainly due to economics and the development of technology.

Complementary to the great expansion of ecclesiastical architecture during the Middle Ages there was the growth of military fortifications demanding more and greater engineering expertise. After the introduction of explosives into Europe a long period followed during which 'engineer' was synonymous with 'military engineer'; indeed, the term 'civil engineer' came into use largely to describe those men whose work was not concerned with military defences.

Until the Middle Ages the most impressive example of military engineering had been executed not in Europe but in China, where the Great Wall stretched with its twists and turns an estimated 1,700 miles (2,736 kilometres), the distance from London to Athens. With its subsidiary defensive off-shoots adding 800 miles (1,287 kilometres), it is claimed to be the only work of man visible from Mars.

Running from Shanhaikwan in the east to a peak overlooking the Great North River far away to the west, the Wall took some twenty years to build in the second century BC. Between 20 feet (6 metres) and 30 feet (9 metres) high, the Great Wall has an upper pavement along which four or five horses can be ridden abreast. Up to a million workers, prisoners of war, and troops were required for the construction work and it is estimated that about 422,400 cubic feet (11,960 cubic metres) of stone or brick was needed for every mile of the Wall. So many died on the job that it has been called the longest cemetery in the world. Protecting the fertile country of southern China from raiders coming in from the more barren north, the Wall was reinforced by towers which were sometimes less than 200 yards (183 metres) apart. Furthermore, to the north of the Wall itself there were isolated watch towers, each stocked against a four-month siege, which

The Great Wall of China, second century BC,
approximately 1,700 miles (2,736 kilometres) long, was restored in the 1950s.

could give a warning of coming attack.

The Great Wall was a more ambitious project than any carried out by the Romans – or their successors, for that matter. But after the withdrawal of the Romans from western Europe and the end of centralized government there came a fresh need for individual defensive castles on a completely new scale.

Initially, a castle might consist merely of a simple timber building standing on top of an earth mound, artificial or natural. This was the 'motte', to which there was concurrently added the 'bailey', an encircling enclosure at the foot of the mound, and if possible a moat. These early motte and bailey castles are described in an account of the area around Calais written about 1130. 'It is the custom of the nobles of that neighourhood to make a mound of earth as high as they can and dig a ditch about it as wide and deep as possible. The space on top of the mound is enclosed by a palisade of very strong hewn logs, strengthened at intervals by as many towers as the means can provide. Inside the enclosure is a citadel, or keep, which commands the whole circuit of the defences. The entrance to the fortress is by means of a bridge which, rising from the outer sides of the moat and supported on posts as it ascends, reaches to the top of the mound.'

This early type of castle changed throughout the years as new methods of attack were devised to overcome the defenders. Scaling ladders long enough to reach the tops of the walls were built; so were battering rams similar to those which had been used in ancient times. The practice of 'sapping' or mining beneath the walls, was improved, while catapults were built capable of throwing over the castle walls stones weighing up to 600 pounds (272 kilograms).

After the return of the first Crusaders in the twelfth century, defences were greatly improved. In the Middle East the men had been faced with the sophisticated fortifications of the Byzantine Empire, many of which were now imitated in Europe. All were designed to make the best use of the land, being built on craggy rocks where possible and sited to bring attackers under fire from slings and arrows.

In more developed castles, a second line was built, inside the protecting outer walls, and sometimes a third. The outer, or curtain, wall was further strengthened by bastions, towers which projected from the walls and enabled the defenders to bring under crossfire any attackers who approached. The bastions, which were less than an arrow-shot apart, were at first square, but were later rounded since this made them less vulnerable to the battering ram.

Entrances and their approaches were made more complicated so that attackers had to run the gauntlet before reaching the main gateway, while most castles now reverted to the use of the portcullis, a heavy gate which could be raised or lowered within vertical guides. 'If a large number of the enemy come in,' rote Tacitus in describing the portcullis in Roman times, 'and you wish to catch them you should have ready above the centre of the gateway a gate of the stoutest possible timber overlaid with iron. Then when you wish to cut off the enemy as they rush in you should let this drop down and the gate itself will not only as it falls destroy some of them, but will also keep the foe from entering, while at the same time the forces on the wall are shooting at the enemy at the gate.'

Although castles on the continent of Europe continued to increase in complexity, it was in Britain, during the reign of Edward I, that the most impressive fortifications of the Middle Ages were built. Edward returned from the Crusades in 1274 and two years later began his conquest of Wales, an operation that was concluded in less than a decade. Conway, Caernarvon, Harlech and Beaumaris on the north coast were among the most impressive of his castles, which had the same function as those of the Norman Conquest – that of holding down a hostile population.

Beaumaris, the last of the Welsh castles, built in the quarter-century between 1295 and 1320, stands on the shores of the Menai Straits which separate the island of Anglesey from the Welsh mainland, and is still an impressive illustration of the military engineer's art. No less than four hundred masons, thirty smiths, and a thousand labourers were employed in building what was to be one of the most formidable links in the English defensive chain. The inner ward or bailey once enclosed the banqueting hall, the state rooms, the domestic apartments and a small chapel, the foundations of which can still be traced. Beyond these is the first line of fortifications, square in shape and defended by six bastions and two strong gatehouses, each with portcullis. Beyond this again is the outer curtain wall, breached by two gateways sited off-centre so that attackers would be under fire from the main bastions. Surrounding the curtain wall is the moat, filled by the sea and broken by a dock that allowed ships to come up to the castle gate.

The struggle between defence and attack, in which every new device used by a castle's defenders would be countered before long with a new method of attack, might have gone on indefinitely but for the introduction of explosives into Europe from the fourteenth century onwards. Explosives – mainly gunpowder – were greatly to reduce the importance of the castle since stone walls could no longer provide a plausible defence against attack by cannon, and brought science and technology on to the battlefield in a way that transformed military engineering.

There is no doubt that gunpowder, a mixture of saltpetre, charcoal and sulphur, was known in ancient times in the Far East, although doubts have been cast on the popular theory that the Chinese were the first to use it. Certainly the Byzantine emperors employed gunpowder to defend Constantinople in the seventh century and a few centuries later it was used by both Moors and Christians during the campaigns waged for the conquest of Spain. The monk and scholar Roger Bacon is generally credited with introducing gunpowder to England early in the thirteenth century, but there were difficulties in properly mixing the three ingredients and it was not until about 1320 that a German monk, Berthold Schwarz, solved the problem and made gunpowder a reliable weapon of war.

Edward III used explosives against the Scots in 1327 and nineteen years later cannon were used against the French in the battle of Crécy. Siege cannon were soon made more mobile, wheeled artillery was in use by 1521, mortars by 1543, and before the end of the century explosives had so altered the possibilities of war that the utility of the castle as a stationary point of defence was disappearing.

These changes in warfare brought about by the introduction of gunpowder took place slowly, over a number of centuries. They were important not only for the practice of engineering, but also for the civilian organization of European states and it is necessary to carry them forward into the seventeenth century before describing the more fundamental changes in thought that had by that time taken place.

Production of gunpowder soon meant the establishment of powder works, and thus the involvement of industry and of Government. The larger numbers of men required to deal with the more complex defences which sprang up as the use of explosives spread tended to bring armies also under increased central control. Moreover, the building of more complicated defences in themselves led on to the more scientific study of surveying, of soil conditions, of the best methods of digging ditches and of the most efficient ways of organizing hundreds of men for the work. All this was knowledge gained for use by the civilian engineers who in the following centuries were to excavate the world's canals and build the world's railways.

Much of this change was due to the genius of Sébastien Le Prestre de Vauban, Marshal of France, who brought to perfection a defensive concept developed over the preceding two centuries. His complex plans began to change the face of France throughout the second half of the seventeenth century, and brought to a climax the long series of defensive developments which followed the introduction of explosives on to the battlefield.

Vauban's work was based on the construction of a continuous polygon of earthworks and ditches which encircled the area to be defended. They were

Bombardment of Tournay, 8th July, 1709, during the
War of the Spanish Succession. In the foreground are star-shaped defences.

The
SIEGE of TOURNAY
July y̆ 8th 1709.

developed into three separate systems, each of which was adapted to suit the terrain since, as Vauban maintained, 'the Art of Fortification does not consist in rules and systems, but solely in common sense and experience.' The first system was largely an extension of earlier defence schemes, consisting of the curtain, or main defensive wall, before which there were demi-lunes, outworks with crescent-shaped necks, and astride which there were defensive bastions. The polygonal or star-shaped layout allowed the approaches to the fortifications to be covered not only by artillery but by musket fire.

This bastion system was used in more than thirty of Vauban's fortifications, but experience revealed one weakness: if even a single bastion were captured by the attackers, then the whole fortification was likely to be taken. He therefore devised a more complex plan in which the bastions were constructed as triangular works separated from the curtain by a deep ditch. Development of defence in depth was further extended in 1698 after France lost the town of Brisach under the Treaty of Ryswick. Commissioned to design a new fortress within French territory, Vauban drew up the plans for Neuf-Brisach whose sixteen-pointed defences enclosed an intricate arrangement of works from which troops could cover all possible approaches with fire.

However, Vauban, who directed the construction of thirty-three fortresses and the improvement of more than a hundred, also developed into the 'parallel lines' method the existing practice of attacking them. In this a series of trenches was dug ever nearer to the defending walls. So successful was the method that Vauban used it effectively in more than fifty sieges during the French wars of the seventeenth century. Quentin Hughes, a writer on fortifications, has given, in his dissertation on *Military Architecture*, a lucid account of how Vauban employed this tactic to destroy the most formidable works of military engineering. 'Starting about 600 metres out from the fortress to be attacked, and working at night, he detailed small detachments of men, 25 and an officer, to open up trenches running parallel to the front he wished to attack. On the second night he constructed batteries whose guns were intended to keep the defenders quiet. Then began zig-zag approaches, each approach concealed from the besieged; by the fourth night the approaches had reached half-way, when a second parallel trench was constructed. More zig-zags followed and by the ninth or tenth night the third parallel was cut at the foot of the glacis [the long gentle slope leading up to the walls]. The artillery next opened up a breach in the ravelin [the triangular earthwork built in the ditch before the curtain] and then in the bastion. The crowning of the covered way was the most dangerous task (at Philipsburgh [in the Low Countries] in 1676 it had cost the lives of 1,200 men) and the work was best done systematically. The attackers then descended into the ditch, taking cover behind the rubbish which had been brought down in the breach of the ravelin and the bastion. In ordinary circumstances the assault on the fortress would commence at dawn on the twentieth day of the investment.'

While it was the series of star-shaped defences for which Vauban was most famous, he also developed and exploited the defensive use of canals and other artificial waterworks which had been originated in the late sixteenth and early seventeenth centuries by Dutch engineers. Thus he worked out an ambitious plan for linking with canals all the Flanders ports and the inland towns on which they relied for supplies.

However, Vauban's most brilliant military engineering project was the transformation of Dunkirk from an ordinary port into a major coastal fortress. When finished, the scheme had surrounded Dunkirk with virtually impregnable defences. Inside the lines, there were storehouses and a huge inner harbour protected by a great lock of Vauban's design. Outside coastal defences completed the transformation of Dunkirk into a bastion which no enemy could secure.

Quite as important as the physical defensive systems in which Vauban utilized the prototypes of civil engineering organization for military purposes, was the setting up in 1675, on his recommendation, of the French Corps des Ingénieurs du Génie militaire. It was followed, in 1689, by the establishment of a corps of naval architects, the Ingénieurs-constructeurs de la Marine and, in 1720, of the Corps des Ingénieurs des Ponts et Chaussées. The first two of these organizations gave military and naval engineers a new independence from the generals and admirals who were now, in practice, forced to take more notice of engineering advice. At the same time Génie officers were recruited almost entirely from those who had received a scientific and mathematical education, thus giving them a higher status. The Ingénieurs des Ponts et Chaussées were organized on military lines even though they were employed entirely on such public works as bridge and road building. In the same way as their military and naval equivalents, they made full use of the advances in scientific knowledge that the eighteenth century was to provide in such volume.

One result of Vauban's initiative was that in the eighteenth century French civil engineers began to occupy a dominating position that they kept for a century. Britain's great nineteenth-century engineers John Rennie and Thomas Telford found it necessary to read in French the works of men who had been encouraged by the French organizations. Bernard Forêt de Bélidor's *La Science des Ingénieurs*, published in 1729, not only brought together a number of writings on structural engineering but was reprinted for more than a hundred years. His *Architecture hydraulique* followed in 1739, and the two books are still considered to have been the first scientific textbooks on engineering.

Bélidor's books, which for many engineers divided the pre-scientific approach from the scientific, came at the end of long and gradual change whose beginnings can be traced back to the early thirteenth century. It was this change of approach which lay beneath the numerous small developments which took place between the thirteenth and seventeenth centuries and which led on to the climactic rise of steam power and the massive engineering and industrial transformations which followed.

These transformations from a world of guesswork and necromancy, intuition and alchemy, to a world of mechanics, chemistry and physics, are often epitomized by the names of Leonardo da Vinci, Copernicus and Galileo. Important as they were, men before and men after also played seminal parts in the story. Among them was Roger Bacon, who taught that experimentation and the expression of results in mathematical form was the road to scientific and engineering advance. He was, moreover, in no doubt about what the outcome of such an approach would be. 'Machines for navigation can be made without rowers so that the largest ships will be moved by a single man in charge', he wrote about 1250. '. . . cars can be made so that without animals they will move with unbelieveable rapidity . . . flying machines can be constructed so that a man sits in the midst of the machine revolving some engine by which artificial wings are made to beat the air like a flying bird . . . machines can be made for walking in the sea and rivers, even to the bottom without danger.' Despite his writings, Bacon's influence was slight and it was only gradually that, during the following century, men making gunpowder, refining metal or tanning leather, began to record the details of what they did and extrapolated from those details the principles by which they could operate more efficiently.

Diffusion of the new approach and the spread of new methods was slow throughout the Middle Ages, one reason being the Black Death which began in 1348, caused 100,000 deaths in London alone, and killed between a half and a third of Britain's population of between three and five million. The Black Death was as devastating elsewhere in Europe, disrupting whole societies and bringing almost to a halt the interchange of ideas. A further handicap to advance was the Hundred Years' War, a series of wars fought intermittently between England and France from 1337 till 1453. Progress would have been even slower had it not been for the invention of printing and the dissemination of writings from earlier times. Thus the manuscript of Vitruvius' main work, found in a Swiss monastery in 1414, was published by Fra Giovanni Giacondo who planned the foundations for the Pont Notre Dame in Paris and used in them the water-resistant pozzuolana cement, now rediscovered with the help of Vitruvius.

However, the change from working by tradition and intuition to working with the help of experimental results eventually made itself evident, and was notable in the activities of Leonardo da Vinci, whose experiments to solve structural problems and to discover the strength of materials were among the first of their kind. Theoretical work represented only one side of Leonardo's efforts. Largely self-taught, he had by the age of thirty achieved fame throughout northern Italy and the catalogue of his credentials which he presented when entering the service of Ludovico Sforza in Milan in 1482 gives an enlightening picture of the qualifications and skills of the fifteenth-century military engineer.

'I have bridges of a sort extremely light and strong,' he said, 'adapted to be most easily carried, and with them you may pursue, and at any time flee from, the enemy; and others, secure and indestructible by fire and battle, easy and convenient to lift and place. Also methods of burning and destroying those of the enemy.

'I know how, when a place is besieged, to take the water out of the trenches, and make endless variety of bridges, and covered ways and ladders, and other machines pertaining to such expeditions.

'Item. If, by reason of the height of the banks, or the strength of the place and its position, it is impossible, when besieging a place, to avail oneself of the plan of bombardment, I have methods of destroying every rock or other fortress, even if it were founded on a rock etc.

'Again I have kinds of mortars; most convenient and easy to carry; and with these can fling small stones almost resembling a storm; and with the smoke of these cause great terror to the enemy, to his great detriment and confusion.'

Leonardo continues with a long list of accomplishments, quoted by Bertrand Gille in *The Renaissance Engineer*, which indicates what was expected of the engineer whose abilities could be adapted to either war or peace. Leonardo's record suggests, moreover, that his claims were no idle boasts, since for nearly forty years he served a succession of rulers in northern Italy, not only carrying out military works but planning canals, devising a scheme for making the River Arno navigable up to Florence and another for clearing the Pontine Marshes,

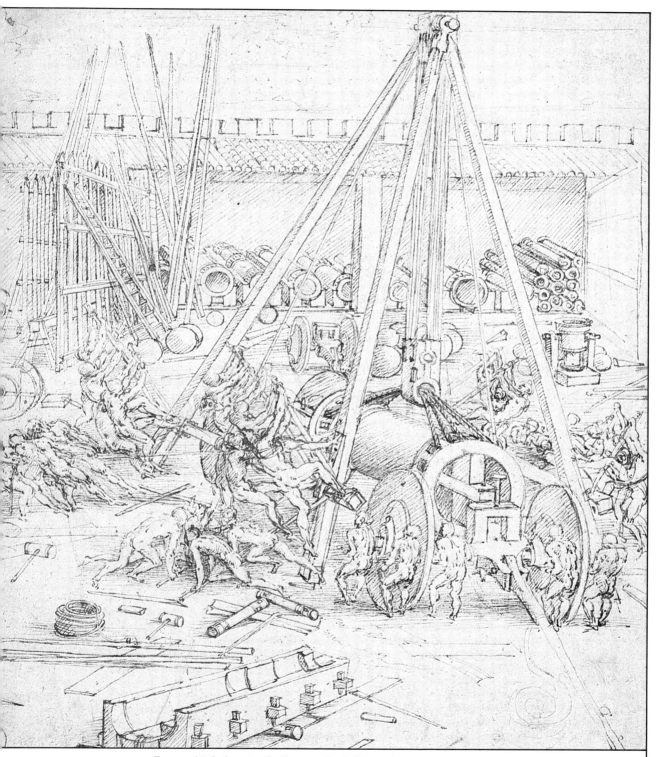

Pen and ink drawing by Leonardo da Vinci about 1487 of the
Courtyard of a Foundry. The barrel of a gun is being raised for mounting on to a gun carriage.

south-east of Rome. Lack of money stopped many of these schemes, but their engineering originality is shown by his notebooks.

There, he recorded many of the experiments which would have been unthinkable a century earlier, but were to become accepted practice little more than a century in the future.

Thus, to discover the tensile strength of wire he fed sand into a basket suspended by a measured length of the wire. When the wire broke, a spring stopped the flow of sand and the weight of sand that had caused the wire to fail could easily be measured. He also demonstrated by experiment what many architects had suspected: that a pillar consisting of a compact bundle of shafts would support a much heavier load than the sum of loads which the shafts could support separately. He established laws for the elasticity of beams, first by calculation, then by experiment, and in his work on hydraulics used streams of differently coloured liquids to find out how water from different sources would mix – research which had a direct impact on the design of mills.

While Leonardo typified the movement by engineers towards the more scientific approach, his notebooks also reveal the growing contemporary interest in machinery. Here are designs for earth-moving equipment and for dredgers, for mechanical equipment which would automatically cut wooden screws, and for the prototypes of the military tank and the flying machine. The notebooks also show that Leonardo experimented with connecting-rod and crank systems for turning continuous circular motion into alternating rectilinear motion. One of these systems consisted of a wheel toothed round only half of its circumference and engaging with two lantern wheels on opposite sides of the main wheel's diameter and fixed to the same axle. Each lantern wheel was thus moved in turn, the continuous circular motion being transformed into an alternating circular motion. Rectangular alternating motion could then be obtained by use of a ratchet.

Most of Leonardo's ideas were recorded in a code that was not to be deciphered until years after his death. His influence on his contemporaries was not, therefore, as great as it might have been, since it was limited to that of the works he actually carried out. Nevertheless, as civil as well as military engineer, he helped to create the age of experiment on which the industrial revolutions of the following centuries were to be based. But other men were also important. Although not themselves engineers, their beliefs and discoveries examined traditional ideas about the natural world; after all, if it were possible to question the way in which the universe operated, it became more plausible to question accepted beliefs about the methods by which buildings and ships were constructed, or water and wind power was utilized.

The list of men who paved the way for the use of steam should, therefore, include not only those who built the first primitive engines but also those who investigated the physics of the world in which they lived, and those who had set that world within a new framework. Among the first of these was Nicolas Copernicus, a Polish astronomer, and a near-contemporary of Leonardo, who showed that the earth was not, as had been taken for granted, the centre of the universe but that, together with the planets, it revolved around the sun. The Copernican view of the universe which he outlined in *De Revolutionibus Orbium Coelestium*, completed in 1530 but published only in 1543, the year of his death, was to be confirmed in the next century by Galileo, whose telescope revealed the phases of Venus, the moon's craters and other astronomical features.

Professor of Physics and Military Engineering at the University of Padua, Galileo spent much time at the arsenal of Venice – the state which controlled Padua University – and his study of how heavenly bodies moved was to be closely linked with the fundamentals of engineering practice. His astronomical investigations showed that an object could move under the influence of two forces operating on it at the same time. This not only enabled him to make a science of gunnery in which shot moved under the double influence of an explosive and of the pull of gravity, but led him to found the science of mechanics, the branch of physical science which deals with the behaviour of matter under the action of force. This led, in 1638, to his *Dialogues on the Two New Sciences*, one being the science of statics, or forces in equilibrium; the other being dynamics which deals with the behaviour of bodies under the influence of forces which produce changes of motion in them. In theory, Galileo outlined the basis of the square-cube law – that volume increases as the cube of linear dimensions while strength increases only as the square; in practice, he began to place the behaviour of loaded structural members on a systematic, empirical and theoretical basis. Thus his work not only lessened the gap between the theoretical scientist and the practical engineer but also provided the basis for the work in the following centuries of Robert Hooke, the English physicist, Leonhard Euler, the Swiss mathematician, and Charles Coulomb, the French physicist.

Galileo's work was evidence of how closely in his day a theoretical study of the heavens could be linked to practical engineering matters, and the same was true of his successor, the Dutch physicist and astronomer Christiaan Huygens. During the sixteenth and seventeenth centuries the experiments carried out by scientists and engineers to test their theories had to be recorded with an accuracy that demanded clocks better than those which existed, and it was Huygens who was mainly responsible for providing them. Finding that his telescopic observations required more exact timekeeping, he recalled that years earlier Galileo had noted that the timing of a chandelier's swing appeared to be constant whatever the amplitude of the swing. The clockmakers of the Middle Ages had built mechanical clocks operated by a falling weight. They were more rugged than the water clocks they displaced but no more accurate and Huygens began to experiment with pendulums. He found that for the swing of a pendulum to be really constant the swing had to be through an arc that was not completely circular. He therefore attached a pendulum to a clock, inserted at its fulcrum a device which gave it the necessary swing, and used weights to keep the pendulum on the move. The result was a greatly more accurate clock, and later improvements kept it correct to the nearest second.

If time was important to experimental results, so was temperature, and the century that saw the first accurate clocks witnessed also the first accurate thermometers. Galileo is credited with having a 'Florentine thermometer' in which the expansion of liquid in a thin tube showed changes of temperature. This was improved by the German physicist Gabriel Daniel Fahrenheit who used mercury in a narrow tube and graduated the tube from 32, the point at which water froze at the normal barometric pressure, to 212, the point at which it boiled. The French physicist René Antoine de Réaumur used alcohol rather than mercury to graduate a scale of 80 degrees between the freezing and boiling points of water, and the Swedish astronomer Anders Celcius used mercury in a tube registering between zero and 100 for the two points (the centigrade scale).

In addition to the more accurate methods of measuring time and temperature – both important as engineers continued to make more use of iron – there was the greater knowledge provided by science of the possibilities which could be exploited by ropes and pulleys, by toothed gears, worm gears, and hoisting devices. Wood was still the material usually used for such aids but despite this it steadily became easier throughout the fifteenth and sixteenth centuries to carry out engineering operations which involved the cutting of watercourses, the construction of mills or the mining of ores from the earth. Simple jib cranes became increasingly more common as well as drums around which rope could be let out or drawn in by men or horses working on the radial handles protruding from the drums. Not only was water power becoming more popular to turn power shafts but the shafts themselves were increasingly used to operate gearing which itself could drive a variety of simple wooden equipment.

Symptomatic of the work now possible for the first time was the successful movement in 1586 of the giant 327-ton Egyptian obelisk which lay on its side before the Vatican in Rome to a new site in the great piazza before St Peter's, and its erection there into the vertical. This was carried out by Domenico Fontana, one of the first men to lay claim to the title of civil engineer. The huge obelisk, brought to Rome from Egypt in ancient times, lay half buried in soil close to the sacristy on the western side of St Peter's. Pope Paul III consulted Michelangelo and Sangallo, but both men said the obelisk would be impossible to move. One of his successors, Pope Sixtus V, turned to Fontana.

The young engineer-architect gathered immense beams of wood, large pieces of iron and what were described as miles of the stoutest rope, as well as windlasses and tackle, scores of horses and nine hundred men. Five levers and forty windlasses would be needed to move the obelisk once it had been surrounded by a timber cage, he decided.

The preparations took seven months and finally, on 30 April 1586, huge crowds gathered to watch the construction of the immense wooden cage round the obelisk, its movement to a new site and then its re-erection. Fontana had worked out the operation in minute detail: a trumpet call announced the operation of the windlasses and a bell announced their stoppage. Strict silence was demanded and the death penalty was ordered for those who broke it. A gallows with attendant hangman had been erected near the site.

The obelisk was successfully half raised. Then it stuck. In the excitement a sailor from San Remo shouted out 'Water, give water to the ropes.' He was marched off to the gallows. But water was poured on the ropes, movement of the obelisk restarted, and the sailor was reprieved and rewarded for his advice by a visit to the Pope. San Remo, moreover, was given the privilege of supplying the palm branches for St Peter's each Palm Sunday.

Fontana who, with Giacomo della Porta, successfully undertook the vaulting of the great cupola surmounting St Peter's in Rome, left unfinished after the death of Michelangelo, typified a new race of architect-engineers whose beliefs rested on the principles of scientific experiment followed earlier by Leonardo. These principles continued to spread, both on the continent of Europe and in Britain where they culminated, in the seventeenth century, in the work of Robert Hooke. As a physicist, Hooke was to be remembered mainly for his law of elasticity, later of immense use to constructional engineers, and his revelation of the principles which underlay the hairspring, work thus making possible the elimination of the pendulum in clocks, the building of ships' chronometers and eventually the wristwatch.

Hooke not only changed Britain's Royal Society 'from a band of scientific virtuosi to a professional body of scientists' but in the process provided the new ground rules on which engineering would in future move forward. Their key lay in experiment, as he made clear in a paper on *Mr. Hooke's Method of Making Experiments*.

'The Reason of making Experiments', this begins, 'is for the Discovery of the Method of Nature, in its Progress and Operations. Whosoever therefore doth rightly make Experiments, doth design to enquire into some of these Operations; and, in order thereunto, doth consider what Circumstances and Effects, in that Experiment, will be material and constructive in that Enquiry, whether for the confirming or destroying of any preconceived Notion, or for the Limitation and Bounding thereof, either to this or that Part of the Hypothesis, by allowing a greater Latitude and Extent to one Part, and by diminishing or restraining another Part, within narrowed Bounds than were at first imagin'd, or hypothetically supposed. The Method therefore of making Experiments by the Royal Society, I conceive, should be this.' Hooke then stresses the care and accuracy with which such experiments should be made and adds, 'After finishing the Experiment, to discourse, argue, defend, and further explain, such Circumstances and Effects in the preceding Experiments, as may seem dubious or difficult: And to propound what new Difficulties and Queries do occur, that require other Trials and Experiments to be made, in order to their clearing and answering: And farther, to raise such Axioms and Propositions, as are thereby plainly demonstrated and proved. Finally, an account of the whole matter should be read for eventual discussion by the Society.'

Although Hooke's work was primarily designed to test scientific theories, its effect frequently had an impact on the day-to-day problems with which engineers and builders were constantly concerned. A printed collection of his papers dealing with experiments, for instance, covers the floating of lead, the pressure of water in pipes, the elasticity of the air, compression, sound, and the expansion of water. There should be no surprise that Hooke the physicist should have been appointed surveyor when much of the City of London had to be rebuilt after the Great Fire of 1666.

His work on springs, and his Law which evolved from it, was part of the scientific investigation of materials, and of their characteristics and strength, which was to divide the old world of engineering from the new. So was the growing understanding of the composition of forces, which lies at the heart of statics, the mathematical and physical study of the way that materials behave under the action of forces when no motion is involved. These and other matters were investigated not by the engineers who were to use the knowledge acquired but by scientists or philosophers dedicated to discovering more about the natural world and the forces that operated in it. Thus to a greater extent than ever before the engineer who built bridges and buildings or built and used the primitive machinery that soon came into existence, depended on scientists. Practice began to depend on theory; hit and miss, trial and error, began to give way in engineering to techniques which rested on mathematics and experiment. All was ready for exploitation of the steam engine which was already entering history with consequences which few had begun to imagine.

THE COMING OF STEAM

Despite the advances of the Renaissance which had made it possible for science to give a newly strengthened backbone to engineering, many old practices continued until the eve of the new age of steam and the Industrial Revolution which it brought about. This was certainly true of the smelting of iron where, as T.S. Ashton makes clear in his survey of *Iron and Steel in the Industrial Revolution* the change from earlier days was limited to the, admittedly important, higher temperatures brought about by the blast furnace. 'The ore', he says 'was first crushed into fragments and mixed with a small quantity of marl and lime which served to bind it together, and the mass was divided into lumps which were placed on a forge and surrounded with charcoal. By means of a blast produced with leathern bellows, worked either by manual labour or by water power, the fire was maintained at a moderate temperature; and the metal was brought to a pasty, rather than a liquid form, the impurities being removed by repeated hammering, "even" says Sturtevant, "as the whey is wrung out by the violence of the Presse, and so the curds are made into a cheese". Several heatings and hammerings were necessary before the ore was finally transformed into a bloom of wrought iron, ready to be worked into implements by the smith. Though the quality of the finished metal was high, the quantity produced at any one forge was necessarily small, and a considerable weight of metallic iron was left behind in the cinders or slag.'

More than improvements in the making of iron were to be needed before the potentialities of steam as a revolutionary source of power could be exploited. First, the physical properties of steam, and the means of controlling them had to be determined. Secondly, technology had to provide steam-tight joints and containers able to withstand the pressures exerted by steam. Neither of these requirements were available during the Middle Ages or in the period immediately afterwards. Leonardo had suggested in 1495 that a projectile might be fired by creating a build-up of steam but there is little indication that this was more than a proposition without practical foundation. However, the seventeenth century, which was to see the acceptance of the experimental method in both science and engineering, witnessed numerous attempts to harness the power of steam. As early as 1606 Giovanni Battista della Porta of Naples stated that as steam condensed inside a closed vessel a vacuum would be created. Della Porta was a Neapolitan natural philosopher. The Italian translation of his *Pneumaticorum libri*, 1601, included a description of a steam engine anticipating that later built by Thomas Savery. He subsequently designed an engine which could raise a column of water in that way. This was the principle which led to the first steam engines even though he wrongly believed that steam was air. It was left to Solomon de Caus, a French landscape gardener, to point out nine years later that steam was evaporated water and that steam pressure would be greater than air pressure.

One of the first proposals in England for using steam as a propelling force was made by Nathaniel Nye, the mathematician, in *The Art of Gunnery*, published in 1647. Here he proposed to 'charge a piece of ordinance without gunpowder' by using water instead of gunpowder, making the cannon airtight with a piece of wood, loading it with shot and then applying a fire to the breach 'till it burst out suddenly'. In Italy Giobattista Branca described how a wheel might be made to revolve by steam jets, and in 1659 a steam-driven pump for water drainage was proposed in *The Elements of Water Drawing* written by R. D'acres.

The first hint of practical use came in 1663 with the publication of the second Marquis of Worcester's *A Century of Inventions*. 'I have discovered', he wrote there, 'an admirable and very powerful means of raising water by means of fire, not by suction, for then, as the philosophers say, one can be limited *intra spheram activitatis*, as suction only operates for a given distance. But there is no limit to my means if the vessel is strong enough. By way of trying it I took a whole cannon, the mouth of which had burst, and three parts filling it with water, I closed the end which had burst and the touch-hole with screws. I kept up a very strong fire inside, and in twenty-four hours the gun broke up with a loud report.'

A good deal of mystery surrounds the Marquis of Worcester's 'water commanding machine' as he called it. He is said to have experimented on the walls of Raglan Castle, and a machine was seen working at Vauxhall in London which it was claimed could raise water 40 feet (12 metres). But there has remained some doubt as to whether its main power really came from steam, and although the inventor was given a ninety-nine year monopoly by Parliament, the machine never seems to have got beyond the trial stages.

However, a decisive move had been made on the Continent in 1643 when the Italian physicist Evangelista Torricelli showed that the atmosphere had a finite weight; the fact that the weight changed from day to day led to the construction of the first barometer. More important was the discovery that if the air could be taken out of a closed container, and kept out, there would be created a zone, known as a vacuum, whose characteristics were totally different from those of the world in which men lived. It was startling enough to learn that man breathed under the pressure of about 15 pounds per square inch, produced by the weight of the air forever pressing down on the earth; it was even more startling to realize that in a vacuum the lack of air caused a flame to flicker out and small animals to die; and it appeared that sound did not pass through a vacuum.

Of more practical interest than these curious characteristics was a demonstration that the weight of air could be used to push a piston down into a cylinder in which a vacuum had been created. Otto von Guericke, a German physicist who was also mayor of the city of Magdeburg, demonstrated in 1654 how powerful was this air pressure when operating against a vacuum. Fifty men pulling on a rope attached to a piston were unable to stop it from moving as a vacuum was created on the other side of the piston. Then metal 'Magdeburg hemispheres' as they were called, fitted together and evacuated of air, were held together by air pressure even when teams of horses tried to pull them apart.

Once the power of air pressure became known, new ways of creating a vacuum were sought, particularly in France where Louis XIV had founded the Académie Royale des Sciences in 1666. Von Guericke had used a simple bellows-operated pump but a French Huguenot, Denis Papin, Professor of Mathematics at Marburg University, tried other methods, including the explosion of gunpowder to drive air from a cylinder. A number of experimentalists, among them Christiaan Huygens, whose pendulum of 1656 had led to improved clocks, and the Abbé Hautefeuille, tried to use this method, but although the explosion would drive air out through a valve in the cylinder it was found impossible to build up the vacuum by successive charges, while other difficulties resulted from the by-products of the explosion. 'Since it is a property of water', Papin wrote in his *Acta Eruditorum Lipsiae* in 1690, 'that a small quantity of it turned into vapour by heat has an elastic force like that of air, but upon cold supervening it again resolves into water, so that no trace of the elastic force remains, I concluded that machines could be constructed wherein water, by the help of no very intense heat, and a little cost, could produce that perfect vacuum which could by no means be obtained by gunpowder.'

Papin had begun by sealing a cooking pot containing water and turning it into what he called a 'Digester or Engine for softening Bones'. When it was heated, the steam accumulating inside created a presssure that raised the boiling point of water and allowed bones to be softened and meat cooked in less time than normal. It was, in fact, the first pressure cooker.

The same year, 1679, Papin heated a cylinder which contained water and was closed at the top with a sliding piston. When the water boiled the steam pushed up the cylinder. When the water was allowed to cool the steam condensed, a vacuum was created in the upper part of the cylinder and atmospheric pressure forced the piston down.

So far, these experiments had been of little more than academic interest. But towards the end of the seventeenth century an economic revolution was starting in Britain: the timber from the great forests which had once covered so much of the country was beginning to run out. What remained was badly needed for shipbuilding, for the 'wooden walls' with which the Navy was to defend the country for another two centuries. In place of wood men began to use coal. And once they started mining deep for coal they were faced with the miner's ubiquitous problem: seepage of water into the workings. First hand pumps, then an ingenious assortment of pumps operated by animals, were for a while sufficient to deal with the problem. But by the start of the eighteenth century some coal mines were 400 feet (120 metres) deep and by mid-century half as deep again.

The first attempt at solving the drainage problem was made by Thomas Savery, a military engineer in London. Nothing is reliably known about the events which led him to construct his first steam pump. But in July 1698 he obtained the grant of a patent for 'A new invention for raiseing water and occasioning motion to all sorts of mill work by the impellent force of fire, which will be of great use and advantage for drayning mines, serveing towns with water, and for the working of all sorts of mills where they have not the benefit of water nor constant windes.'

This predecessor of the engines which were to turn the wheels of the Industrial Revolution was a comparatively primitive affair and it was used for a comparatively mundane task. Water was heated in a boiler and the steam created then led through a tap into a container from the bottom of which a suction pipe led down, past a non-return valve, into the water which was to be raised. A tap was then turned to allow cold water to fall on the container of steam, the steam condensed, a vacuum was created, and water was drawn up the suction pipe. At the next cycle of operations the water was driven out of the apparatus via a second non-return valve.

Savery demonstrated his engine before the Royal Society in June 1699. Although its principles appeared to be sound, it was clear that poor workmanship limited its practical efficiency. However, Savery was not despondent and three years later was advertising his 'miners' friend' in glowing terms. 'Captain Savery's engines' he claimed, 'which raise water by the force of fire in any reasonable quantities and to any height, being now brought to perfection and ready for public use: these are to give notice to all proprietors of Mines and Collieries which are encumbered with water, that they may be furnished with engines to drain the same, at his workhouse in Salisbury Court, London, against the Old Playhouse, where it may be seen working on Wednesdays and Saturdays in every week from 3 to 6 in the afternoon, when they may be satisfied of the performance thereof with no less expense, than any other force or horse or hands, and less subject to repair.'

The response from mine and colliery proprietors was poor, but the engines were used for drainage purposes in a number of private estates. Among them were the grounds of Campden House, Kensington, and an account in *New Improvements of Planting and Gardening* maintained that it had 'succeeded so well that there has not been any want of water since it was built.' Suction and delivery pipes were 3 inches (7.6 centimetres) in diameter, cost of the installation was £50, representing a vastly greater sum in today's money, and about 52 gallons (236 litres) could be raised every minute. A second engine for the same purpose was installed in the grounds of Sion House, in west London. Others may have been used at mines, although the engine was inefficient for a number of reasons. The steam cock and the condensing water cock had to be turned on and off by hand. And since water could only be drawn up for a limited height, a number of engines would, in a deep mine, have to be installed at different levels.

Such limitations tended to delay the introduction of steam and until mid-century conventional horse pumps continued to be built. Just what they involved is made clear in the proposals of John Smeaton for an engine to be worked by 'one large or two light Horses' and designed to pump water to a lake in the Dowager Princess of Wales' gardens at Kew. Smeaton, born in 1724, and reported to be the first self-styled British civil engineer, had by 1750 begun to acquire the reputation which led to his designing pumping engines, bridges, canals and harbour works as well as the famous Eddystone Lighthouse.

'The following things are supposed to be done at the expense of her Royal Highness', Smeaton wrote in his proposals for the Kew pumping engine. 'The drain or gutter, leading from the engine to the Chinese house to be lowered four feet; the ground to be cleared and levelled for the engine to stand upon; the earth to be wharfed up with brick-work; the well cleaned, and the brick-work repaired, where necessary; and, in case it is thought necessary to underpin the groundsill of the shade with brick, the materials to be led from the river to the place, and the labourers to assist in lifting and digging what may be required during the setting up. The engine to work with an Archimedes screw, 2 feet 8 inches diameter, and 24 feet long; the shade for the horses to be 30 feet from out to out, and the mean diameter of the horse track 24 feet.

'This engine to raise 1200 hogsheads in four hours, with one large horse, or two light ones, such as have heretofore been used.

'A small pump, to be worked occasionally by the engine, for raising water from the gutter to the cistern, in the kitchen garden.

'The whole of the machinery and frame-work thereof, with the shade for the horses' walk, and a cover from the screw, with a cistern and sluice at the foot thereof, to be completed in the most substantial and workmanlike manner, for the sum of one hundred and fifty pounds.'

This reliance on horses, constantly hoofing their circular walk, was already lessening in Smeaton's day, first under the pressure of Savery's invention and then of its successors. Savery utilized the force of the atmosphere to operate his engine's working stroke and his invention is more accurately called an atmospheric than a steam engine. Nevertheless while it used a vacuum to draw up water, it was the pressure of steam which forced the drawn-up water out of the engine, and this naturally limited its use. It was followed by the brainchild of Thomas Newcomen, not only very different but basically simpler and more efficient; so much so that Newcomen engines remained in use for more than a century.

Thomas Newcomen came from Dartmouth in Devon and was well aware that in Cornwall to the west the pumping-out of tin mines, as well as of coal mines, was a growing problem. He solved it with his Newcomen engine, the first of which was erected not in Cornwall but above a colliery near Dudley, Staffordshire, in 1712. So successful was it that before the end of the century more than sixty Newcomen engines were working in Cornwall and some hundreds elsewhere in Britain.

Outwardly, the main innovation appeared to be a massive horizontal rocking beam. The operation of the engine is graphically described today by R. J. Law of the Science Museum which has a unique collection of early steam engines. 'A vertical cylinder open at the top was supplied with steam from a boiler beneath, which resembled a brewer's copper', he has written. 'The piston, packed with leather and sealed with a layer of water on top, was hung by a chain from the arch head of a rocking beam. From the other end of the beam the pump rods were suspended. When steam at slightly above atmospheric pressure was admitted into the cylinder, the piston was drawn up by the weight of the pump rods and any air or water blown out of the cylinder through water-sealed non-return valves. After the steam valve was closed, the steam in the cylinder was rapidly condensed by a jet of cold water. The unbalanced atmospheric pressure drove the piston down, raising the pump rods and making the working stroke. The cycle was then repeated, the steam valve and injection cock being opened and closed by a plug rod hung from the beam. The Dudley Castle engine had a cylinder of 19 inches internal diameter and made a stroke of about 6 feet. At each stroke it raised 10 gallons of water 51 yards and at 12 strokes per minute developed about 5½ horse power.'

Soon after the Dudley installation, improvements began to be made. A rod called the plug-frame, which carried a number of projections, was fitted to the beam; these projections opened and shut the valves controlling the steam and the condensing water, and operation of the engine thus became automatic. Then John Smeaton, the engineering all-rounder, redesigned individual parts of the engine and further improved its overall operating efficiency.

From the start, the Newcomen engine was more economical than earlier pumping machines and when the second engine was installed at Griff Colliery, near Coventry, the cost of running it was reported to be £750 a year less than the money spent on maintaining the horses it displaced.

The story was to be the same elsewhere, with the result that steam engines soon became an item in the British export trade. The first to work abroad was installed for mine drainage at Chemnitz (now Karl Marx Stadt in East Germany) in 1729, an account of its savings being given by Marten Triewald, a Swedish engineer, who quoted the Austrian Imperial authorities. 'The horse-whim or artifice has drawn 25 feet water per minute by means of two sets of pumps 8 inches in diameter, with a lift of 2½ feet, which makes altogether 5 feet, and five lifts in a minute', he said. 'Even if one allows the fire machine only 13 strokes or lifts of 6 feet, it will be easily seen how, by subtracting the 25 feet of water which the horse-artifice is lifting in the same time as the fire machine lifting 78 feet, the fire machine has actually been lifting 53 feet of water more than the horse artifice. As to the cost, the horses with their apparatus have been drawing 900 imperial gulden a month. Whilst the fire machine, with all its staff and firewood has drawn only 400 imperial gulden a month.'

A major initial expense in the early engines was the cylinder, which was at first cast in brass. Within a few years, however, the cost was reduced by nine-tenths as it was found that cast-iron cylinders could be produced with sufficient accuracy. The ability of the expanding iron industry to turn out cylinders accurate enough to satisfy the demands of the steam engine builders illustrates the close way in which the stories of steam and iron were to be intimately linked as Britain moved towards the Industrial Revolution and the engineering opportunities which it was to offer.

Steam engines required iron, not only for the boilers in which steam was created but for the pistons which moved up and down in cylinders and for the various connecting rods and ancillary equipment essential to the finished engine. But iron smelting still needed charcoal, the charcoal made from the wood of the great forests from which, it was estimated, 2,000 fully grown oaks were required for a single 1,350-ton warship. Coal could be used as a substitute for charcoal for certain purposes such as heating. But in the business of iron smelting coal was for long considered useless, and for reasons which only slowly came to be understood. Eventually it was realized that coal contains sulphur, an element which unites easily with iron, as well as other impurities. Coal-smelted iron, therefore, contained unwanted contaminations which prevented its use for many purposes.

With coal ruled out as an iron-smelting fuel, and the demands on charcoal continually rising with the need for more wooden warships, efforts were soon being made to solve the problem. Dud Dudley, a natural son of Edward Dudley, 5th Baron, took out a number of patents early in the seventeenth century, the second being for 'the Makeing of Iron with Sea or Pitt Coals, Peats, or Turffs, as Abovesaid, and with the same to Rost, Melt, or Refyne all Metalls of what Nature soever.' Dudley claimed in his *Mettalum Martis*, published in 1665, to have made 'much good iron', but the claim is not substantiated and at the start of the eighteenth century there was only one way of dealing with the problem, and that not a very satisfactory one. This was to use a reverberatory furnace in which flames were forced down, or reverberated, from the roof, there thus being no contact between the metal and the fuel. The method had its limitations but the variety of objects that could be made by it is illustrated by an advertisement in *The Postman* for 24-6 December 1700. 'At Mr. Stringer's Iron Foundry and Refinery in Blackfryers near Ludgate', it says, 'are cast without Wood, Charcole or Bellows, Cannons, Bombs, Shot Sheels, etc. Bells of any size or tone, Potts and Kettles,

Pit head of a coal mine, location
unknown, showing steam winding gear about 1820.

hollow Rolls, Stoves, Cockles and Bars for Sugar-Works, solid large Rolls for flatting of Iron, Brass, Copper or Lead, Rolls for Mints, Stoves, Backs and Hearths for Chymneys, Flower Pots, and Balconies, and Hatter Basons, Plates for Packers and Hotpresses, very large plates for looking-glass Grinders, Cylinders for Water-works, various things for Millwork, Boxes for Coaches, Carts and Drays, Anvils for Smiths and Forges. All sorts of Chymical Vessels that can be made in Iron or in Stone Glass . . .'

While Mr Stringer was producing what he could from his reverberatory furnace, and while Savery was experimenting with his atmospheric engine, the real break-through in the production of iron was made by Abraham Darby of Coalbrookdale, the first of a long line of iron founders.

Coalbrookdale, where Darby set up business in 1709, lies on the River Severn, eleven miles from Shrewsbury, and can without exaggeration be called the real birthplace of Britain's Industrial Revolution. The engineering sites which quickly grew up in and around it soon acquired a romantic character very different from that which today surrounds much industrial activity, an attitude exemplified by a description in a Shropshire gazetteer published in 1851. 'Coalbrook Dale, a winding glen two miles from Madeley, hemmed in by lofty hills & hanging woods, is celebrated for the most considerable iron works in England:' it says, 'the forges, mills and steam engines, with all their vast machinery the flaming furnaces & smoking chimneys and handsome residences nestling under the cliffs of the hills, have altogether a most romantic and singular appearance, and perhaps in no part of the globe are features of so diversified and wonderful a character brought together within so limited a compass – here art has triumphed over nature, and the

A view of Coalbrookdale, on the River Severn
near Shrewsbury, in 1788. This engraving was coloured by hand.

barren wilderness has been converted into one of the most animating abodes of commerce, and being studded with residences of taste and elegance, it gives the whole a very interesting appearance.'

The area had been a centre of the iron-founding industry since Tudor times, having the three basic needs of accessible iron ore, nearby forests for production of charcoal, and water power to provide the blast for the blast furnaces. But nearby there was also coal, and Darby decided to find some way of using it. His solution was extraordinarily simple. Charcoal was made by partly burning wood in open stacks. Why not do the same with coal and thereby turn it into coke from which the sulphur would have been driven off?

Abraham Darby's experiments were successful and, by the time of his death in 1717, he had succeeded in making coke-produced iron suitable for a wide range of tasks.

His innovations and those of his son, Abraham Darby II, who brought to Coalbrookdale the steam engine for pumping water in summer droughts, a method of making wrought iron from coke-produced pig iron, and the use of rails for transport, are vividly described in a letter written about 1775 by Abraham Darby II's wife Abiah. Her father-in-law, she wrote, 'suggested the thought that it might be practable [sic] to smelt the Iron from the ore in the Blast Furnace with Pit Coal. Upon this he first try's with raw coal that came out of the Mines, but it did not answer. He not discouraged, had the Coal Coak'd into Cynder, as is done for drying Malt, and it then succeeded to his satisfaction. But he found that only one sort of pit coal would suit best for the purpose of making good iron. These were beneficial discoveries, for the Moulding and Casting in Sand instead of Loam was of great service, both in respect to expence and expedition, and if we may compare little things with great – as the invention of printing was to writing, so was the moulding and casting in Sand to Loam.'

During the summer, water turning the wheels which operated the furnaces dropped too low to be efficient. 'But my husband', Abiah continued, 'proposed the errecting [sic] of a Fire Engine to draw up the water from the lower Works and convey it back into the upper pools, that by continual rotation of the water the furnaces might be plentifully supplied; which answered exceeding well to these Works, and others have followed the example . . .

'But all this time the making of Barr Iron at Forges from Pit Coal pigs was not thought of. About 26 years ago my Husband conceived this happy thought – that it might be possible to make bar from pit coal pigs. Upon this he sent some of our pigs to be tryed at the Forges, and that no prejudice might arise against them, he did not discover from whence they came, or of what quality they were. And a good account being given of their working, he erected Blast Furnaces for pig iron for Forges . . . Many other improvements he was the author of; one of service to these Works here, they used to carry all their mine and coal upon horses backs, but he got roads made and laid with sleepers and rails, as they have them in the North of England for carrying them to the Rivers, and bring them to the Furnaces in Waggons. And one waggon with three horses will bring as much as twenty horses used to bring on horses backs. But this laying the roads with wood begot a scarcity and rose the price of it, so that of late years the laying of rails of Cast Iron was substituted, which, although expensive, answers well for ware and duration. We have in the different Works near twenty miles of this road, which cost upwards of eight hundred pounds a mile . . .'

Iron foundries were no longer limited to afforested areas but could be set up wherever coal was available. Even greater freedom came when 'Iron-mad Jack', John Wilkinson, also working in the area, used steam power to produce the air blast for the blast furnaces, a move which meant that foundries need not be sited near a river. It was Wilkinson who, in the second half of the eighteenth century, greatly increased the efficiency of the steam engine by building his boring mill which could provide engines with accurately machined bores. Until this invention, bores had been finished by a cutter head mounted on a rotating pole. Wilkinson's alternative was to devise a system for using a rotating bar which had bearings at either end and along which the cutting head could be moved as required. Initially built to produce more accurately bored cannon, the method was adapted to boring steam cylinders and James Watt, who was to revolutionize the steam engine in the latter part of the eighteenth century, stipulated that the cylinders for all his engines should be made on the new boring mill.

Wilkinson, one of the first iron founders to use steam and to encourage others to do so, was a key figure in the rise of the iron industry whose needs were more and more served by steam. In 1770 he visited France where he set up a foundry and advised French engineers on supplying Paris with water from the Seine, a project for which he eventually provided 40 miles (64 kilometres) of cast iron piping. He stayed at Le Creusot for three years, advised the local iron founders and remodelled the iron centre in the town. In Britain he built the first iron boat in 1787 and successfully launched it into the Severn, much to the disappointment of a large crowd who expected it to sink.

Throughout the eighteenth century, the use of iron from the coke-fired blast furnaces was considerably extended. In 1722 patents were granted for making smoothing irons of cast iron and later in the century gun carriages were cast, as were coffin nails and tacks. Cast-iron hinges and cast-iron buttons became fashionable. Richard Dearman developed a process for making cast-iron hoes for the West Indian Plantations, while John Turton of Bristol was making cast-iron manacles, armlets worn by the slaves who used the hoes.

Meanwhile the need for iron rose following the outbreak of the war with the American Colonies in 1776 and the consequent rearmament in Europe. 'Their great guns', says Macpherson of the firm of Carron in 1777, 'which were cast solid and bored by a drill worked by the whole force of the River Carron, were exported to Russia, Denmark, Spain etc.; and the quantities were so considerable that the Government was unwilling to let them be carried in ordinary ships, lest they should fall into the hands of the American cruisers. The company thereupon fitted out a stout ship of their own, properly armed and manned, for the purpose of carrying to Spain 300 iron guns from three to twenty-four pounders.'

While Darby and his successors had solved the problem of producing cast iron without using wood for their furnaces, difficulties still remained in turning the coke-smelted iron into satisfactory wrought iron, since the result tended to be so brittle that it disintegrated under the hammer blows which should have transformed it. The problem was eventually solved by Henry Cort of Fontley Forge, Hampshire, who in 1784 patented a process whereby the molten pig iron was kept hot in a reverberatory furnace in which the fuel was kept separate from the pig iron. The reaction was helped by 'puddling', a stirring of the molten mass by long iron bars, described in homely terms by one of Cort's contemporaries. 'As the stirring of cream, instead of mixing and uniting the whole together, separates like particles to like, so it is with the Iron,' he said; 'what was at first melted comes out of the furnace in clotted lumps, about as soft as welding heat, with metallic parts and dross mixed together but not incorporated. These "great cinders of iron" were put under the forge hammer and then passed through the rollers, "and by this simple process all the earthy particles are pressed out."'

The importance of Cort's process can hardly be over-estimated since it not only determined the course that much civil engineering was to take but directly affected the geography of the Industrial Revolution. 'His discovery', it was stated in 1948 by T.S. Ashton in his *The Industrial Revolution, 1760-1830*, 'was one of the outstanding events in the history of technology. It had the effect of freeing the forge-masters from their dependence on the woodlands, just as Darby's discovery had freed the furnace owners. It liberated England from the necessity of importing large amounts of charcoal-iron from the Baltic, at a time when political relations with Sweden and Russia were liable to rupture. It drew the forge part of the industry from its scattered haunts to the coalfields, where the making of finished iron could be carried on in close proximity to the furnaces. And it led to the growth of great integrated establishments, in which all processes, from the mining of ore and coal to the slitting of rods, were controlled by a single group of proprietors. Within a relatively short space of time the industry came to be concentrated in four main areas, and new types of communities, with dense populations, grew up around the pithills and slagheaps of Staffordshire, South Yorkshire, the Clyde and South Wales. The output of iron increased vastly; metal came to take the place of timber and stone in works of construction; the hardware industries expanded the range of their products and there was hardly an activity – from agriculture to ship-building, from engineering to weaving – that did not experience the animating effects of cheap iron.'

During the later part of the eighteenth century civil engineering was thus offered a material which while not new was now available at a cost much lower than had previously been considered possible. Not suprisingly, the ironmasters were adept at finding fresh uses for their products. Iron chairs and iron pipes became common. Iron rails – an omen for the future – increasingly replaced the wooden rails on which horse-drawn carts carried coal from the collieries.

'Iron-mad Jack' even designed an iron coffin in which he was to be buried. More significantly, the Severn was spanned in 1779 at Coalbrookdale by the world's first iron bridge.

By modern standards the best cast iron, although weak in tension, was strong in compression, and the arch bridge was, therefore, one obvious use. The first attempt to build one was made at Lyons, France, in 1755, and one arch was actually assembled in a builder's yard before the scheme was considered too costly and iron was abandoned for wood. The first designs for Coalbrookdale were made by Thomas Farnolls Pritchard of Shrewsbury who discussed the bridge with John Wilkinson and Abraham Darby III. But Pritchard died in 1777. He left the project in Darby's hands, but who was mainly responsible for the final design remains uncertain.

In many ways it replicated the traditional stone bridges – much as the first 'horseless carriages' were built on the lines of coaches, even though they were powered by engines rather than hooves. The arch of a stone bridge consists of wedge-shaped hand-cut voussoirs and Darby's iron bridge incorporated voussoirs of cast iron which fitted into and supported each other, no bolts or rivets being

used. With a span of 100 feet 6 inches (30.6 metres), a total length of 196 feet (60 metres) and a rise of 50 feet (15 metres), it consisted of five main iron ribs, each cast in two pieces, weighing 5¾ tons (5.84 metric tons) each and supporting a 24 foot (7.3 metre) carriageway carried on sand-cast iron plates 2½ inches (6 centimetres) thick. The complete bridge, using 378½ tons of iron was built in three months. The bridge is still standing, and if it had a fault it was that of being too light by comparison with stone bridges. The inward thrusts of the approaches on which it rested were not fully counteracted by the weight of the iron and had to be replaced by cast-iron subsidiary arches. 'This must', it has been said, 'have been an early case of that difficulty of putting new wine into old bottles which constantly troubles materials engineers.'

Many other iron bridges followed before the end of the century. One was designed by the reformer Tom Paine after he had gone to America in the latter half of the eighteenth century. It was to have been erected across the River Schuylkill in Pennsylvania, but when the project was abandoned, Paine had the plans executed by the Rotherham Ironworks in Britain. The bridge, assembled for demonstration purposes in Paddington, London, was finally erected across the River Wear at Sunderland in 1795.

While Paine's design was a 'one-off', the engineer Thomas Telford, who was concerned in more than thirty canal-building schemes, as well as in the building of many roads, designed dozens of iron bridges and viaducts during his career. When county engineer for Shropshire, he built in 1796 the iron bridge over the Severn at Buildwas, and later the famous Waterloo Bridge at Bettwys-y-Coed in North Wales, and the Craigellachie Bridge over the Spey in Northern Scotland. Iron figures prominently in the two revolutionary aqueducts with which he carried waters of the Ellesmere Canal System over the Dee at Pont-Cysyllte and over the Ceriog Valley at Chirk, and in 1801 he designed a single-span iron bridge of 600 feet (183 metres) which was to cross the Thames in London. The plans for what would have been an extraordinarily beautiful structure were abandoned only because of the difficulties created by the approaches.

On the Continent the first iron bridge was a footbridge, only 40 feet (12 metres) long, built in 1797 by a wealthy estate owner in Lower Silesia. It was soon followed by something far more ambitious – the Pont des Arts across the Seine in Paris in 1803, and the Pont d'Austerlitz which was started the next year.

While the use of iron had thus been developing, many engineers had been improving the performance of the Newcomen steam engine by making its separate components more efficient. Prominent among them was John Smeaton who tackled the task scientifically, studying boilers, cylinders and valve gear one by one and recording the improved efficiency – in other words, the reduction of fuel consumption – which improvements in each item could bring about. The improvements were cumulative and when they were built into an engine at Long Benton Colliery in 1772 it was found that to do certain work the fuel needed was only half what it had been a few years earlier.

Thus throughout the second half of the eighteenth century the steam engine and the iron industry moved forward on parallel lines, the advance of each

---

The world's first iron bridge, spanning the River Severn at Coalbrookdale.
The complete bridge, using 378½ tons of iron, was built in three months in 1779.

supplementing the advance of the other. This all paved the way for the vast extension of steam which the next few decades were to witness. While the first engines had been built to drain the coal mines, their successors carried out a growing number of other tasks. Steam was used to power the machinery as Europe became industrialized; it was used to move the horseless carriages of the railway age and eventually it was to replace sail, first on small river craft and then on vessels crossing the Atlantic, or carrying passengers and goods to other distant parts of the world.

These developments, each of which led to the creation of a new department of specialist engineering, would have been impossible without the work of James Watt, a man who is often considered to be the inventor of the steam engine. Although the line in reality goes back through Newcomen and Savery to those physicists who first discovered how steam could be used, the mistake is understandable. During the last third of the eighteenth century Watt so transformed the potentialities of steam that his engines differed from their predecessors not so much in quality as in kind. He introduced a separate condenser, the closed-top cylinder in which steam instead of the atmosphere pressed down the piston; the use of steam alternately on either side of the piston; the expansive working of steam; the idea of parallel motion in which a series of links copied and enlarged the reciprocating motion of one point on the engine to another point; and a method by which a steam engine could perform rotary motion. These improvements progressively decreased the amount of fuel an engine needed, at the same time increased the number of different tasks it could perform, and with the introduction of rotary motion helped steam turn the factory wheels of the Industrial Revolution. If any one man can be named as responsible for the age of steam, it was James Watt.

Born in 1736, the son of a Greenock shipwright, Watt was sent to London to train as an instrument maker, returned to Scotland, and in 1757, at the age of twenty-one, opened an instrument shop at Glasgow University. He had become interested in the possibilities of steam, then the subject of perpetual discussion among engineers. He considered the idea of driving road vehicles by steam and even made a few elementary models. Then, in 1763, there came the opportunity which was to change his life. A model of Newcomen's engine belonging to the Natural Philosophy class needed repair, and Watt was given the task. 'I set about repairing it as a mere mechanician', he later wrote, 'and when that was done, and it was set to work, I was surprised to find that its boiler could not supply it with steam, though apparently quite large enough, (the cylinder of the model being two inches in diameter, and six inches stroke, and the boiler about nine inches diameter).'

Watt soon saw the cause of the problem. The steam chamber of the engine had to be cooled to condense the steam and produce a vacuum. It was then filled with steam once again but, since it had been cooled, a great deal of steam was required just to heat up the chamber. Since this steam performed no useful work it was thus lost for practical purposes. 'I perceived', wrote Watt later on, 'that, in order to make the best use of steam, it was necessary first, that the cylinder should be maintained always as hot as the steam which entered it; and, secondly, that when the steam was condensed, the water of which it was composed, and the injection itself, should be cooled down to 100° [F.], or lower, where that was possible.'

It was only in May 1765 that he saw how it could be done. 'It was *in the Green of Glasgow*', he later told Robert Hart, a Glasgow engineer, 'I had gone to take a walk on a fine Sabbath afternoon. I had entered the Green by the gate at the foot of Charlotte Street – had passed the old washing House. I was thinking upon the engine . . . and gone as far as the Herd's House, when *the idea came into my mind that as steam was an elastic body it would rush into a vacuum, and if a communication was made between the cylinder and an exhausted vessel, it would rush into it and might there be condensed without cooling the cylinder.*'

Thus was born the idea for Watt's revolutionary separate condenser. He decided to keep the condenser clear of air and water by means of a pump, and to keep the cylinder constantly hot by enclosing it in an insulated steam jacket. And to help maintain the heat he would use steam instead of atmospheric air to push down the piston.

'When once the idea of the separate condensation was started,' he subsequently wrote, 'all these improvements followed as corollaries in quick succession, so that in the course of one or two days the invention was thus far complete in my mind, and I immediately set about an experiment to verify it practically. I took a large brass syringe, $1\frac{3}{4}$ inches diameter and 10 inches long, made a cover and bottom to it of tin-plate, with a pipe to convey steam to both ends of the cylinder from the boiler; another pipe to convey steam from the upper end to the condenser (for, to save apparatus, I inverted the cylinder); I drilled a hole longitudinally through the axis of the stem of the piston, and fixed a valve at its lower end, to permit the water, which was produced by the condensed steam on first filling the cylinder, to issue. The condenser used upon this occasion consisted of two pipes of thin tin-plate, ten or twelve inches long, and about one-sixth inch diameter, standing perpendicular, and communicating at top with a short horizontal pipe of large diameter, having an aperture on its upper side, which was shut by a valve opening upwards. These pipes were joined at bottom to another perpendicular pipe of about an inch diameter, which served for the air and water-pump; and both the condensing pipes and the air-pump were placed in a small cistern filled with cold water.'

When the miniature prototype was coupled up to a small boiler and set in motion, the savings of steam and fuel were as Watt had expected. 'A large model, with an outer cylinder and wooden case, was immediately constructed', he wrote, 'and the experiments made with it served to verify the expectations I had formed, and to place the advantage of the invention beyond the reach of doubt.'

In 1769 he took out a patent for 'A New Method of Lessening the Consumption of Steam and Fuel in Fire Engines' and set up another experimental engine, this one with a 5 foot stroke. But he was now married, and to provide for a family he started in business as a civil engineer, carrying out surveys for the canals which were beginning to criss-cross the country. There was little time for continuing work on the new engine and it was not until 1773, when he had gone into partnership with Matthew Boulton, a leading Birmingham manufacturer, that work on the separate condenser was seriously recommenced. Watt's good fortune in linking his progress to Matthew Boulton has rarely been explained better than by Samuel Smiles, the famous nineteenth-century English writer. 'Many distinguished inventors are found comparatively helpless in the conduct of business, which demands the exercise of different qualities', he wrote, ' – the power of organising the labor of large numbers of men, promptitude of action in emergencies, and sagacious dealing with the practical affairs of life. Thus Watt hated that jostling with the world, and contact with men of many classes, which are usually encountered in the conduct of any extensive industrial operation. He declared that he would rather face a loaded cannon than settle an account or make a bargain; and there is every probability that he would have derived no pecuniary advantage from his great invention [i.e. the steam engine], or been able to defend it against the mechanical pirates who fell upon him in Cornwall, London, and Lancashire, had he not been so fortunate as to meet, at the great crisis of his career, with the illustrious Matthew Boulton, "the father of Birmingham".'

Despite his reluctance to deal with the business sharks of his time, Watt exuded a confidence that was remarked upon by no less an acute observer than Sir Walter Scott, the famous novelist. 'Amidst this company stood Mr. Watt, the man whose genius discovered the means of multiplying our national resources to a degree perhaps even beyond his own stupendous powers of calculation,' he wrote. 'This potent commander of the elements – this abridger of time and space – this magician, whose cloudy machinery has produced a change in the world, the effects of which, extraordinary as they are, are perhaps only now beginning to be felt, was not only the most profound man of science, was not only one of the most generally well informed, but one of the kindest of human beings.'

Watt's partnership with Boulton transformed the prospects for the steam engine. Boulton, who was to tell James Boswell, Dr Johnson's biographer, from his Soho works, 'I sell here what all the world desires – power', was primarily a businessman, although one with a knowledge of science, and thus a perfect foil for Watt.

With Boulton's encouragement two full-scale engines were built at Boulton's Soho Works in Birmingham. The cylinders were made from cast iron, a considerable improvement, but teething troubles abounded. 'The engine goes marvellously bad', Boulton wrote to Watt in 1775. 'It made eight strokes per minute; but upon Joseph endeavouring to mend it, it stood still. Nor do I at present see sufficient cause for its dullness. I have a few minutes ago had the top taken out, and find that I can pump down the piston & although I hear ye air pass by it into the cylinder, yet the error is not sufficient to account for its bad going.'

The difficulties were eventually overcome and the following year the two engines were completed and installed, one of them at the Bloomfield Colliery at Tipton in Staffordshire, the other at the New Willey, Shropshire, blast furnaces of the John Wilkinson who had cast the cylinders. Both engines were spectacularly successful since they required only a third of the fuel used in a comparable Newcomen engine. The new Boulton & Watt engines speedily took over from their predecessors and the Soho factory in Birmingham was from now on busy with orders.

The conditions on which the engines were sold were unusual. The firm insisted that the main components should be made by the people they named: thus the cylinders had to be made by John Wilkinson, whose boring machine for cannon-making gave him a superiority over all rivals, while castings for the pumps had to come from the Darbys of Coalbrookdale. Boulton & Watt supplied some of

A typical Boulton & Watt steam engine from the
collection of 15,000 such coloured drawings in Birmingham Public Library.

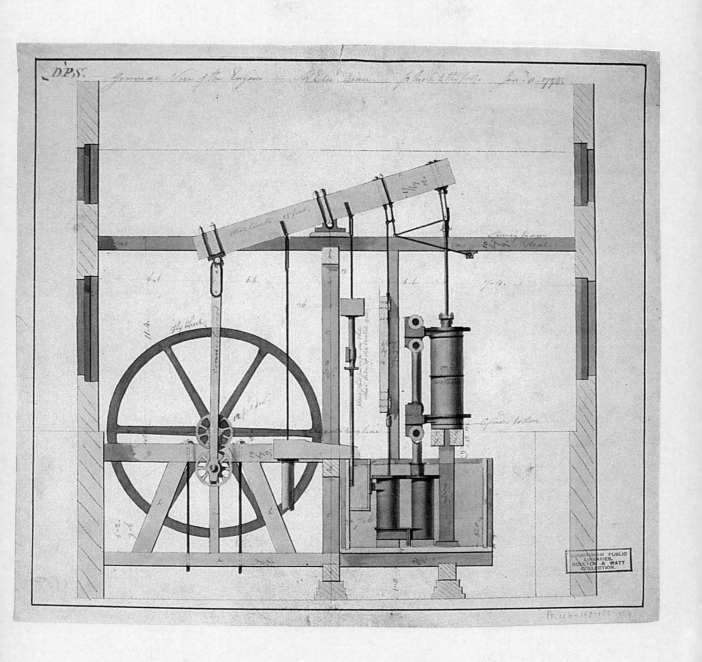

the components and supervised the erection of the engine on site. Payment was made by the user who was charged one-third of the money he had saved on fuel by installing the engine instead of using a Newcomen model. The saving was calculated with the help of a sealed meter which Watt designed and had fixed to the engine beam. This gave the number of engine strokes which, with details of the pump capacity, allowed the saving to be worked out with considerable accuracy.

It was soon clear that the Boulton & Watt engine was far more efficient than the Newcomen. Surely it could soon be used for tasks other than pumping? Boulton and Watt both thought so, and in 1775 Watt succeeded in the difficult legal task of having his patent extended to the end of the century, a personal success but one which for years exercised a stranglehold on development of the steam engine.

Neither man was of the type to rest on his laurels, and during the early 1780s Watt, spurred on by Boulton, lodged more patents which between them transformed the utility of the steam engine – just in time for the Industrial Revolution which was about to sweep first Britain and then the rest of Europe.

The first of the patents, lodged in 1781, dealt with a subject that had rarely been far from Watt's mind since he had made the steam engine a practical proposition – that of finding some way of making the engine produce rotary instead of reciprocating motion. The obvious method was to incorporate a crank of the kind which had been found for years on wood-turning lathes. Watt had himself built a model incorporating a crank, but believed since other men had fitted cranks for other purposes, there was no need to patent it. James Pickard of Birmingham, whom Watt always believed had been given details of the Soho Works machine by one of his workmen, thought otherwise. Pickard patented the use of the crank to produce rotary motion in a steam engine and Watt, rather than challenge the patent, looked about for other means.

His patent was for 'certain new methods of applying the vibrating or reciprocating motion of steam or fire engines, to produce a continued rotative or circular motion round an axis or centre, and thereby to give motion to the wheels of mills or other machines.' Although five different methods of obtaining such motion were described, the most important was that using a connecting rod which drove a flywheel shaft through a sun-and-planet gear. The planet wheel, rigidly fixed to the connecting rod, revolved around the perimeter of the sun wheel which was keyed to the driven shaft, and when both wheels were the same size the driven shaft would make two revolutions each time the engine made a double stroke.

Within two years Watt had erected an engine with sun-and-planet gear for John Wilkinson at whose Bradley Forge works it operated a tilt hammer. Other ironmasters soon followed suit.

Watt followed the sun-and-planet gear by patenting in 1782 a double-acting engine. Until then all engines had delivered power only on the down-stroke and had thus required a flywheel to carry on the engine motion for the rest of the cycle until the next downstroke was reached. Watt removed the disadvantage by arranging for the connections to be reversed when the piston reached the bottom of the cylinder. Steam had been entering at the top and the condenser had been connected to the bottom. But now the position was reversed so that steam pressure acted on the up- as well as the down-stroke.

Since the main beam had to be pushed up as well as pulled down, it was necessary to replace with a rigid connection the chain which had been satisfactory for pulling down the beam. From this requirement came Watt's 'Parallel motion', a Z-shaped linkage of rods which enabled the double-acting engine to provide rotary motion without vibration. 'I am more proud of the parallel motion than of any other invention I have ever made', Watt is reported to have said.

An early rotative engine was bought by the indefatigable John Wilkinson and another by Josiah Wedgwood, the most famous potter of his age, for use in his pottery. With cotton mills growing in numbers and factories of multitudinous varieties anxious to power their machinery by steam, only one more advance was needed: a method of keeping the speed of the engine constant even when it was subjected to varying loads. Watt provided the answer with an adaptation of the centrifugal governor, already in use on windmills.

Watt's governor consisted of an upright spindle at the bottom of which was a pulley and at the top of which were two pivoted levers each carrying a heavy cast iron ball. The pulley was made to revolve by the engine and at normal running speed turned the spindle just fast enough for centrifugal force to push the iron balls slightly outward. If the running speed increased, the centrifugal force also increased, the iron balls flew out farther, and a linkage at their attachment-point to the spindle operated a throttle which decreased the amount of steam going into the engine. Conversely, a drop in speed opened the throttle. The governor not only controlled the speed of the steam engine but could be set to restrict the speed between any chosen upper and lower limits.

With the increasing use of the Boulton & Watt engines for tasks other than raising water it was found necessary to settle one important point. Previously, the efficiency of the engines could be stated simply by announcing that so many tons of coal would raise so much water a height of so many feet. But with the spread of the rotary engine and its use for many differing jobs some universally applicable standard had to be found. In an age when horses were still the main source of energy, the solution seemed obvious. It was given in one of Boulton & Watt's account books for August 1782. 'Mr. Worthington of Manchester wants a mill to grind and rasp logwood and to drive a calendar,' it says. 'The power for all which is computed to be about that of 12 horses. Mr. Wriggley, his millwright, says a mill-horse walks in 24 ft. dia^r and makes 2½ turns p. minute. 2½ turns = 60 y^{ds} p. minute, say at the rate of 180^{lb} p. horse.' Watt then proceeds to work out the size of the engine: '60 yds × 3 = 180 × 180 pounds = 32,400 ÷ 120 feet of piston's motion = 270 lbs × 12 horses = 3240^{lbs} load of cyl^r, which at 5^{lb} p. inch = 29 inch cyl^r, 6 feet stroke, 20 p. minute.'

The figure of a horsepower as being 32,400 pounds (14,696 kilograms) raised one foot high per minute was later changed to 33,000 pounds (14.968 kilograms), but it soon became the practice for the makers to refer to their engines as being rated at so many horsepower, a practice which spread not only throughout Britain but wherever steam was used.

The use of steam, developing steadily throughout the second half of the eighteenth century, was the main factor separating the pre-industrial age from the industrial. It also helped to bring into existence the concept of the civil engineer as a fully-fledged professional capable of handling the multiplicity of different engineering tasks which the flourishing age required. John Smeaton, who made numerous improvements to the Newcomen engine amid a crowded lifetime of other work, was the prototype of this new breed of man. He won the Royal

Society's Copley Medal for improvements on mill work. He built Ramsgate Harbour, the Forth and Clyde Canal, and important bridges in Perth, Banff and Coldstream. He was constantly being consulted by those who wanted waterways improved or swamps drained, and when, in 1769, London Bridge appeared likely to collapse it was John Smeaton, at work in Yorkshire, who was urgently sent for. Two arches had been thrown into one, with the removal of a large pier, and many Londoners were apprehensive of passing over or under the bridge. 'The committee of Common Council adopted [Smeaton's] advice', says an account of his work; 'which was, to repurchase the stones of all the city gates, then lately pulled down and lying in Moorfields, and to throw them pell-mell (or *pierre perdu*) into the water, to guard these sterlings [the starlings or bastions protecting the piers], preserve the bottom from further corrosion, raise the floor under the arch, and restore the head of water necessary for [restoring] the waterworks to its original power; and this was a practice he had before and afterwards adopted on other occasions. Nothing shows the apprehension of the bridge falling, more than the alacrity with which his advice was pursued: the stones were re-purchased that day; horses, carts and barges, were got ready, and the work instantly begun, though it was Sunday morning. Thus Mr. Smeaton, in all human probability, saved London bridge from falling, and secured it till more effectual methods could be taken.'

It was naturally to John Smeaton that, in the early weeks of 1771, another engineer proposed that they and others in the same profession might hold occasional meetings, 'where they might shake hands together, and be personally known to one another; that thus, the sharp edges of their minds might be rubbed off, as it were, by a closer communication of ideas, nowise naturally hostile; might promote the true end of the public business upon which they should happen to meet in the course of their employment; without jostling one another with rudeness too common in the unworthy part of the advocates of the bar, whose interest it might be to push them on perhaps too far in discussing points in contest.'

Smeaton supported the proposal and 'in March, 1771, a small meeting was first established on Friday evenings, after the labours of the day were over, at the Queen's Head Tavern, Holborn.' This was the Smeatonian Society from which, some two decades later, the Institution of Civil Engineers was to emerge.

The need for such an organization had become obvious well before the end of the eighteenth century since there had been numerous and varied repercussions to the new availability of power through the efforts of Boulton & Watt. Their subsequent encouragement of gas lighting soon created, as will be seen, a world of new conditions which transformed life not only in factories but in the home. The growing demands of engineers brought into existence the machine tool industry.

And the availability of steam power, and of engineers who could provide equipment to utilize it, made possible the growth of the textile industry on which so much of Britain's prosperity was to depend for many decades. The processing of wool had a history going back to earliest times, a history of hand-processing later aided by the power of the water mill. But steam power became widely available just as the American colonies won independence and found themselves able to export to Britain large quantities of cotton. As was so often the case, demand stimulated ingenious ways of satisfying it, and a long series of inventions transformed what had for generations been a cottage occupation employing thousands of women at home into a factory industry where the maximum use could be made not only of women but also of children. Employment of children under nine in textile factories remained legal until 1883; indeed, employment of older children spending half of their schooldays in the mill, was forbidden only in 1918.

Even though the mills of the Industrial Revolution employed workers by the thousand, the engineers whose inventions made the mills possible were frequently attacked. Antagonists claimed that work would be taken away from the workers and more than one textile engineer was driven from his home by the predecessors of the Luddites who, a few years later, were to destroy machinery wherever destruction was possible.

Spinning fibrous substances such as cotton or wool into the rounded strands of yarn that can be woven involves two separate operations. The fibre must be drawn out in a continuous manner, as from a spindle, and the material must be twisted in order to give it coherence and strain-resisting power. For centuries this had been done by hand, using either a spindle from which the mass of fibres was unwound, or a spinning wheel.

It was these simple hand operations which were mechanized by a succession of devices in the second half of the eighteenth century. Among the first was the spinning jenny of James Hargreaves, who realized that if a number of spindles were arranged vertically side by side it might be possible for several threads to be spun at once. His first machine, of 1764, incorporated eight spindles, progressively increased to 120, the operator being able to control the spinning by means of a single horizontal clamp which he directed with one hand while using the other to turn the equivalent of the spinning wheel.

Steam power being used for carding,
the preparation of cotton or wool fibres so that they can be used for spinning.

Hargreaves' spinning jenny produced soft yarn suitable for the weft of a fabric – the crosswise threads woven under and over the lengthwise warp – and it was with weft in mind that Richard Arkwright in 1769 patented what came to be known as his water-frame spinning machine since it was first used at a mill at Cromford, near Matlock, powered by a water mill. He had invented, he said in his petition to George III, 'A new piece of machinery never before found out, practised or used, for the making of weft or yarn from cotton, flax and wool, which would be of great utility to a great many manufacturers in this his kingdom of England, as well as to his subjects in general, by employing a great number of poor people in working the said machinery, and by making the said weft or yarn much superior in quality to any ever heretofore manufactured or made.' However, as Arkwright was to state five years later, his machinery was found to be 'peculiarly adapted for spinning Cotton Yarn for Warps, and is principally used for that Purpose.'

One more thing was needed. This was the 'mule' of Samuel Crompton, invented 1779, so called because it was a hybrid, incorporating features from both Hargreaves' spinning jenny and from Arkwright's water-frame machine. The speed of its spindles, of the rollers by which the characteristics of the thread could be controlled, and of its carriage, could all be governed independently, with the result that the quality of the yarn could be varied as required and produced suitable for use as weft or warp.

While the spinning jenny was simple and small enough to be used in cottage homes, this was not true of Arkwright's or Crompton's machines or of Edmund Cartwright's power loom which he worked on from 1785 to 1792 and which gave a further boost to steam. The turning point came in 1785 when a Boulton & Watt steam engine was installed at a Papplewick, Nottinghamshire, mill. From that date onwards steam increasingly powered the multiplicity of devices which year by year improved the efficiency of the textile industry, not only for spinning but for weaving and the other ancillary processes such as carding, willowing and batting. Women worked in steadily smaller numbers at their cottage doors; instead, they, and their children, moved for work into the mills and factories which no longer had to be sited beside running water but could be built wherever a steam engine could be installed.

There were two ways of looking at the changes which the engineers had wrought. The hard labour was certainly taken from many industrial processes, as Edward Baines, editor of the *Leeds Mercury* (and not to be confused with his father Edward Baines), has emphasized in his *History of the Cotton Manufacture in Great Britain* . 'All are moving at once,' he has written, '– the operations chasing each other; and all derive their motion from the mighty engine, which, firmly seated in the lower part of the building and constantly fed with water and fuel toils through the day with the strength of perhaps a hundred horses. Men, in the meanwhile, have merely to attend on this wonderful series of mechanism, to supply it with work, to oil its joints, and to check its slight and infrequent irregularities.'

But if the engineer was lightening labour he was also making much of it unnecessary. In 1783 Thomas Bell invented 'cylinder printing' for fabrics which enabled one man and one boy to do the work previously done by a hundred men and a hundred boys. In the same year oxymuriatic acid was introduced to the bleaching industry, thus allowing work which had previously taken several months to be finished in a matter of days.

It was, moreover, not only the threat of unemployment which arrived with the machines. There was also the change in the tempo of life. If Britain was becoming 'the workshop of the world', its inhabitants sometimes had to pay a heavy price for the success. 'The steam-engine was quite tireless', D.S.L. Cardwell has pointed out in his *Steam Power in the Eighteenth Century: A Case study in the Application of Science*, 'it worked as hard on Saturday night as it did on Monday morning and the men, women and children who tended the machines which it drove were required to keep up the pace which it set. They could not take it easy for a day or two, if they felt like it, and make up the time a little later, as a self-employed man could, for the machine never felt like taking it easy. The pace, then, of life was – and still is – set by the machine.'

It was not only the pace of life but its environment which had been transformed by steam, as Charles Dickens pointed out in his novel, *Hard Times*, published in 1854, when describing the new industrial centre of Coketown. 'It was a town of machinery and tall chimneys out of which interminable serpents of smoke trailed themselves for ever and ever and never got uncoiled', he wrote. 'It had a black canal in it and a river that ran purple with evil smelling dye, and vast piles of buildings full of windows where there was a rattling and a trembling all day long, and where the piston of the steam engine worked monotonously up and down, like the head of an elephant in a state of melancholy madness.'

Dickens' imaginary town typified the scores of new towns which sprang up wherever opportunity offered once industry ceased to be tied to the charcoal-producing forested areas or places where water power was available. These factories were filled with machines operated by steam, water or other mechanical power, – machines which increased in complexity as power was used for more and more tasks that had previously been done by hand. The machines were made by the machine tool builders, a new race of men whose work increased in importance as metal took over from wood and as the rough and ready accuracy adequate for previous generations became useless in an age of pistons, cylinders, and wheels which revolved in bearings at speeds which would have been unthinkable a few decades earlier.

The transformation took place in less than a century. Boulton was overjoyed when in 1776 he found that John Wilkinson was able to bore him a 50 inch (127 centimetre) cylinder that 'doth not err the thickness of an old shilling in no part'; eighty years later Joseph Whitworth, responsible for standardizing screw threads, was using in his workshop a machine capable of measuring to within one millionth of an inch. The vast increase in accuracy which was made possible over the years by the machine tool engineers not only made steam engines steadily more efficient, so that they produced more power from the same amount of fuel, but also reduced maintenance, breakdowns, and similar trials and tribulations which colliery and factory owners were apt to set against the advantages of the new source of power. Quite as important was the fact that metal parts made to the same standards in the same factory could now be interchangeable. This applied not only to major items such as pistons and cylinders for steam engines but to the humble screw, necessary in quantity as machines increased in numbers and variety. Until almost the end of the eighteenth century every screw had to be finished by hand. The result was an imperfect article, and every nut had to be made for its own particular screw if trouble were to be avoided.

Machine tools of a limited nature had been made both on the Continent and in Britain in the mid-eighteenth century for delicate and specialized tasks like watchmaking, but Joseph Bramah, who became famous for his design of the 'unpickable' lock and of the water closet, was the first engineer whose work was seriously to affect the manufacture of industrial tools and equipment.

Bramah, sometimes considered the most versatile of the late eighteenth-century engineers and a man who in the words of his biographer was 'granted no less than eighteen patents for such widely different inventions as water closets and locks, fountain pens and fire engines, printing machines and hydraulic presses, carriage brakes and suspension', was to become almost as famous for the men he hired and trained as for his own work. In 1789 he employed Henry Maudslay, a young man who had been employed in Woolwich Arsenal as a powder boy filling cartridges. Maudslay turned out to be a natural-born engineer, and is thought to have played a part in more than one of Bramah's achievements. Before he left Bramah in 1797 to found his own business, Maudslay had with his employer perfected the slide rest, sometimes called 'one of the great inventions of history'. This was a device which eliminated the need for a lathe operator to hold metal-cutting tools in his hands while operating the lathe. Until its invention an operator had turned his lathe by means of a pole or treadle, and had held the cutting tool up to the metal gripped by the lathe. Maudslay clamped the tool in a rest which could move along, or across, the bed of the lathe, the result being greater accuracy and more speedy operation.

Soon after he left Bramah, Maudslay produced his screw-cutting lathe which, like the slide rest, enabled products to be turned out with accuracy of a totally new order and put an end to the necessity of fitting an individual nut to its

A bench lathe, 1810, made by Henry Maudslay who, at the end of the eighteenth century, began to produce screw-cutting lathes of a new order of accuracy.

THE COMING OF STEAM

individual screw. Similar lathes had been produced outside England but Maudslay's was simple to operate and his continual advocacy for ever greater accuracy encouraged its use. Thus for bench work he devised a screw micrometer accurate to one ten-thousandth of an inch. He called it 'The Lord Chancellor' – the Court of Final Appeal.

Maudslay was a firm supporter of the scientific approach which was to become increasingly evident in the engineering world as the nineteenth century progressed, but he had also a down-to-earth attitude which endeared him to other engineers. 'Keep a sharp look-out upon your material,' he would exhort them; 'get rid of every pound of material you can do without; put yourself the question "What business has it to be there?" avoid complexities, and make everything as simple as possible.'

Maudslay not only produced the lathes, the planing, drilling, and milling machines which themselves produced the equipment to fill the burgeoning factories of the Midlands and the North during the first half of the nineteenth century, but trained many of those who carried out his work. His personal assistant in old age was James Nasmyth whose steam hammer was to revolutionize much engineering in the 1850s. Richard Roberts, best known for his self-acting spinning mule, but also the man who was to solve Robert Stephenson's problem of making rivet holes in his famous Britannia Bridge across the Menai Straits, had worked for Maudslay. So had Joseph Whitworth, on whose efforts and imagination so much of mid-nineteenth century engineering depended.

Whitworth, whose dedication to engineering accuracy had been consolidated by his years with Maudslay, was responsible for an innovation whose effects were crucial to engineering success throughout the later decades of the nineteenth century. This was the standardization of screw threads. Until Whitworth's time, screws had been made by engineers virtually according to individual choice and taste, a freedom which led to chaos when machines, or parts of machines, from different makers were being assembled together. Whitworth made a large collection of the screws then in use and analyzed their numbers of threads per inch, the amplitude of the threads and the shape of the threads. His results were described in a famous paper *On a Uniform System of Screw Threads*, published in 1841. His proposals for standardization to produce maximum efficiency were quickly adopted and 'Whitworth threads' were soon being used throughout British engineering, a practice which continued until 1905 when the British Standard Fine Thread was introduced; even so the Whitworth thread continued to be used for another forty-four years.

The engineering revolution which took place during the final decades of the eighteenth century and the first of the nineteenth was accompanied by two other innovations which helped to shape the future in both Europe and the United States. One was the development by Joseph Bramah into practical equipment, of the principle behind the hydraulic press; the other, a direct outcome of encouragement by Boulton & Watt, was the introduction of gas lighting, an event which lifted the world from the rush and oil-lamp era which had changed little since the dawn of history.

In the middle of the sixteenth century Simon Stevinus, a Belgian-Dutch mathematician who introduced decimal fractions, showed that the pressure of a liquid on a surface depends only on the area of the surface and the height of the liquid above that surface. In the following century Blaise Pascal, the French mathematician who invented a calculator that was an ancestor of the nineteenth-century cash register, not only verified this experimentally but pointed out that 'all these examples show that a fine thread of water can balance a heavy weight.'

Neither Stevinus nor Pascal investigated the industrial implications. But in the 1790s Bramah realized that use could be made of the fact that if a small piston was pushed down into a closed container, then a larger piston in the same container would be pushed up. The smaller piston would have to move down for a greater distance than the large one was pushed up – the relationship being the same as that of their cross-sections – but it would still mean that a small pressure could sustain a heavy weight.

Bramah turned the principle to good use in a variety of ways – in the hydraulic press which he introduced to the woollen industry, and in cranes which he recommended for use in the rapidly expanding docks of London and Dublin. 'I have also now supplied it [the hydraulic principle] with the most surprising effect to every sort of crane for raising and lowering goods in and out of warehouses', he wrote in 1802, according to Ian McNeil in *Joseph Bramah: A Century of Invention, 1785-1851*. 'So complete is the device, that I will engage to erect a steam engine in any part of Dublin, and from it convey motion and power to all the cranes on the keys and elsewhere, by which goods of any weight may be raised at one third the usual cost. This I do by the simple communication of a pipe, just the same as I should do to supply such premises with water.'

It was, in fact, to be some years before hydraulic cranes became widespread, but by the middle of the nineteenth century hydraulic power, based on Bramah's demonstration that its use was a practical proposition, was in operation on dock gates and lock gates, swing bridges and lifts in tall buildings.

In contrast to the use of hydraulic power, which only slowly permeated engineering and whose influence was mainly limited to professional activity, there was the introduction of gas lighting which transformed not only factory life but domestic life and the day-to-day running of Britain's cities. Until the end of the eighteenth century, life after dark went on by courtesy of candles, rush lights and crude oil lamps. Fires were endemic; so much so that it was rarely thought wise to operate factories between dusk and dawn. Travel after dark was invariably dangerous while such occupations as reading or sewing were strictly limited. The coming of gas light, therefore, altered the quality of life in a myriad ways.

J.B. Helmont, a Flemish chemist, had at the end of the sixteenth century investigated the gas produced by burning coal, but his researches were not followed up, possibly due to his preoccupation with mysticism and such subjects as 'the philosopher's stone'. Only in the late eighteenth century were a number of tentative efforts made to utilize the flame which burned when coal gas was lit. With one exception, these were still only patchily explained and it was left to Phillipe Lebon, an engineer from the Service des Ponts et Chaussées, to light and partially heat his house in Paris with gas during the first years of the nineteenth century. Lebon died in 1804. Once again, the possibilities of gas lighting might have died at birth.

By this time, however, something more promising had been started by William Murdock, Boulton & Watt's chief engine-erector in Cornwall. It is symbolic of the inter-relationship of industrial and engineering ideas that Murdock was led on to the miracle-making gaslight by his efforts to find a substance for preserving ships' hulls. Tar made by the distillation of coal was one

A Hesketh steam winding engine at Chatterley Whitfield Colliery, Tunstall.

material with which he experimented. It was a short step for him to start lighting the rooms of his home in Redruth, Cornwall, with gas produced from coal. Britain then had what must have appeared to be limitless coal, and her mill and factory owners were eager to light their new buildings which were effectively closed during the hours of darkness.

In 1798 Murdock returned to Boulton's Soho Works as manager. He at once set about lighting the entire works by gas, generating it by burning coal to red-heat in cast-iron retorts, and leading it to various points in the factory where holes allowed the gas to escape and be lit. When the Peace of Amiens was celebrated in 1802, Murdock lit the front of the Soho factory with gas flares, an event which created great interest throughout the Midlands. Two years later Boulton & Watt were canvassing for orders. The first came from Phillips & Lee, the cotton spinners of Salford, and here, in the mill which Boulton & Watt had built for them in 1801, there was installed a battery of six retorts which provided the gas for 900 burners which lit the mill, a private road and a private house.

Murdock continued to make improvements to the gas-generating apparatus which removed some of its impurities as well as the worst of its smell. However, leadership in gas lighting soon passed – possibly due to the death of Boulton in 1809 – to Samuel Clegg, a former Boulton employee who set up in business on his own. Clegg soon pulled off a major coup by installing eighty gas lights in the London home and premises of Rudolph Ackermann, the London art publisher. His main importance in the story of gas lighting, however, was his co-operation with a German immigrant F.A. Winzer – later changed to Winsor – who inaugurated a public gas supply from a central generating station. The station was used to light Pall Mall and while plagued with early difficulties led on to the London Gas Light & Coke Company in 1812 and the spread of gas lighting throughout London and eventually elsewhere. The parish of St Margaret's, Westminster was lit by gas in 1814. London had 26 miles (42 kilometres) of gas mains by the end of 1816, 122 miles (196 kilometres) by 1823 and 600 miles (965 kilometres) by 1834.

Progress in engineering as the nineteenth century got under way – the gas-lit streets, the steam engines not only pumping coal mines dry but turning the wheels in factories which turned out ever more variegated articles – was not confined to Britain. A whole series of circumstances, among them the huge supply of native coal which initially encouraged the steam engine and the genius of Watt which made steam power practicable, had given Britain a head start in the new industrial age. But on the continent of Europe similar advances were soon being made while across the Atlantic a new nation was flexing its muscular independence. There the great expanses of territory, unmapped and unexplored yet ripe for development, offered opportunities different but quite as great as those of Europe. As it has been written of steam engines in the United States, 'To a new nation they brought locomotives crashing up the Sierra canyons above Sacramento, steam powered ore crushers thundering in Virginia City, factory whistles shrieking over Gastonia, train whistles hooting down the Hudson River valley, side wheelers blowing for their Mississippi landings – all sounds of manifest destiny, of a bold young America that never could have happened without the machine called the steam engine and its millions of good peaceable horses.'

ENGINEERING ARTIFICIAL WATERWAYS

With the expansion of coal mining, helped by steam-driven pumping machines, industry in Britain tended to spring up as near as possible to the coalfields. Yet coal in continually greater quantities had to be moved about the country, while its export became more important throughout the eighteenth century. Britain's road network was poor, the roads themselves being covered in dust-clouds during the summer months and deep in mud during the winter. There was little experience of moving bulk loads by carts, while a packhorse would carry only an eighth of a ton. On a soft road a horse might be able to pull ⅝ of a ton. But if the load were carried by a barge on a waterway, then up to 30 tons could be drawn by the same horse. Yet even the most useable of the country's rivers tended to flood disastrously during the winter; during the summer they might in places be seriously short of water. The result could have been forecast. The demand created the opportunity and during the last half of the eighteenth century and the first decades of the nineteenth canals proliferated, linking the coalfields with the main ports, crossing Britain north and south, east and west, and providing a transport network that was not to be equalled until the growth of the rail and road systems.

The movement of coal to the ports by canal was to have a significant effect on Britain's economic health, but this was only among the more obvious of results. There were to be wide social repercussions of which Thomas Pennant has given an embracing account. 'The cottage', he wrote after one of his famous tours of Britain in the 1780s, 'instead of being half-covered with miserable thatch, is now covered with a substantial covering of tile or slates, brought from the distant hills of Wales or Cumberland. The fields, which before were barren, are now drained, and, by the assistance of manure, conveyed on the canal toll-free, are cloathed with beautiful verdure. Places which rarely knew the use of coal, are plentifully supplied with that essential article, upon reasonable terms; and, what is of still greater public utility, the monopolizers of corn are prevented from exercising their infamous trade; for, the communication being opened between Liverpool, Bristol and Hull, and the line of canal being through countries abundant in grain, it affords a conveyance of corn unknown to past ages. At present, nothing but a general dearth can create a scarcity in any part adjacent to this extensive work.'

If the canals were thus a civilizing influence throughout the country, they had another effect that can easily be overlooked. For the first time in Britain – at least since prehistoric times when slaves were marshalled in some numbers – it was necessary to gather and organize large bodies of civilians. They were the 'navigators' or 'navvies' who dug and embanked the canals, and their organization paved the way both for the large labouring gangs that a few decades later were needed to build the railways, and for the armies of men and women who were to work in William Blake's 'dark Satanic mills' of the Industrial Revolution.

Canals had been built since ancient times. In 910 BC Darius I had dug the 37 mile (59 kilometre) Grand Canal linking the Nile and the Red Sea. Some 40 feet (12 metres) deep and 100 feet (30 metres) wide – giving it roughly the same cross-section as the *Queen Elizabeth* – the Grand Canal was a major achievement and the most formidable of the scores of smaller canals which later appeared in Upper and Lower Egypt. The Chinese were building canals at least from the third century BC and in 215 BC Ling Ch'u is recorded as having constructed a 90 mile (145 kilometre) canal from the Han capital of Ch'ang-an to the Yellow River. Others followed and in AD 70 the first sections were cut of the 600 mile (965 kilometre) Grand Canal of China, its route followed also by an Imperial Highway, which was eventually carrying two million tons of goods annually. The Chinese were also responsible for at least two canal 'firsts'. One came in AD 984 when a part of the Grand Canal was rebuilt and a double slipway over which barges were hauled was replaced by gates only 250 feet (76 metres) apart. This was in practice the first canal lock. Earlier there had been single gates which were used to regulate water levels but these took a long time to operate and usually resulted in the loss of large amounts of water. The lock, which was to become such a familiar feature of waterways during the canal age of the eighteenth century, eliminated both disadvantages.

The second Chinese record was made in 1283 when a 700 mile (1,126 kilometre) northern branch of the Grand Canal from Huaian to Peking was taken across a watershed in the Shanting hills. As Europeans were to discover five hundred years later, crossing a watershed presents numerous problems, notably that of providing water on the summit stretch to make up for that lost in the locks that are necessary either side of it. The Chinese, like most Europeans later, solved the problem by diverting minor rivers into the upper stretches of the canal.

The Romans dug canals as part of their water supply systems in Italy, and in Britain cut the Fossdyke from Lincoln to the River Trent. In 1564 Exeter Corporation in South Devon built a 1½ mile (2.4 kilometre) canal to by-pass a portion of the River Exe which had become unnavigable, and this was followed by a number of schemes which were river improvements rather than canals in the conventional sense.

More like the genuine canal was the New River, an artificial watercourse nearly 40 miles (64 kilometres) long which was excavated in the early years of the seventeenth century to bring water to London from springs at Chadwell and Amwell in Hertfordshire. Until the last years of the sixteenth century, London had been provided with water from wells which was distributed by conduits to various parts of the city. As the city grew, supplies became less adequate and in 1582 Peter Morice, a Dutchman, installed a pumping engine in one of the arches of old London Bridge which had been built more than three and a half centuries earlier. The engine was worked by water wheels driven by the force of the Thames tides which rushed with great velocity through narrow openings in the bridge. The water was then distributed, partly by hollowed-out wooden logs, partly by lead conduits, to houses which were not too far from the river.

With the continued growth of London, Morice's pump, and a second one which he soon installed, failed to satisfy the demand, and towards the end of Queen Elizabeth's reign the Corporation of the City obtained an act empowering them to cut a watercourse from any part of Middlesex or Hertfordshire where a number of springs were known to exist. Sufficient money was for long lacking, and it was not until 1609 that the Corporation took up the offer of Hugh Myddelton, a London goldsmith, to build and finance such a scheme himself. As was to happen during the canal age a century and more later, there were objections that the artificial river would flood good farming land, that the Church would be robbed of its tithes, and that both water rights and the landowner's privacy would be lost.

The objections were overcome and Myddelton's New River began to be cut. As many as six hundred men were employed at the same time on the task of digging the 10 foot (3 metre) wide trench into which the spring waters were eventually diverted. Myddelton's aim was to plan the watercourse so that it followed the contours of the land as far as possible. This meant that the line was by no means straight, and although the springs were only some 20 miles (32 kilometres) from London, the New River was almost 40 miles (64 kilometres) long. At places timber troughs lined with lead were constructed to carry the water over roads, and about 160 over-bridges or under-bridges were built. The average drop from the springs in Hertfordshire to New River Head in London was 2 inches (5 centimetres) per mile, but sometimes more and where the fall would have been too rapid, 3 foot (0.914 metre) perpendicular stop-gates were erected.

Construction continued without major trouble, and on Michaelmas Day, 1613, an opening ceremony was organized at New River Head in the presence of the Lord Mayor, the Aldermen and the Common Council, and hundreds of spectators. Sixty workmen carrying spades and other tools marched round the reservoir to the sound of drums and trumpets, and a ceremonial speech was read: 'At the conclusion of the recitation, the flood-gates were thrown open and the stream of pure water rushed into the cistern amidst loud huzzas, the firing of mortars, the pealing of bells, and the triumphant welcome of drums and trumpets.' London's new supply of water was to be distributed by no less than 400 miles (640 kilometres) of pipes, at first of elm but later replaced with cast iron.

Little of the experience from these earlier times appears to have been used by Britain's first canal builders. What they did have, however, was a good knowledge of water control, built up during the centuries when watermills were a main source of power. Also, the Duke of Bridgewater, so-called 'father of inland navigation', had during his Grand Tour of the Continent learned much when inspecting the Languedoc Canal linking the Mediterranean to the Atlantic.

In the 1720s Liverpool merchants had financed a number of short canals which cut the cost of coal transport by linking the rivers Mersey, Ribble and Irwell. They were followed in 1757 by what is sometimes described as England's first real canal, linking the Mersey, and thus the port of Liverpool, with the Lancashire coalfields. But it is with the Bridgewater Canal that the canal age was really launched, inspired by the Duke's impressions of the Languedoc Canal, today called the Canal du Midi. This was, in fact, the second of two great achievements which in the seventeenth century made France the most successful of European canal builders. The first was the Briare Canal linking the Loire and the Seine, planned in 1603 and opened in 1642 after a series of delays caused by changes of management after the assassination of Henry IV, the canal's royal patron, in 1610. Following the River Trezee from Briare, the canal rose 128 feet (39 metres) to the watershed then dropped 266 feet (81 metres) to the Loing at Montargis whence barges continued north to the Seine. After a second canal had been dug from the Loire at Orléans to Montargis, traffic on the Loing became so great that the original Briare Canal was extended to the Seine.

Fine an engineering achievement as this was, it was overshadowed at the end of the century by the splendid Canal du Midi, initially the brainchild of King Francis I who, in 1516, had discussed possible routes with Leonardo da Vinci. Nothing came of these early proposals, and a century and a half was to pass before Louis XIV ordered the start of the work under the engineer Pierre-Paul Riquet and the statesman Jean Baptiste Colbert.

The canal which so impressed the Duke of Bridgewater ran 180 miles (290 kilometres) from Sete on the Mediterranean to the city of Toulouse where it linked up with the River Garonne flowing north to Bordeaux and the Atlantic. The Canal du Midi was a complicated engineering enterprise. Many locks were needed on the ascent from the Mediterranean to Toulouse, including a staircase of eight near Beziers. There were numerous aqueducts carrying the canal across rivers and valleys, and a 180 yard (165 metre) tunnel. Riquet had, moreover, dealt dramatically with the problem of providing an adequate water supply where the canal crossed the watershed between the Mediterranean and the Atlantic. At lower levels, water loss in canals due to seepage or the operation of locks can usually be made good by taking in water from streams and rivers. At higher levels this can be a difficulty. Riquet dealt with it by digging a 27 mile (43 kilometre) feeder canal which brought water from the River Sor at the Montagne Noire and a shorter feeder bringing water to a reservoir specially built for the purpose.

The Canal du Midi was a success from its earliest days and the Duke of Bridgewater was not slow to grasp how something similar could help distribute the coal from his collieries at Worsley, north of Manchester. In 1759 he hired James Brindley, an engineer who had already shown his ingenuity in the use of steam engines for draining coal mines, to undertake the work for him. Between Worsley and Manchester there runs the River Irwell and the Duke's first plan was to take his canal down to the level of the Irwell on the north side in a flight of locks, and to build a second flight on the south side to take it up on to higher ground again. Brindley proposed a far more audacious solution. Why not, he asked, build an embankment on the north side of the Irwell to take the canal, then carry it across the river on an aqueduct? The idea of carrying barges on a channel of water high above the Irwell was considered impractical by many, but the Duke supported Brindley and on 17 July 1761 the first barge passed over the viaduct at Barton. The viaduct was about 200 yards (183 metres) long, 12 yards (11 metres) wide, and consisted of three central semicircular arches carrying the canal 39 feet (12 metres) above the river. 'The effect of coming at once on to Barton Bridge, and looking down upon a large river, with barges of great burthen sailing on it; and up to another river hung in the air, with barges towing along it, form altogether a scenery somewhat like enchantment,' Arthur Young, the late eighteenth-century agriculturalist and author, was to write, 'and exhibit at once a view that must give you an idea of prodigious labour.'

At both ends of the canal Brindley incorporated ingenious engineering constructions. At Worsley he drove the canal into the mines themselves, so that coal could be loaded directly on to the barges; at first there was only 2 miles (3 kilometres) of underground canal, then extensions brought the total underground network to more than 40 miles (64 kilometres) as the Duke enlarged his mines. At Manchester it was first intended that the canal should end at the foot of Castle Hill. However, Brindley drove into the hill, then sank a vertical shaft from the ground to the canal terminus. 'The barges', said Brindley's biographer, Samuel

Smiles, 'having made their way to the foot of this shaft, the boxes of coal were hoisted to the surface by a crane, worked by a box waterwheel of 30 foot diameter and 4 foot 4 inch wide driven by the waterwheel of the river Medlock. In this contrivance Brindley was only adopting a modification of the losing and gaining bucket, moved on a vertical pillar, which he had before successfully employed in drawing water out of coal-mines.'

At Worsley, Brindley built a number of other devices, including a steam engine which drained those galleries of the mine below canal level. At one mine entrance he erected a water bellows which forced fresh air into the workings; and where the canal entered the workings he built an overshot mill of a new kind driven by a 24 foot (7 metre) diameter wheel which operated three pairs of stones for grinding corn, a dressing or boulting mill and a machine for sifting sand and mixing mortar.

Brindley's canal, the first major construction of its kind in Britain, halved the cost of bringing coal from the Worsley mines to Manchester, and by the time that the first barge passed over the Barton Viaduct Brindley was already planning the Duke's next project. It was to be an extension of the canal from Manchester to the Mersey at Runcorn, which would give access for barges from Manchester to the open sea at Liverpool. This time objections were raised not only by landowners whose property would be crossed – a frequent cause of complaint which made it necessary for a Canal Bill, granting powers of compulsory purchase, to be passed by Parliament before canal work could begin – but by others who feared that a breach of the canal embankments would cause disastrous flooding. Brindley guarded against the latter possibility by dividing the canal into separate sectors with floodgates which would automatically be brought into operation if a breach in the banks occurred.

With the opening of the Manchester-Runcorn canal the economic benefits of canals became indisputable and Brindley was soon afterwards planning the details of an ambitious scheme thought up by the Duke and Josiah Wedgwood the potter. This was nothing less than a canal almost 140 miles (225 kilometres) long which would link the Mersey at Runcorn with the River Trent, and both with the River Severn, thus creating a water-borne connection between the ports of Liverpool, Bristol and Hull. Furthermore, it would provide the right sort of transport for the fragile goods of the Potteries whose output was fast beginning to expand.

The Grand Trunk Canal, as it was called, cut across a major watershed and its problems can be judged from the fact that it required thirty-five locks to take it up to the watershed on the north and forty more to take it down on the south side. It crossed the River Dove by a twenty-three arched aqueduct. It crossed the River Trent at four places and it needed about 160 minor aqueducts, 109 road-bridges and five tunnels including the 2,880 yard (2,633 metre) Harecastle Tunnel, which was rightly counted as one of the great engineering achievements of the age. 'Gentlemen come to view our eighth wonder of the world, the subterraneous navigation, which is cutting by the great Mr. Brindley, who handles the rocks as easily as you would plum-pies, and makes the four elements subservient to his will,' a local writer said of the Harecastle Tunnel in 1767. 'He is as plain a looking man as one of the boors of the Peak, or as one of his own carters; but when he speaks, all ears listen, and every mind is filled with wonder at the things he pronounces to be practicable. He has cut a mile through bogs, which he binds up, embanking them with stones which he gets out of other parts of the navigation, besides about a quarter of a mile onto the hill Yelden, on the side of which he has a pump worked by water, and a stove, the fire of which sucks through a pipe the damps that would annoy the men who are cutting towards the centre of the hill. The clay he cuts out serves for bricks to arch the subterraneous part, which we heartily wish to see finished to Wilden Ferry, when we shall be able to send Coals and Pots to London, and to different parts of the globe.'

The great Harecastle Tunnel was not completed until 1777 but soon afterwards the Grand Trunk Canal was opened and within a few years was more than justifying the expectations of the Duke and Wedgwood. 'From a half-savage, thinly-peopled district of some 7,000 persons in 1760, partially employed and ill-remunerated,' Samuel Smiles wrote of the Potteries, in *Lives of the Engineers*, 'we find them increased, in the course of some 25 years, to about treble the population, abundantly employed, prosperous and comfortable.'

The Grand Trunk Canal, linking the Trent and the Mersey, was the first leg of Brindley's 'Cross', the second being provided by the Staffordshire and Worcestershire Canal which ran to the Severn at Stourport and the Coventry and Oxford Canal which gave a link with the Thames and whose tortuous 11 mile (18 kilometre) summit level took its barges a distance of only 4½ miles (7 kilometres) in a straight line. These connections between the four main rivers of southern England were followed by a network of shorter waterways in the Midlands which brought about a demand for something more direct than the Coventry-to-Oxford link with the Thames. The answer was the Grand Junction Canal which ran from the Midlands to the Thames at Brentford on the outskirts of London; off-shoots led to the River Nene at Northampton and to the Trent via the River Soar.

A lithograph by Thomas Shepherd of the Regent's Canal, London.

Brindley did not live to see the completion of the 'Cross', but he had in Thomas Telford a worthy successor. Telford came just those few years later that were needed if an engineer were to utilize the benefits of cast iron. Whereas Brindley often relied on brick culverts, Telford used iron – notably in his 1,000 foot (305 metre) long aqueduct which carried his Ellesmere Canal across the River Dee. He also benefited from the experience of his predecessors and when traffic on the Trent and Mersey Canal demanded something bigger than Brindley's Harecastle Tunnel, Telford built a large one alongside it in less than three years compared with the eleven years that the first tunnel had taken.

In the north, Scotland was to benefit from the Forth and Clyde Canal, which bisected the country at its narrowest, and by two canals built partly for non-commercial reasons. One was the 9 mile (14 kilometre) Crinan Canal which, with the aid of fifteen locks, cut through the Mull of Kintyre from Crinan to Ardrishaig and was dug to help develop western Scotland; the second was the Caledonian Canal which ran from Fort William on the west coast to Inverness on the east and enabled warships to avoid the long and dangerous voyage round the north of Scotland. Both, however, were also used for commerce.

Many of the minor canals which are no more than a footnote to the history of the era were yet of much importance to local inhabitants, as is indicated by one account of the opening of the Bude Canal in Cornwall. 'The Committee of Management,' it says, 'supported by the neighbouring gentry, on the arrival of the loaded boats at the point of debarkation, marched into the town of Holsworthy in procession, the band playing "See the Conquering Hero Comes", and hailed by the acclamations of the populace . . . the dinner, provided at the Stanhope Arms, was composed of the choicest viands; and the hilarity, happiness and unanimity of all present, were most auspicious.'

Although different in detail, most canals presented certain similar problems to the engineers. Most of them were built before the foundation of the Ordnance Survey and while in many cases local maps were available they could rarely be relied upon; and there was in existence nothing comparable to current topographical maps. The line of the canal, therefore, had to be surveyed from scratch. In order to keep down the cost, difficult engineering works were avoided wherever possible. Tunnels, embankments and aqueducts all increased the expense and the ideal was for a canal route to contour round hillsides rather than cut through them. A further complication was that the Act of Parliament necessary before work on a canal could begin, invariably laid down that it should not pass nearer than a stipulated distance from certain villages or houses. Above all, the engineer had to ensure that there would be enough water to keep his canal full.

Once the line was decided upon, stone or bricks had to be acquired for the inevitable bridges and sometimes for tunnels or aqueducts if they were to be lined. Although steam-powered excavators were eventually used, all the early canals were dug by the hundreds-strong teams of navigation cutters, navigators or navvies who moved across country as the canal progressed. A towpath had to be built beside the canal for the horses who would eventually tow the boats, and part of the canal engineer's art lay in so plotting the route that the excavated soil could be used to build up the towpath embankments.

Where a canal passed through clay or similarly watertight soil little further work was called for. Elsewhere, a lining of puddle was necessary, puddle being, according to Abraham Rees in *The Cyclopaedia, or Universal Dictionary of Arts,*

Sciences and Literature, 'a mass of earth reduced to a semi-fluid state by working and chopping it about with a spade, while water just in the proper quantity is applied, until the mass is rendered homogeneous, and so much condensed, that water cannot afterwards pass through it, or but very slowly. The best puddling-stuff is rather a lightish loam, with a mixture of coarse sand or fine gravel in it; very strong clay is unfit for it, on account of the great quantity of water which it will hold, and its disposition to shrink and crack as this escapes.'

However well the canal banks were made, they were always in danger of damage from vessels that travelled too fast or otherwise set up unusual currents. Thus when the canal connecting Edinburgh to the Forth and Clyde Canal was completed, it was banned to the new paddle steamers then being tentatively introduced. James Nasmyth, the English engineer, proposed, instead, that a chain should be laid along the bottom of the canal. It would enter the canal vessel at the bows, pass over a notched pulley, and leave the vessel at the stern. A small steam engine on the barge would rotate the pulley and thus pull the vessel through the water. But so great was the prejudice against steam that even this system was banned – even though it was used on many continental canals. Years later Nasmyth watched it in efficient operation for pulling vessels across the River Hamoaze outside Devonport where he was enlarging the naval dockyard.

The stone or brick aqueducts presented their own problems but these became less important after 1793 when a cast-iron trough supported by nineteen arches on 126 foot (38 metre) masonry piers was used successfully to carry the Ellesmere Canal over the River Dee at Pont Cysyllte. Other aqueducts lined with cast iron soon followed.

Locks are used when it is necessary for a canal to continue at a different level. The history of the lock is shrouded in as much mystery as surrounds the start of most technological innovations. Few reliable details of how the Chinese worked their locks are available. Credit is given by the engineering historian Cresy to Pietro and Dionysius Domenico of Viterbo for building a lock in northern Italy in 1481. Another is reputed to have been built in 1488 on the River Brenta near Padua, and in 1497 Leonardo da Vinci supervised the building of six in a scheme which united Milan's two existing canals. These six locks were operated by what were known as mitre gates which closed or opened horizontally and were thus different from the first Italian locks. These involved the raising or lowering of a solid wooden portcullis which stopped the canal waters or let them through. Leonardo's sketchbook drawings of the San Marco lock are the oldest that are known of what is basically the model for contemporary canal locks. However, there is a strong tradition that Dutch engineers, building their canals a century and more earlier, had used the lock in a number of places.

The truth is probably that the ingenious Italians of the Renaissance made their locks more watertight and, therefore, more efficient than their predecessors. The principle of operation is simple enough, since the lock is merely a watertight chamber, closed by gates at each end and situated between two stretches of a river or canal which are at different levels. By opening a sluice closed against the upper water level, the lock is filled to this level, the gate is opened and into the lock there move vessels on their way down. The gate is then closed, a sluice at the other end is opened, and the water in the lock drops to the lower water level on to which any vessels in the lock can move out after the lower gates are opened. The process is reversed if vessels are to be moved from lower to higher level. If the difference in height is considerable a series of locks can be placed end to end, forming what is called a flight of locks, among the most famous being the engineer John Rennie's flight of twenty-nine locks near Devizes on the Kennet and Avon canal.

Before the canal age, locks had been used for many years on river navigations and William Vallans in 1577 in his *Tale of Two Swannes*, one of the earliest examples of blank verse outside the drama, describes one on the River Lea at Waltham Abbey on the outskirts of London. Of those who passed it, he said, '. . . a rare device they see, / But newly-made, a waterworke; the locke / Through which the boates of Ware doe passe with malt. / This locke contains two double dores of wood, / Within the same a cesterne all of Plancke, / Which onely fils when boates come there to passe / By opening of these mightie dores.'

When a flight of locks was impossible, or would have been too costly or too slow in operation, one of two other engineering devices could be used. One was the incline, the second the lift. Primitive inclines, in which barges were hauled up a paved incline by rope and capstan, were used in ancient China, but the first in Great Britain were built in the late 1760s on the Tyrone Canal in Ireland. They were unsatisfactory and were followed in 1788, by an incline on the Ketley Canal in Shropshire which used counterbalancing cars or cradles into which the boats were manoeuvred, before being lifted 73 feet (22 metres). Similar inclines quickly followed on the neighbouring Shropshire Canal and others in the West Country.

Canal lifts consisted of two counterbalancing water-filled cast-iron troughs. They were among the later of the canal engineer's devices and could be extremely effective. A lift on the Grand Western Canal, erected in 1838, could raise an 8 ton boat 46 feet (14 metres) in three minutes. Hydraulic lifts eventually replaced the simple kind, the first in Britain being that at Alderton which in 1875 connected the River Weaver with the Trent and Mersey Canal more than 50 feet (15 metres) apart in height. Here, two troughs 75 feet by 15½ feet (23 metres by 4.7 metres), containing 5 feet (1.5 metres) of water and supported in the centre by a hydraulic ram were used. When 6 inches (15 centimetres) of water was drawn off from the lower trough it rose more than 45 feet (13.7 metres); then the force of the hydraulic ram took over and brought the trough to the upper level. A 100 ton barge could thus be raised in two and a half minutes compared with the hour needed to raise it to the same height in a series of locks. So successful was the Alderton lift that the consulting engineers who installed it, Clark and Sandeman, were asked to build a similar installation in Belgium for lifting 500-ton barges.

The height of 'canal mania' was reached in Britain during the last decade of the eighteenth century when canal mileage increased by nearly a third from 2,200 in 1790 to 3,000 in 1800 – roughly the same amount that was added during the next fifty years. The traditional canals, initially dependent on the movement of coal from the coalfields to the towns where it would be used, or the ports from which it would be exported, were superseded by the railways on which the first mobile steam engines ran in increasing numbers from the 1820s onwards.

In the United States, where canals were also to be displaced by railways, the story was rather different for two reasons. There, the canal system began to be built much later and was not fully established when the railways began to run; secondly, the financial and geographical backgrounds were in strong contrast to those in Britain.

Although Cadwallader Colden, Surveyor of the Province of New York, proposed as early as 1724 that a canal should be dug between the Hudson River at

Albany and Lake Erie, the project had to wait for a century. And although Colden's friend, Benjamin Franklin, wrote to the Mayor of Philadelphia in 1772 saying, 'Rivers are ungovernable things, especially in Hilly countries, Canals are quiet and very manageable,' it was another twenty-eight years before America's first canal was opened. This was the 22 mile Santee Canal linking the Santee River with the headwaters of the Cooper River and thus with Charleston Harbour. Built by Christopher Senf who had served with Hessian troops of the British Army which surrendered when General Burgoyne laid down his arms at Saratoga in 1777, the Santee Canal required twelve locks and eight masonry aqueducts. Successful though it was, another seventeen years were to pass before De Witt Clinton, Mayor of New York, authorized the Erie Canal which was to demonstrate the immense transformation in trade and social conditions which a single canal could bring about.

One reason for America's somewhat reluctant start in canal building was the lack of rich financiers. Such projects were even more speculative in the United States than they were in Europe, and more than one American engineer lamented the fact that America had no 'spirited noblemen' to finance their schemes. As justification of their complaints is the fact that up to the outbreak of the Civil War in 1861 more than half of the entire canal investment came from the public works programmes of New York, Pennsylvania and Ohio. There was some reason for private caution since conditions were so much less promising than in Britain. There the bulk of canals was within a 180 mile (290 kilometre) sided square based on Liverpool, Hull, London and Bristol, and containing large areas of population. In the United States far greater distances had to be traversed, high mountain ranges had to be crossed or circumvented, and the population was more widely dispersed. The only counterbalancing factor was that the Americans had the advantage of extensive slave labour.

The dramatic changes that canals could bring about would not have been realized even when they were but for Clinton's determination in pushing through the building of the Erie Canal. Eventually started at Albany in 1817, it was dug for the 363 miles (584 kilometres) to Buffalo by 1824. Half of the route lay through forests and swamp, and eighty-two locks were required to cope with a difference of 571 feet (174 metres) in height between the two ends. In the early stages of the work special cement was imported from England. Then Benjamin Wright, senior engineer of the Erie Canal, discovered in New York State a deposit of limestone from which the necessary quality of cement could be made, and imports ceased. The opening of the canal had immense repercussions since it provided a link between the Hudson Valley and the Great Lakes Basin and encouraged trade between the territories on either side of the Appalachians. Significantly, too, it helped make New York City, instead of Philadelphia, the most important port in the United States.

But the new water route served the double task of drastically reducing freight prices for materials being sent west and making the way easier for the thousands of immigrants thronging New York. It encouraged the building of other canals which within a few decades transformed the life of whole states. 'Canals were intended [in the United States]' Robert Payne has pointed out in his survey *The Canal Builders*, 'to bring remote country districts close to the ocean, but their main effect at the beginning was to bring about intercourse between the big cities and the interior. Canals introduced a new way of life, warm, exotic, audacious and uncomfortable. Showboats tied up at the wharves, and soon the songs of New York were being sung on the remote plains of Ohio, Indiana, and Michigan. Canal barges crowded with merchandise became huge floating emporiums, laden with the latest dresses, whalebone corsets, jimcrack jewelry, and agricultural implements. Captains of barges saluted each other with musical blasts from their horns while their wives hung out washing on clothes-lines and looked after the pigs and chickens on board.'

The Erie Canal was only two years old when Pennsylvania decided to tackle an even more ambitious scheme. This was nothing less than the building of a canal from Philadelphia to Pittsburgh, nearly 400 miles (640 kilometres) away, across country that included a mountain ridge whose lowest crossing point was at some 2,300 feet (701 metres). The justification for attempting such a scheme was given by Governor John Andrew Schultze in his Annual Message to the State in January 1826. 'There can be no doubt of the superiority of transportation by water,' he said. 'It brings the articles and produce so much nearer to market, that it gives a value to what would otherwise have rotted on the surface, or lain neglected in the bowels of the earth. It increases the value of his labor to the farmer, by lessening the charge of conveyance to market; and for the same reasons enables him to get his returns at a cheaper rate. It raises the price level, creates improvements, and by the consumption it occasions, and by the mills and manufactures erected, establishes a market at home; the best of all possible markets.'

With this in mind the Pennsylvanians began their enterprise. It took eight years to complete and when finished included a 172 mile (277 kilometre) stretch that required a hundred and eight locks, a second 104 mile (167 kilometre) stretch needing sixty-six locks, and a crossing of the Alleghanies in which the boats were drawn up inclined planes for 1,400 feet (427 metres) on one slope and let down 1,170 feet (357 metres) on the other, steam engines being used to operate the cables. The Pennsylvania Canal was never a financial success despite being something of an engineering wonder; nevertheless by 1840, 1,000 miles (1,600 kilometres) of canal had been dug in Pennsylvania.

There was also the Delaware-Hudson Canal which had one similarity with the first British canals: they had been built mainly to move pit coal while the Delaware-Hudson link was needed to move the huge anthracite deposits of north-east Pennsylvania to the Hudson and thence downstream to the coast. During the year following its opening in 1829 some 7,000 tons of anthracite were carried; by 1872 the annual movement had risen to 29 million tons. For the first few years of operation, barges were locked down into a still pool created on the Delaware by damming, then taken across the river by ferry rope before being locked up on the far side of the river. This process caused so much inconvenience to Delaware traffic that John August Roebling, later famous for his Brooklyn Bridge, built a suspension viaduct across the river, one of four with which he improved the canal between 1847 and 1850.

Short canals began to link towns along the eastern seaboard and then in 1848 there came the Illinois-Michigan canal which gave a water route between Lake Michigan and the Gulf of Mexico and was to be an important factor in the growth of Chicago. The peak of the American canal age was reached in the early 1850s by which time 4,500 miles (7,200 kilometres) had been built. Then, as had already happened in Great Britain, railways began to replace canals and within a few years useable mileage had been halved. From this point the story began to take a shape different from Britain's. There the decline of the canals continued, so that in the twentieth century they have become very largely holiday highways for those who enjoy the quiet of transport by narrow boat. In the United States, on the contrary, the initial setbacks were followed by an expansion of useable mileage to 22,000 (35,400 kilometres).

In the United States the rise of the railways was parallelled by the rise of the steamboat, pioneered by such men as Fulton, so that the main routes for inland water transport switched from canals to improved rivers. Towards the end of the nineteenth century the railways became predominant, the eclipse of the steamboat being accelerated by the Civil War. Nevertheless, as late as World War I the US Federal Government used money in an attempt to revitalize America's inland waterways industry. In the twentieth century, however, little genuine canal transport has survived. The vast majority of the 22,000 miles (35,400 kilometres) forming the present system is made up of the Mississippi system, the sea-level 'intra-coastal' waterways, short river navigations and those in the Great Lakes region. There is almost no direct comparison with the genuine canals of Europe which link navigable rivers.

Yet if the conventional canal was in some countries to be overshadowed by the railways from the second half of the nineteenth century onwards, there was another and more adventurous kind of canal which soon came into existence. In Europe and America man-made water communications had helped to change the physical conditions of life, to make available over large areas the fuel, the building materials, the food, which had previously been procurable only to a much more limited extent. But if engineers were able to work this transformation – and to do so, it must be pointed out, at considerable profit to the businessmen who supported them – surely canals might cope with the awkwardnesses which geography had thrown up on a major scale in some parts of the world? Why should it not be possible to emulate Louis XIV who had joined the Mediterranean and the Atlantic with the Canal du Midi?

A number of such canals were now to be built, the first being the Gotha Canal linking the Baltic with the North Sea through Sweden. Utilizing Lake Wenern and a number of smaller lakes, it was 120 miles (193 kilometres) in length, 55 miles (88 kilometres) being actual canal. The Corinth Canal and the Kiel Canal followed, but by far the most ambitious were to be the Suez and Panama Canals, both presenting massive problems to their engineers. In addition, the Suez Canal, the 100 mile (160 kilometre) link between the Mediterranean and the Red Sea, raised intricacies of politics and organization even more complicated than its engineering difficulties, great as these were. 'I believe', Lord Palmerston said of the canal in Parliament on 7 July 1857, 'that it is physically impracticable, except at an expense which would be far too great to warrant the expectation of any returns.'

Although the builders of the ancient world had not contemplated a link such as the canal, which was to run in an almost straight line between the Mediterranean at Port Said in the north and the newly created town of Suez in the south, there had been many attempts to keep in being the canal which the Pharaoh Sesosteris I had dug about 2000 BC between the Red Sea and the Nile, which itself flowed north into the Mediterranean. After the canal had fallen into disuse, apparently due to silting by desert sands, Pharaoh Nech began to reopen it about

A convoy of southbound ships in the Suez Canal, Bellah Loop Bypass, waiting for northbound ships to pass.

600 BC but gave up the attempt because of engineering difficulties. The same thing happened when Darius resumed the task almost a century later, although one reason for abandonment in this case was a belief that the level of the Red Sea was many feet higher than that of the Mediterranean, and that a project to link the two could be disastrous.

The conviction lasted into the eighteenth century and played its part in stopping in their tracks various Suez canal proposals. Even in 1798 Napoleon – whose orders had included a command 'to cut a canal through the Isthmus of Suez, and to take all necessary steps to ensure the free and exclusive use of the Red Sea by French vessels' – was told by his engineer, Le Père, that at high tide there was a 30 foot (9 metre) difference in the Mediterranean and Red Sea levels.

Only in the mid-nineteenth century did the two requirements for progress, a new need and the appearance of a man dedicated to fulfil it, radically alter the prospects. The need followed the great growth of sea-borne trade which the steamship had encouraged. By the mid-1850s the thousands of extra miles followed by ships as they steamed down the west coast of Africa and rounded the Cape of Good Hope to reach the Indian Ocean on their way to the Far East were a growing burden on a growing trade. The benefits of a Suez Canal became more obvious with the years.

With the need there arrived the man: Ferdinand de Lesseps, an unlikely figure to find among engineering heroes. De Lesseps had spent the first part of his working life in the French Diplomatic Service, much of it in Egypt, where he became friendly with Said Pasha, soon to be ruler of the country. And in Alexandria he became obsessed with the idea of building a canal between the Mediterranean and Suez, 'an enterprise', he later wrote 'which had taken possession of my imagination after reading the memoirs of Le Père, the head engineer on the expedition of General Bonaparte.' By de Lesseps' day British surveyors had shown that there was no difference in water levels between the Mediterranean and the Red Sea. Thus the success of a canal was not automatically ruled out and from then on it became the obsession of de Lesseps' life to build it.

His opportunity arrived in 1854, after he had retired from French service. Said Pasha succeeded to the Viceroyship of Egypt. De Lesseps left France to congratulate the new Viceroy personally and on 15 November 1854 was granted a concession to build the canal. Two years later the concession was ratified by the Turks who still controlled Egypt. The Suez Canal Company was formed, and the money for the project quickly raised, half of it in Europe, where France was the biggest contributor, half provided by Said Pasha.

On 25 April 1859 de Lesseps turned the first spadeful of earth at Port Said and work began along the 100 mile (160 kilometre) route to the south. The line chosen crossed Lake Manzala for 28 miles (45 kilometres) and south of this made use of Lake Bellah, 8 miles (12.9 kilometres) long; Lake Timsah, 5 miles (8 kilometres); and the 23 miles (37 kilometres) of the Bitter Lakes.

While the use of these existing waterways reduced the need for excavation it created its own problems. Lake Manzala, for instance, was only 5 foot (1.5 metres) deep, with a bed of shifting sand, and the minimum depth of water in the canal was to be 26 feet (8 metres). The problem was solved by using local labourers to scoop up the mud, squeeze the water from it and then compact it into artificial walls which were eventually built higher than the lake surface. This stage was followed by mechanical dredging which dug down to solid clay and deposited the clay on the newly built walls.

The dredgers formed part of a fleet of sixty vessels, including some specially designed for work on the canal which could cut rock under water. Steam-driven, the vessels produced 10,000 horsepower and could move 6 million cubic feet (0.17 million cubic metres) of material a month. Many others were also tailor-made for specific jobs, since the route taken by the canal varied from area to area. The Jisr Ridge, for instance, was made up of rock covered with loose sand which tended to fill up excavations as soon as they were dug. South of the Bitter Lakes, on the last stretch to Suez, there lay the Shallufa Ridge where 67,000 cubic yards (52,000 cubic metres) of rock had to be moved by blasting.

To the natural difficulties of the ground there was added the intense heat of the climate in which the thousands of labourers had to toil, and the problem of providing drinking water for such a large work force. It had been stipulated in de Lesseps' concession that he should build not only the navigation canal but also a sweet water canal which would bring water eastwards from the Nile to the Canal Zone. But until 1863 when this was completed as far as Ismailia, half way down the navigation canal, condensers had to be constructed.

The same year Said Pasha died. His successor, Ismail Pasha, was less well disposed towards the enterprise and for a while it appeared that operations might be stopped. Neither Britain nor Turkey was happy at the prospect of what they saw as a French-controlled canal governing an all-important short cut to the Far East, and it was only after complicated negotiations that the work was resumed.

By the late summer of 1869 the canal was nearing completion, the 100 mile (160 kilometre) waterway having a surface width of from 150 to 300 feet (45 to 90 metres), a minimum depth of 26 feet (8 metres) and a bottom width of 72 feet (30 metres). At 6-mile (9.6-kilometre) intervals side-basins had been excavated so that vessels could pass one another. At Port Said two breakwaters each more than a mile (1.6 kilometres) long had been built out into the Mediterranean while a substantial mole had been constructed at Suez. By mid-August only two dykes at the northern end of the Bitter Lakes prevented the waters of the Red Sea from meeting those of the Mediterranean. On the 15th de Lesseps, standing by the southern of the two dykes, dramatically anounced: 'Thirty-five centuries ago Moses ordered the waters of the Red Sea to draw back, and they obeyed him. Today the Sovereign of Egypt orders them to return and once more they obey.'

As an explosion breached the dam, the surge of water pouring through was so great that workmen had urgently to strengthen the second obstruction. If the waters had broken through before settling, further and expensive dredging would have been necessary. But the northern dam held and the following day the waters from the two seas were allowed to meet.

In 1870, 486 vessels passed through the Suez Canal. By 1875 the number had risen to 1,264 and by 1880 had reached 2,026. The majority of the ships were British and in 1875 the British Government bought from the Egyptians 176,602 of the Suez Canal Company's 400,000 shares. Its anxiety to control the Suez Canal was not based only on considerations of trade. Any water link between the Mediterranean and the Red Sea had strategic implications as General Gordon was to stress a decade after the Suez Canal was opened. Working in Jerusalem in 1883, he speculated on the building of a canal from Haifa to the Gulf of Akaba, and wrote to a friend, 'Such a canal would close all attacks from Russia upon Palestine except upon the line between Haifa and Zerin, and strangely enough would force her to attack on Megiddo (Armageddon). It would prepare the way for a United Europe to put this thus isolated Palestine under a common ruler, and would bring about the true prophecy of the Scriptures. All nations would come here and colonise.'

Nothing was to come of Gordon's proposal, and the world had to be content with de Lesseps' canal. It was an unqualified success and it led to the building of the more ambitious Panama Canal although only after de Lesseps himself had suffered a humiliating defeat. For the Panama project presented engineering problems far greater than those which had been faced in Egypt.

The benefits to be gained from a waterway crossing the 50 mile (80 kilometre) isthmus linking north and south America had been obvious from about 1530 when it was finally realized that no natural waterway existed between the Caribbean and the Pacific. Until a canal was cut, ships sailing from Europe to the Pacific would be forced to pass through the Straits of Magellan – or traverse the

A freighter in the Miraflores Lock, Panama Canal, which leads 8 miles (13 kilometres) to the Bay of Panama, providing a 54-foot (16.5-metres) two-step drop from lake level to a sea-level channel.

North-West Passage if that conjectured route were ever found to exist. Yet for some centuries the prospects for a canal looked remote. The backbone of the Cordilleras which stretched from the Rockies to the Andes would have to be cut through. The area was one of the rainiest in the world, with an annual fall of 140 inches (355 centimetres) on the northern side, and certainly one of the most unhealthy. Rivers which could rise 30 feet (9 metres) in a day crossed the isthmus and complicated all plans for cutting through it.

Not until 1875, six years after the Suez Canal had begun to operate successfully, did Lucien Napoleon Bonaparte Wyse found 'La Société Civile Internationale du Canal Interocéanique' and obtain a concession from the Colombian Government to build a canal across the isthmus. Nothing came of the project and it was left to de Lesseps to put forward his own scheme at an International Engineering Congress held in Paris four years later. This was for a 28 feet (8.53 metre) deep canal which would be cut through Panama, at the narrowest part of the isthmus, at a cost of £26,400,000.

De Lesseps' name was enough to attract money to the Compagnie Universelle du Canal Interocéanique de Panama which was set up in 1881. Yet neither his personal experience nor that of his engineers was sufficient to deal with the problems of the enterprise. Quite apart from the engineering difficulties, yellow fever and malaria were endemic and as many as 500,000 men are claimed to have died from disease while building the canal.

In 1888, with the work only two-fifths finished, de Lesseps' company went bankrupt and for almost a decade it appeared that the project had been abandoned for good. Then the facts of geography and the exigencies of war prodded it alive. The Americans had for long appreciated the advantage which a Panama Canal would give to their own policies and strategies, a point brought home during the Spanish-American War of 1898 when a US battleship, anchored off San Francisco, was urgently needed in the West Indies. Instead of the 4,600 mile (7,400 kilometre) voyage eventually made possible by the Panama Canal, the ship had to make the 13,400 mile (21,565 kilometre) journey around Cape Horn.

The Americans at once began diplomatic moves and in 1904 Panama granted the United States a concession for the canal. The next, and most important, move was the engineering decision on which of two plans should be followed. One, which had been adopted by the ill-fated de Lesseps project, was for a sea-level canal to be built from the Atlantic near the port of Colón to Panama on the Pacific. The alternative was for a high-level canal. Vessels would pass up into this near the northern end of the route and after crossing the isthmus would descend in one or more locks to the level of the estuary leading to Panama.

Both presented great, but different, difficulties. The sea-level route would require excavations 45 feet (13.7 metres) deeper than the higher one; but the higher route would need extensive locks. The US Board controlling the operation recommended by eight votes to five that the sea-level route be followed. It was eventually overruled by Congress which, in June 1906, decided on the high-level route.

The earlier company had excavated 548,000 cubic yards (419,000 cubic metres) of earth, and much of its work could be incorporated in the new scheme, which was as well since the problems facing the Americans were even more formidable than expected. When work started it was thought that 103 million cubic yards (79 million cubic metres) would have to be excavated; the real figure turned out to be 195 million (149 million); a figure perhaps justifying the comment by the famous scholar and statesman, Lord Bryce, that the canal was 'the greatest liberty man has ever taken with nature.'

Among the northern part of the route there ran the River Chagres, so turbulent that there was no possibility of canalizing it for navigation. Instead, its waters were dammed so that behind it there was formed the artificial Lake Gatún, 164 square miles (425 square kilometres) in area, twice the extent of Lake Maggiore and four-fifths the size of the Lake of Geneva. Upstream from Lake Gatún the inflow of the Chagres and its tributaries was controlled by diversionary channels, turning the lake into a vast reservoir.

The dam hemming in Lake Gatún was itself a major engineering feat. The ground here was made up of alluvial deposits of gravel firmly cemented with mud and clay. It was watertight but soft, extended downwards for 280 feet (85 metres) before solid rock was reached, and would spread under the weight of a concrete dam. An earth dam 2,100 feet (640 metres) wide at its base and with a centre spillway through which 137,000 cubic feet (39,008 cubic metres) of water could be released, was therefore built across the mile and a half wide valley floor. At each end of the dam, rock bulwarks were built up of material brought from as far as 26 miles (42 kilometres) away. Between these two bulwarks silt was deposited and the water in it removed with the help of hydraulic presses. The silt was then packed down to form an 'hydraulic fill'.

While the work of preparing Lake Gatún was going on, engineers were building to the north of it the lock which was to bring vessels from the Colón level up to that of the lake, 85 feet (26 metres) above. The biggest of its three steps was 32 feet (9.7 metres) high and the locks were able to handle two ships at a time.

South of the lake there came the most formidable obstacle: the barrier of hills more than 300 feet (90 metres) high which linked the main mountain chains of north and south America. They were breached by the Culebra Cut, 9 miles (14 kilometres) long, of which Lord Bryce was to write, 'Never before on our planet have so much labour, so much scientific knowledge, and so much executive skill been concentrated on a work designed to bring the nations nearer to one another and serve the interests of all mankind.'

As work in the Cut progressed, 200 miles (320 kilometres) of railway line were laid down, tier above tier, to move the excavated soil. Scores of steam shovels and thousands of men were employed to keep the movement going and special equipment was designed and built to help them. There was, for instance, the Ledgerwood Unloader. 'Railway trucks provided with flaps were used, these flaps making a single platform of the whole train,' it has been written of the contraption. 'At the rear of the train was a plough which could be drawn by a wire rope attached to a drum carrier on a special car in the forepart of the train. When the car arrived at the dumping ground the drum was started, and the plough, advancing from the rear, swept the 320 cubic yards of rock from the sixteen cars in seven minutes.' Progress was also helped by the 'track shifter', invented by one of the canal workers, which raised track and ties clear of the ground and laid them down up to nine feet away.

Eventually the Cut, 272 feet (83 metres) deep in places, was completed. During the final stages other work was concentrated on a single steplock at Pedro Miguel and, further south, a double flight lock at Miraflores. These were major undertakings using huge masses of material, broken stone brought from Porto

Bello, 20 miles (32 kilometres) east of Colón, sand from Nombre de Dios and cement from the United States. Finally, in 1914, the first ships passed through the 50 mile (80 kilometre) Canal, 300 feet (91 metres) wide with a minimum depth of 41 feet (12.5 metres), and uniting two oceans.

In addition to canals of the magnitude of Suez and Panama which linked different seas and oceans, the closing years of the nineteenth century witnessed a unique British canal-building achievement: the Manchester Ship Canal whose 35 miles (56 kilometres) brought ocean-going vessels into the centre of the city of Manchester. A plan for linking this heart of the cotton trade with the open sea had first been proposed in 1824 when it was suggested that a canal could be dug between the city and the Dee estuary. Nothing came of this, partly due to opposition from Liverpool whose inhabitants believed that its dock trade would be wrecked, and it was only in 1885 that the Manchester Ship Canal Act was passed – at the third attempt and after nearly £350,000 had been spent on Parliamentary and legal fees.

The canal as envisaged in the Act was an almost staggeringly ambitious project, even though only 35 miles (56 kilometres) long. It was to start in the west at Eastham on the Mersey Estuary, and for its first 8 miles (13 kilometres) which took it to Runcorn, was to be built out from the southern bank of the Mersey. Thence a new cut was to take it to Manchester. The canal, which could accommodate 15,000-ton vessels, rose to Manchester in four sets of locks, was crossed by nine main road bridges – seven of them being swing bridges – and five railway bridges or viaducts. The set piece engineering feature of the canal, however, was the Barton swing bridge which replaced Brindley's old aqueduct and carried the Bridgewater Canal over the Ship Canal. The bridge, operating on a pivot pier in the Ship Canal, consisted of a 234 foot (71 metre) long trough 19 feet (5.8 metres) wide with a water depth of 6 feet (1.8 metres), and with watertight gates at each end.

The engineering work demanded not only 17,000 men, but the largest concentration of mechanical power assembled in the country for a single task. It included 60 Ruston & Proctor steam excavators, 231 steam engines for pumping, 194 cranes, 59 pile-drivers, 173 locomotives, 6,300 wagons and 230 miles (370 kilometres) of railway line. Some 70 million bricks were used and 1.25 million cubic yards (0.95 million cubic metres) of concrete.

Despite financial problems, the death of the contractor, and serious storms and floods, the whole complex enterprise was completed as planned and on 1 January 1894, seventy-one merchant ships steamed through the canal from the open sea into the new Port of Manchester.

In scope and implications, the Manchester Ship Canal was to be equalled as a commercial seaway only by the St Lawrence Seaway almost half a century later. One of the greatest engineering marvels of the twentieth century, the Seaway today allows 20,000 ton ocean-going ships to bring their cargoes 2,300 miles (3,700 kilometres) from the Atlantic Ocean into the heart of the North American continent.

More than four hundred years ago the French explorer Jacques Cartier, sailing up the St Lawrence, found his way barred by impassable rapids above what is now Montreal. Optimistically, he named them Lachine, believing that beyond them lay the route to the Far East. In fact, they blocked the water route to the Great Lakes – Ontario, Erie, Huron, Michigan and Superior – the inland sea of

North America as it has been called. During the following centuries many efforts were made to link the individual lakes through which, with the exception of Michigan, the frontier between Canada and the United States was to run. Yet only with the coming of the twentieth century was it possible to envisage a great navigation system linking the Great Lakes basin with the open sea; and not until 1959, when the Seaway was formally opened by Queen Elizabeth II and President Eisenhower, was there a successful end to years of daring construction work and to half a century of argument. Those who built the Seaway had to overcome not nature alone but disagreements between Canada and the United States. 'Few projects have been so bitterly opposed or inspired so many opinions, arguments, legal battles, treaties and inter-government memoranda,' says the first President of the St Lawrence Seaway Authority, Lionel Chevrier. '. . . The seaway story is a chronicle of men fighting for self interest against nations fighting for national interest.'

The St Lawrence, draining 700,000 square miles (1,813,000 square kilometres) – an area more than three times the size of France – is known throughout its 2,000 miles (3,200 kilometres) by a variety of names: the St Mary's River, the St Clair River and the Detroit River among others. When the Seaway was planned the river was still barred by dangerous rapids while its waters were with minor exceptions flowing to the ocean unused. Today the Seaway traffic is measured in millions of tons while its power development project shared by Canada and the United States, taps a potential of more than 2 million horsepower of electricity.

The Lachine Rapids which stopped Jacques Cartier are today avoided by the Côte Ste Catherine Lock in which ships are raised more than 30 feet (9 metres) to the level of Lake St Louis above the rapids. More than 18 million gallons (82 million litres) of water are needed to fill the 768 foot (234 metre) long lock and the job is done in seven minutes. At the western end of Lake St Francis the two Beauharnois locks raise ships a total of some 80 feet (24 metres) after which they pass through the Beauharnois canal and out into Lake St Francis.

Up to this point, all the locks and canals were built by Canada's St Lawrence Seaway Authority but above Lake St Francis there begins an international zone containing a complex of locks, canals, control dams and power stations. Once through this, ships enter the waters of Lake Ontario and, after passing the inland ports of Kingston, Toronto and Hamilton, enter the Welland Canal.

The first Welland Canal was completed in 1829 but it was shallow, badly conceived and as originally planned would have raised vessels the 326 feet (99 metres) from the level of Lake Ontario to that of Lake Erie by means of no less than forty locks. Today only eight locks have to be negotiated, three of them allowing passage of ships in both directions simultaneously. Beyond the Lake Erie ports, and those of Lake Huron, lies a last group of locks from which ships emerge into Lake Superior for the final run to Thunder Bay, 2,300 miles (3,700 kilometres) away from the Atlantic.

The navigation season on the Seaway runs from 1 April to 15 December from which date the lower stretches are normally blocked by ice. During the eight

A vessel outward bound in the
Manchester Ship Canal approaching Eastham Locks.

and a half months that the Seaway is open each year more than 21 million tons of grain from the Prairie Provinces and the American mid-western states are carried to the grain elevators on the lower reaches of the St Lawrence and to markets abroad. The United States has recently proposed an eleven-month season but Canada has so far been reluctant to agree since it appears that while the cost of keeping the Seaway open would be borne by Canada the benefits would be gained by the United States.

Ability to divert and control waters to make fresh trade routes reached new heights of efficiency with the Panama and St Lawrence schemes of the twentieth century. Long before this, however, the power of steam had helped to revolutionize man's control of water in other ways. One of them was to make practicable drainage schemes previously beyond man's power to operate but which now began to change the face of the land. It was particularly true of the Netherlands whose low-lying acres had been under constant threat from the sea from the start of recorded history. On the night of 14 December 1287, an inundation had drowned 50,000 people and thirteen years later a similar catastrophe formed the Zuyder Zee. On the night of 18 November 1421 sixty-five villages were inundated and some 10,000 people drowned.

The Dutch had responded, first by building artificial mounds or islands, known as 'terps', along the sea-coast and linking them with dykes, but by the fourteenth century they were developing three different kinds of embankment to hold back the sea. One was the *slikkerdijk* which had an earth core and slopes reinforced with osiers or straw, and plastered with clay. A second was the *wierdijk* made up of seaweed instead of osiers, while the third, the *rietdijk*, used bundles of reeds, tied together, instead of the seaweed. Later, embankments were replaced with *krebbingen*, structures consisting of two rows of piles with bundles of faggots placed in between them. In 1408 the first windmill was installed for pumping out water from inland areas and many others were built after the disaster of 1421. Ingenious sluices were devised for use in the dykes; these opened to allow the waters to drain out to sea at ebb tide, and were closed at the flood. Prominent in this work was Jan Leeghwater, mill constructor and engineer who not only began to put the country's water defences on a scientific basis, but helped her military defence when the Prince of Orange was besieging the town of 's-Hertogenbosch in which the invading Spanish were defending themselves. The Spanish had reinforced their defences by flooding the surrounding country. Leeghwater drained it with windmills and the town was recovered by the Dutch.

In 1500 Andries Vierlingh found that sea dykes could be strengthened if certain plants were encouraged to grow on the sand of which the dykes were formed; and the following century the mayor of Amsterdam began the scientific study of flood control by recording the height of the sea at eight places near the city every half hour.

Special dredgers were soon being built to scour river beds so that inland water would drain out to sea more easily. There followed the construction of tidal reservoirs which could be filled at high water, then emptied at low water to create an artificial scour. The number of windmills grew from scores to hundreds, and the Dutch for more than a century became the leaders of drainage engineering throughout Europe. Samuel Smiles, recording British achievements, had to recall that not only did Cornelius Vermuyden drain the Great Level of the British Fens

The twin flight locks at Thorold on the Welland Canal, Canada, built for
two-way traffic. Three of the eight locks allow passage of ships in both directions simultaneously.

but that another Dutchman, Freestone, reclaimed the marsh near Wells in Norfolk. Canvey Island, off the mouth of the River Thames, was embanked by Joas Croppenburgh and Dutch workmen, while Joan Johnson, also a Dutchman, planned and built a new haven at Great Yarmouth. When the River Witham burst its banks at Boston, it was Matthew Hake who was sent for from Flanders to deal with the emergency.

One of the few areas of Europe where the Dutch hydraulic engineers were unsuccessful was the Pontine Marshes, the 300 square miles (777 square kilometres) of country lying south-east of Rome between the Lepini Mountains and the Tyrrhenian Sea. The area, within the territory of the Papal State, had been malaria-infested marsh since earliest times, and although it had been partially drained at the height of the Roman Empire, when it became possible to build a section of the Appian Way along its eastern boundary, the engineering work had fallen into decay long before the beginning of the sixteenth century when Pope Leo X commissioned a series of drainage works. Pope Leo died before the project was seriously started and it was only in the 1580s that Pope Sixtus V inaugurated a scheme on which 2,000 men were employed. Once again the death of the Pope brought the operation to an end.

In 1623 the Netherlander Nicolaas Corneliszoon de Witt arrived in Rome with plans to resurrect the scheme, now to be financed by Dutch and Italian merchants. This time it was the engineer's sudden death which stopped work and it was only in 1676 that another Dutchman, Cornelius Janszoon Meijer, started it up again. After his death, the drainage scheme was carried on by his son with some engineering success. However, opposition from local inhabitants who did not wish to have their way of life altered, prevented the plans from being fully carried out.

These unsuccessful schemes rested entirely on the construction of drainage canals to take the run-off from the Lepini Mountains and lead it into a major new canal which would then take the water into the sea. This was never sufficient, and the Pontine Marshes were only finally cleared and transformed when in the 1930s the Fascist goverment launched a major scheme not only to divert the contemporary run-off, but to drain the whole area and build in it a number of new towns in which they settled some hundreds of families.

The Pontine Marshes was almost alone among Dutch drainage failures and one of their greatest engineers, Cornelius Vermuyden, even had a partial success in dealing with the British fenlands, one of the most intractable areas in the country. A region of 306 square miles (792 square kilometres), the Fens had been a problem since the end of the Roman occupation. During that occupation much of the fenland was cultivated but in the following decades, for reasons that are still not clear, it degenerated into sparsely occupied marshland. In the Middle Ages attempts were made to keep in repair the embankments of the river which lazily traversed the country on their way to the North sea, but they were only half effective and in 1600 there was passed 'An Act for the recovering of many hundred thousand Acres of Marshes'.

As a result, Vermuyden was hired to drain the Fens. His plan was to cut new channels to carry water at a greater speed than it could acquire in the winding rivers. Notable among the cuts was the 70 foot (21 metre) wide, 21 mile (34 kilometre) long Old Bedford River, to be followed some years later by the New Bedford River which ran parallel to it. For a while the scheme worked. But as the drainage succeeded, the peat which covered the area steadily shrank. Its level sank until by the beginning of the eighteenth century it was lower than the cuts into which its water should have been draining. The result was that when Daniel Defoe, the famous English writer, looked from the Gogmagog Hills in the early 1720s what he saw was, 'Fen country almost all covered with water like a sea; the Michaelmas Rains having been very great that Year, they had sent great floods of Water from the Upland Countries, and these Fenns being as may be very properly said, the sink of no less than thirteen Countries . . . they are often overflowed.'

The next experiment was with the windmill. But as John Rennie put it, 'windmills can only work when the wind blows, and it is frequently calm when most wanted.' Thus the windmill was eventually replaced by steam-driven pumping machines, and from 1820 onwards they appeared in ever greater numbers throughout the Fens.

However, steam proved to be an answer to drainage problems not only in England but also in Holland. The first Boulton & Watt steam engine reached Rotterdam in 1785 to drain the polder of Blydorp and Kool. Another followed five years later, and four more were ordered in the early years of the nineteenth century. The draining of the Dutch polders by steam was so successful during the nineteenth century that plans were made to build a dam across the Zuyder Zee and then drain it by steam pumps, a task now being accomplished.

Great as was the project to drain the Zuyder Zee, it has been overshadowed by the Delta Plan, the most ambitious and complex of all the world's drainage schemes. This was started after a disaster on the night of 31 January 1953 when the sea broke through the dams protecting southern Holland and drowned nearly 2,000 people. The Delta Plan, now almost completed, has involved the building of storm barriers on the lower reaches of the rivers Rhine and Scheldt and of dams to join the islands lying at the mouths of the river estuaries, the most important of these being 8 miles (13 kilometres) long.

Such a vast engineering operation would have been impossible without the development of new equipment. Important among the innovations had been a redesigned form of the huge concrete caissons which were part of the artificial Mulberry harbours used by the Allies during the invasion of Normandy in 1944. Dozens of these were built, towed into position and sunk to provide foundations for the new dams. Nylon mattresses were devised to protect the sea bed from damage, and special vessels were designed for laying underwater stone and for underwater asphalting. The opening and closing of sluices in the new defences is dependent on changing water levels, on wind direction and on wave height, figures for which are constantly fed into a computer which automatically operates the sluices at the right moment.

The use of computers in the Delta Plan's huge engineering water control schemes typifies the inter-relationship of different engineering disciplines which is a prominent feature of the twentieth century. But that inter-relationship has always been present to some degree; indeed, the experiences gained by the canal builders were to be found invaluable when the steam engine, translated into a viable mobile power unit, brought in the railway age which was to relegate many canals to the dustbin of industrial history.

THE COMING OF THE IRON HORSE

The works of the canal builders, whether those of Thomas Telford in the 1800s or the twentieth-century achievements of the St Lawrence Seaway and the Dutch Delta Plan, are still among the most impressive illustrations of man's mastery of water and his diversion of it to his own ends. Yet while the canal age was still flourishing a new use for steam was beginning to create its own corps of professional specialists: the railway engineers whose work was to transform the countries they laboured in and the lives and social attitudes of the inhabitants.

Telford himself foresaw the coming change as early as 1800 when he wrote, according to J. Plymley in his *General View of the Agriculture of Shropshire*, 'Since the year 1797 . . . another mode of conveyance has frequently been adopted in this county [Shropshire] to a considerable extent; I mean that of forming roads with iron rails laid along them, upon which the articles are conveyed on waggons containing from six to thirty cwt.; experience has now convinced us that in rugged country or where there is difficulty to obtain lockage, or where the weight of the articles of produce is great in comparison with its bulk, or where they are mostly to be conveyed from a higher to a lower level, in these cases iron railways are in general preferable to canal navigation.'

Tracing back the origin of almost any such technological development is like opening a succession of Chinese boxes: another is to be found inside each, so that it seems that the last will never be reached. With the mobile steam engine, however, the engine designed to move on either road or rail, pulling carts or coaches without the aid of horses, ideas changed almost suddenly into hardware with the emergence of Richard Trevithick, a Cornishman who developed the high-pressure steam engine that Watt had always thought too dangerous. Trevithick had all the heartiness of his age, little regard for danger, and a boundless imagination. Davies Gilbert, born David Giddy, later to become President of the Royal Society, recalled that Trevithick, 'a Tall and strong young man had made appearance among Engineers, and that on more than one occasion he had threatened some people who contradicted him, to fling them into the Engine Shaft.' Confident that undreamed of steam applications plus the use of iron would help change the world, Trevithick and a collaborator filed in a single year patents for floating docks, iron ships, iron masts, methods of bending timber, diagonal framing for ships, iron buoys, rowing trunks and steam cooking.

His predecessors in plans for using steam to move vehicles included Nicholas Joseph Cugnot, a French military engineer who in 1769 built a steam carriage which could carry four people at 2.25 mph (3.6 kph). Although the steam ran out after about a quarter of an hour, the French military authorities encouraged Cugnot to build a steam-operated vehicle for drawing artillery. The result was a three-wheeled carriage, two of the wheels being at the rear and the third, a driving wheel more than 4 feet (1.2 metres) in diameter, being mounted on a pivoted iron forecarriage which could be turned by gears to steer the vehicle. The engine, mounted on top of the forecarriage, had two single-acting inverted cylinders, connected by a rocking beam, the motion being transmitted to the driving axle by pawls acting on two reversible ratchet wheels. The boiler was a spherical copper vessel below which the fire was held in a copper container. A driver's seat and a crude brake acting on the driving wheel completed the apparatus.

Cugnot's second machine, like his first, had certain built-in disadvantages. The three-wheeled construction made the vehicle unstable while the driver had to stop regularly to top up the boiler with water and to add fuel to it. However, these problems could probably have been removed if Cugnot had been given the opportunity. Unfortunately, his main backer was Etienne François, Duc de Choiseul-Amboise, who fell from his position in the Government in France before further work could be authorized. Choiseul's successor was uninterested in steam engines, even for drawing artillery, and the world's first steam-powered road vehicle remained in the Paris Arsenal until it was removed to the Conservatoire National des Arts et Métiers in Paris at the end of the eighteenth century.

Cugnot was ahead of his time, and the same was true of William Murdock, one of Boulton & Watt's employees who did so much to introduce gas lighting into Britain. A working model of Murdock's steam carriage was built between 1781 and 1784 at Redruth in Cornwall. It was not clear how closely, if at all, he worked with Watt but in April 1784 Watt lodged a patent covering various devices, including steam engines 'to give motion to wheel carriages'. He failed to go ahead with the idea. So did Murdock, and the way was left open for Richard Trevithick, whose interest in steam originated not so much from Watt's engines as from the limitations which Watt put on them with his patents. In this, Watt was little different from many other engineers and inventors who try to keep competitors out of the field. The practice, however, tends to encourage the more ambitious rivals to circumvent the patents. In the case of steam, Richard Trevithick was just such a man, one who had grown up in a county where the earlier Newcomen engines were steadily being replaced by Boulton and Watt's. Before Trevithick reached his teens there were more than a score of them working in Cornwall and he was accustomed to hearing the mine owners complaining about the burden of Watt's patent payments.

Determined to break Watt's monopoly, Trevithick decided to do away with the condenser that was now considered an essential of any successful steam engine, and to compensate for the portion of an atmosphere thus lost by using steam at a pressure of several atmospheres.

The first engineer to propose using steam at more than atmospheric pressure had been Jacob Leupold of Leipzig who in 1725 planned a pumping engine which would have discharged the steam into the air after it had done its job. But the engineers of the day could not produce boilers capable of working under pressure for any length of time and the same was true almost half a century later when Watt was issued his first patent. Although that patent covered the use of high-pressure steam, it was an idea that Watt never translated into practice. He believed, probably with good reason, that boilers could not yet be made to withstand the pressure indefinitely and he was fearful of accidents which would discredit the use of steam altogether.

At the end of the century Trevithick was more confident. In 1801 he supplied a Cornish mine with the first engine to use steam satisfactorily at a pressure of about 100 pounds per square inch (6.89 bars), and the following year supplied another which worked with steam at a pressure of 145 pounds (9.99 bars). He had been correct in believing that such engines would be safe if properly used but in 1803 a mechanic inadvertently tied down the safety valve of an engine and the explosion of the boiler killed four men. Despite this setback Trevithick continued to supply 'Cornish engines' as they were known. They were not only more economical than the Boulton & Watt engines which had preceded them, but they were, when generating the same horsepower, both lighter and smaller. This in itself helped to popularize the use of steam just as the factories of the Industrial Revolution were demanding small, adaptable power sources that could be used for a variety of tasks.

Much the same process was continuing in America where Oliver Evans took the lead in producing engines working at about 100 pounds to the square inch (6.89 bars). American technology still lagged behind that of Britain and Evans had to face considerable difficulties. However, he was among the first to see the value of steam for propelling river craft, and built an amphibious steam dredger, the *Orukter Amphibolus*. Although the 15½-ton craft came to grief on the roads of Philadelphia, Evans was to design more than one engine for use in the paddle steamers which in the next decades were to help open up the American West.

However, it was in England, with Trevithick's encouragement, that steam was eventually to start a transport revolution. His vision was of his engine as a prime mover which could supplant the horse not only from the plodding routine of operating pumping engines and the like but as a means of transport. With the growth of the iron-smelting industry the wooden rails along which horses pulled waggons from collieries to works had, first in Coalbrookdale and then in other places, been reinforced with nailed-on cast-iron plates. These had then been replaced by flat iron rails which in turn were replaced in the 1770s with flanged metal rails. Would it not be possible, Trevithick argued, for his engine to be mounted on wheels which would without the use of horses be propelled along iron tracks? Might it not, indeed, be possible for his steam engine to run on the ordinary roads, without rails at all?

The road locomotive or steam carriage was first tried following experiments – these now being the order of the day – to discover what would happen if the axles of a vehicle were turned by a power pack mounted on it, rather than by forward movement as when the vehicle was pulled by a horse. Would there be sufficient friction between wheel and road surface? After successful trials near the Cornish town of Camborne, Trevithick set about building what was, with the exception of Cugnot's experimental machines, the first horseless carriage in Europe.

Models were made, and their efficiency tested, in the homely atmosphere of the family kitchen. 'A boiler something like a strong iron kettle was placed on the fire,' wrote Trevithick's son Francis, 'Davies Gilbert was stoker and blew the bellows: Lady Dedunstanville was engine man and turned the cock for the admission of steam and was "charmed to see the wheels go round". . . . Shortly afterwards another model was made which ran round the table, or the room. The boiler and the engine of this second model were in one piece; hot water was poured into the boiler and a red-hot iron put into an interior tube, just like the hot iron in tea-urns.'

Little is known about the full-scale vehicle which was eventually built but, judging by the drawings of Trevithick's son, it consisted of a horizontal boiler in which was inserted a cylinder whose piston moved a crank which itself turned a rear axle supporting two of the four wheels. After a debut on Christmas Eve 1801, the vehicle suffered an accident four days later when it was being driven by Trevithick's cousin, Captain Andrew Vivian. According to Davies Gilbert, 'the carriage was forced under some shelter and the Parties adjourned to the Hotel, & Comforted their Hearts with a Roast Goose and proper drink, when, forgetful of the engine, its Water boiled away, the Iron became red hot, and nothing that was combustible remained either of the engine or of the house.'

For some inexplicable reason there was now a hiatus of about two decades and it was only in the 1820s that the first steam coaches began to run on Britain's roads. Among the first was that of Goldsworthy Gurney which in 1829 ran at 15 miles (24 kilometres) an hour from London to Bath and back again. A few years later Walter Hancock's London and Paddington Steam Carriage Company was running 10-passenger steam coaches between the capital and Surrey and Kent, and was claimed to have carried more than 12,000 passengers by 1836.

Apart from taking the initial steps at the beginning of the century, Trevithick had no part in all this and it was for the construction of the first steam locomotive to run on rails that he achieved fame. The vehicle was built after an iron-master had bet a neighbour 500 guineas that a steam engine could not haul 10 tons of iron on the metal rails – known as a tramway – from Penydarren in South Wales, to Abercynon, some 10 miles (16 kilometres) away. This Trevithick successfully accomplished on 21 February 1804. 'In the present instance,' wrote a *Cambrian* reporter three days later, 'the novel application of Steam, by means of this truly valuable machine, was made use of to convey along the Tram-road ten tons long weight of bar iron from Penydarren Iron Works to the place where it joins the Glamorganshire Canal, upwards of nine miles distance and it is necessary to observe that the weight of the load was soon increased from ten to fifteen tons, by about seventy persons riding on the trams, who, drawn thither (as well as many others) by invincible curiosity, were eager to ride at the expence of this first display of the patentee's abilities in this country. . . . It is not doubted but that the number of horses in the kingdom will be very considerably reduced, and the machine in the hands of the present proprietors, will be made use of in a thousand instances never yet thought of for an engine.'

Yet the steam tramway engine failed to catch on and Trevithick began to experiment with steam to drive vessels on the canals which by now criss-crossed southern England and the Midlands. He was drawn to the idea following a proposal that his engines might be converted to pull fire-boats across the English Channel and into the port of Boulogne where Napoleon was believed to be concentrating his ships for an invasion of Britain. Trevithick at once carried out a trial. 'There was an engine, a 10 In. Cylinder put in to a barge to be carried to Macclesfield for a cotton factory, and I tryed it to work on board,' Trevithick told Gilbert. 'We had a fly wheel on each side [of the] barge and a crankshaft that was thrown across the deck. The wheels had flat boards of 2 ft. 2 in. long and 14 In. Deep, six on each wheel, like an under-shut water-wheel. The extremity of the wheels went abt. 15 miles pr. hour. The barge was between 60 and 70 Tons burthon. It wod go in still water abt 7 Miles p. Hour. This was don onely to try what effect it wod have. As we had all the apparatus of old mat[eria]ls at the Dale, it cost little or nothing to put together. I think it wod have drove it much faster with sweeps.'

Another use for steam emerged when Trevithick designed and built a dredger for work on the Thames, a gun brig on to which was mounted a high-pressure Trevithick engine powering an endless chain of buckets. However, so costly was the machine that the authorities decided to continue with manual labour.

Trevithick was typical of those eighteenth-century engineers who were unwilling to admit that anything was impossible. After the failure of his first steam-driven road vehicles, he wrote to Sir John Sinclair of the Board of Agriculture saying, 'the first and only self-moving machine that ever was made to travel on a road with 25 tons at four miles per hour and completely manageable by only one man, I think ought not to be dropped without further experiments.' Then, in 1808, he returned to the locomotive, which he appears to have envisaged for use on either road or rail.

It was natural that the horseless carriage should be displayed before connoisseurs of horseflesh, and on Sunday, 17 July 1808, the following notice appeared in the *Observer*: 'The most astonishing machine ever invented is a steam engine with four wheels so constructed that she will with ease and without any other aid, gallop from 15 to 20 miles an hour on any circle. She weighs 8 tons and is matched for the next Newmarket meeting against three horses to run 24 hours starting the same time. She is now in training on Lady Southampton's estate adjoining the New-road near Bedford Nursery, St. Pancras. We understand she will be exposed for public inspection on Tuesday next.'

This was 'Trevithicks Portable Steam Engine', advertised with the words 'Mechanical Power Subduing Animal Speed' and 'Catch me who can'. While it apparently never put in an appearance at Newmarket, it was demonstrated on a circular track on the ground now occupied by Euston Square in central London. But interest was less than expected; Trevithick once again began to concentrate on stationary engines, and the way was left open for the development of the railway by George Stephenson two decades later.

However, Trevithick's imagination never flagged and at the age of sixty-one, following the passage of the Reform Bill, he planned an extraordinary construction in celebration of the Bill – a column that would have been more than twice as high as St Paul's. 'Design and specification for erecting a gilded conical cast-iron monument', ran his description. 'Scale, 40 feet to the inch of 1000 feet in height, 100 feet diameter at the base, and 12 feet diameter at the top; 2 inch thick, in 1500 pieces of 10 feet square, with an opening in the centre of each piece 6 feet diameter, also in each corner of 18 inches diameter, for the purpose of lessening the resistance of the wind, and lightening the structure; with flanges on every edge of their inside to screw them together; seated on a circular stone foundation of 6 feet wide, with an ornamental base column of 60 feet high; and a capital with 50 feet diameter platform, and figure on the top of 40 feet high; with a cylinder of 10 feet diameter in the centre of the cone, the whole height, for the accommodation of persons ascending to the top. Each cast-iron square would weigh about 3 tons, to be all screwed together, with sheet lead between every joint. The whole weight would be about 6000 tons. The proportions of this cone to its height would be about the same as the general shape of spires in England.

'A steam-engine of 20-horse power is sufficient for lifting one square of iron

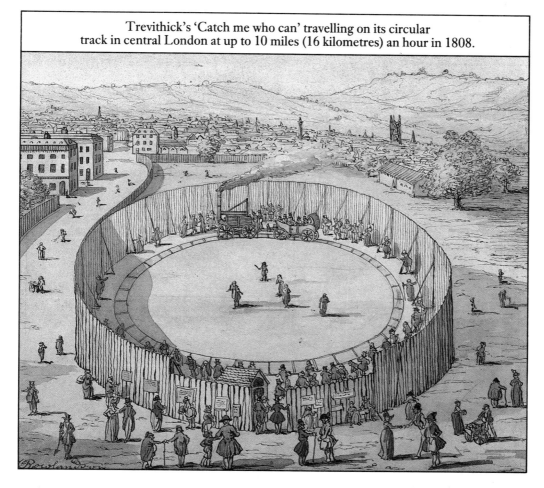

Trevithick's 'Catch me who can' travelling on its circular
track in central London at up to 10 miles (16 kilometres) an hour in 1808.

to the top in ten minutes, and as any number of men might work at the same time, screwing them together, one square could easily be fixed every hour; 1500 squares requiring less than six months for the completion of the cone. A proposal has been made by iron founders to deliver these castings on the spot at 71s. a ton; at this rate the whole expense of completing this national monument would not exceed 80,0001 [£80,000].'

Trevithick's plans for his giant tower were as unsuccessful as his attempts to harness steam to vehicle on rails. Yet the increasing use of iron instead of wood for the tracks used by horse-drawn vehicles leaving the collieries was a pointer to the way things were going. So was the move towards public use of metal tracks on which horses pulled loads which in turn led on to steam railways. As early as 1799 a scheme was promoted for building a horse tramway between London and Portsmouth, and four years later the first section was opened between Wandsworth and Croydon. Two years after that, in 1805, a second section was opened as far as Merstham in Surrey. About twenty similar horse-operated trackways were opened in the early years of the nineteenth century, most, such as the line at Coalbrookdale on the Severn, serving collieries or ironworks.

However, all these experiments which either petered out or, at least, were not followed up. What was needed after Trevithick's imaginative if unsuccessful enterprises was a man with the required dedication and the technological genius to produce a steam engine small enough to be mobile and efficient enough to produce sufficient power to draw useful loads.

The man was to be George Stephenson, born into a life of poverty outside Newcastle in 1781. Working through a series of humble jobs in the local collieries, Stephenson devoted himself to study of the steam engine with a passion that other men grant to science, the law, an over-weening hobby, or the making of money. He was certainly to become rich as 'the father of the railways' but it was not the driving force behind his work. For him – as for his son Robert – the success of railway transport was the be-all and end-all of existence. He worked his way up from fireman to enginewright. He dismantled and then put together more than one colliery engine. From his scanty earnings he paid to be taught to read and write, and then began to study scientifically how steam could be used. In 1814 he built his first engine. It pulled 30 tons at Killingworth Colliery, where he worked, faster than horses could move the load. Five years later, when the Hetton Colliery Company built an 8 mile (13 kilometre) wagon-way to the River Wear, Stephenson was appointed engineer. In his first post as civil engineer he used various methods, as a report in the *Newcastle Chronicle* of 23 November 1822 made clear. 'On Monday the 18th inst. the Hetton Coal Company effected the first shipment of their coals at their newly created staith,' said the account. '. . . Five of Mr. George Stephenson's patent travelling engines, two 60 horsepower reciprocating engines and five self-acting inclined planes, all under the direction of Mr. Robert Stephenson . . . simultaneously performing their various and complicated offices . . . exhibited a spectacle . . . interesting to science and encouraging to commerce.'

Stephenson's first engines suffered from numerous minor troubles. He and his son Robert dealt with them throughout the next few years, finally producing a prototype of what was, in essentials, to be the railway engine for the next few decades. An important improvement was the use of waste steam which he directed as a blast to stimulate combustion. With it there went a completely redesigned boiler in which the earlier single fire tube was replaced by a number of small tubes connected to a water-jacketed firebox. The heating surface was thus greatly increased which meant that more power could be produced from every pound (0.45 kilograms) of fuel. He also provided a physical link between the cylinders and the wheels, and at the same time improved the joint adhesion of the wheels to the track by the use of horizontal connecting rods.

As he built in these improvements, Stephenson produced increasingly efficient engines for drawing coal from the local collieries. So far, however, this was the only job which his mobile steam engines were designed to undertake. The crucial advance came in 1825.

As far back as 1810 there had been proposals for a link between the south Durham coalfields and the port of Stockton, at the mouth of the Tees, from which the coal was shipped. Some favoured a canal, others a land link, and when the Bill authorizing a railway between Stockton and Darlington was passed in 1821 it was left open as to whether a land link should be powered by horses or by steam. Stephenson had by now built more than a dozen engines, and in 1822 was appointed engineer to the company which was to build the new line. He at once surveyed the route, then successfully advised that a new Bill be presented which would permit the carriage of passengers as well as goods and the use of both stationary and mobile steam engines. It was a turning point in Stephenson's career and soon afterwards he opened his own engine works at Newcastle with Robert in charge.

After the route between Stockton and Darlington had been surveyed, Stephenson himself supervised the laying of the track. It was of wrought-iron rails which though dearer than the cast-iron track previously used elsewhere did not fracture so easily. The distance between the two rails was 4 feet 8½ inches (1.43 metres) as recommended by the company, a gauge which was to become the standard, first in Britain and then throughout much of the world. It was the gauge of the stagecoach wheels, and, according to some authorities, identical with that of Roman chariot wheels. But the question mark that Trevithick thought he had removed still hung over the problem of traction. Surely, it was argued, the wheels would merely slip round on the iron track? Stephenson, correctly, believed otherwise.

Everything went according to plan on 27 June 1825, the date of the railway's opening run. With George Stephenson at the controls, 'Locomotion No. 1' drew out from Darlington pulling six wagons laden with coals and merchandise, the committee of the company and local notables seated in a coach, six wagons filled with passengers, fourteen wagons with places reserved for workmen, and six wagons carrying coal which were unhitched en route. Some six hundred passengers were carried on the inaugural run.

While steam continued to be used, the line was for a while leased out to coach proprietors who employed horse-drawn coaches with wheels modified to fit the line. But in 1833 the practice came to an end and the line was kept for steam traffic only.

A more ambitious proposal was now being considered: nothing less than a railway between Liverpool and Manchester. It was not the length of the line – nearly 40 miles (64 kilometres) – which made the project seem so daring but the problems of the country it would have to cross. No less than sixty-three bridges would be required, as well as a nine-arch viaduct and a lengthy cutting 100 feet (30 metres) deep. In addition there would be the 4 mile (6.4 kilometre) crossing of Chat Moss, a stretch of swampy ground long considered impassable.

These problems were made to sound almost insuperable when the Liverpool and Manchester Railway Bill was debated in Parliament. Stephenson, who had been commissioned for the work and had to defend the scheme against objectors, found that in some quarters the feeling against railways was fanatical. Coach builders and all others whose living depended on horse transport, protested strongly. So did many landowners. Even the more conservative engineers of the period used such words as 'impractical'.

After an initial rejection the Bill was amended and finally approved. Building of the line went ahead. The crossing of Chat Moss was successfully accomplished. A long tunnel was built at the Liverpool end to avoid interference with main city streets, the Sankey Valley was crossed by the nine spans of the Sankey Viaduct and outside Liverpool the 2 mile (3.2 kilometre) Olive Mount Cutting was dug through the red sandstone. But there was still debate as to whether an engine pulled by steam rather than horses would be able to do the job. The scheme's financial supporters remained nervous, and to calm their nerves the directors made a daring proposal. There should, they announced, be a competition so that the best engine could be chosen.

The site selected for the trial was Rainhill, almost mid-way between Liverpool and Warrington, and here a 3½ mile (5.6 kilometre) track was laid out. Six tons was the weight limit for the competing engines which were required to draw three times their own weight at 10 miles (16 kilometres) an hour. Boiler presssure was to be no more than 60 pounds per square inch (4.14 bars), and the height of the engine no more than 12 feet (3.7 metres).

On the morning of 6 October 1829 five steam engines arrived for the contest. 'For the accommodation of the ladies who might visit the course (to use the language of the turf) a booth was erected on the south side of the railroad, equidistant from the extremities of the trial ground,' *The Times* reported on 8 October.

'Here a band of music was stationed and amused the company during the day by playing pleasing and favourite airs.

'The directors, each of whom wore a white riband in his button-hole, arrived on the course shortly after 10 o'clock in the forenoon, having come from Huyton on cars drawn by Mr. R. Stephenson's locomotive steam carriage, which moved up the inclined plane from thence with considerable velocity. Meanwhile ladies and gentlemen in great numbers arrived from Liverpool and Warrington, St. Helens and Manchester, as well as from the surrounding country, in vehicles of every description. Indeed, all the roads presented on this occasion scenes similar to those which roads leading to race-courses usually present during days of sport. The pedestrians were extremely numerous, and crowded all roads which conducted to the race-ground.'

During the afternoon, as three hundred special constables tried to keep the crowd of 15,000 away from the course, the competitors steamed up and down the track more for amusement than for experiment. 'Mr. Robert Stephenson's carriage attracted the most attention during the early part of the afternoon,' *The Times* continued. 'It can without any weight being attached to it, travel at the rate of twenty-four miles in the hour, shooting past the spectators with amazing velocity, emitting very little smoke but dropping its red-hot cinders as it proceeded. Cars containing stones were then attached to it, weighing, together with its own weight, upward of seventeen tons, preparatory to the trial of the speed being made. This trial occupied, with stoppages, seventy-one minutes, and proved that the carriage can, drawing three times its own weight, run at the rate of more than ten miles an hour.'

The trials were spread over three days, during which the paces of five competitors were tested. Two of them, the 'Cycloped' and the 'Perseverance', failed to reach the required speed. The 'Sans Pareil' broke down. The fourth engine, the 'Novelty', had been made in great haste by John Ericsson, a Swede who had settled in Britain and was to have a distinguished engineering career in the United States. 'It was the lightest and most elegant carriage on the road yesterday,' *The Times* declared, 'and the velocity with which it moved surprised and amazed every beholder. It shot along the line at the amazing rate of thirty miles an hour! It seemed, indeed, to fly, presenting one of the most sublime spectacles of human ingenuity and human daring the world ever beheld.'

But before the trials were finished, the 'Novelty' over-reached itself. Her boiler exploded and she had to be withdrawn from the contest.

Finally there came Stephenson's 'Rocket', with George Stephenson on the footplate. The 9,500 pound (4,309 kilogram) engine working at 50 pounds to the

A detail of Stephenson's
'Rocket', now in the Science Museum, London.

square inch (3.45 bars) and driving a pair of wheels by 8 inch × 16 ½ inch (20 by 42 centimetres) cylinders not only performed without effort but, when the wagons were disconnected, reached 30 mph (48 kph) without trouble. The directors of the company lost no time in ordering eight 'Rockets'.

George Stephenson and his son were not only the leading railway engineers of the day. They were also adept at public relations. They made certain that the success of the railway engine would be known and a few weeks before the official opening of the Liverpool and Manchester line invited Fanny Kemble, the actress who had made a sensational debut the previous year, to travel on the line.

'We were introduced to the little engine which was to drag us along the rails,' she later wrote. 'She (for they make these curious little fire horses all mares) consisted of a boiler, a stove, a small platform, a bench, and behind the bench a barrel containing enough water to prevent her being thirsty for fifteen miles – the whole machine not bigger than a common fire engine. She goes upon ten wheels, which are her feet, and are moved by bright steel legs called pistons: these are propelled by steam, and in proportion as more steam is applied to the upper extremities (the hip joints, I suppose) of these pistons, the faster they move the wheels; and when it is desirable to diminish the speed the steam, which unless suffered to escape, would burst the boiler, evaporates through a safety valve into the air.

'The reins, bit and bridle of this wonderful beast is a small steel handle, which applies or withdraws the steam from the legs or pistons so that a child might manage it . . .'

Then the speed rose to 30 mph (48 kph). 'You cannot conceive what that sensation of cutting the air was; the motion is as smooth as possible too. I could either have read or written; and as it was I stood up and with my bonnet off drank the air before me ... When I closed my eyes this sensation of flying was quite delightful, and strange beyond description ...'

The Liverpool and Manchester line was officially opened on 15 September 1830, the guest of honour being the Duke of Wellington. Most things went well, the event being marred only by the death of William Huskisson, Member of Parliament for Liverpool, who at a halt en route stepped down from the directors' carriage into the path of an engine approaching on the second of the two parallel tracks which had been laid.

The success of the new line signalled the start of the 'railway mania' whose intensity more than equalled that of the great era of canal building at the end of the previous century. When a Bill was presented in 1832 for a London-Birmingham line its advocates pointed out that the speed of the Liverpool-Manchester line was double that of the fastest stagecoach, that the cost of the journey had been halved and that of the 700,000 travellers who had so far been carried, only one had met with a fatal accident, that one being Huskisson. The record thus gave a great impetus to the railway movement, and 970 miles (1,561 kilometres) of line had been authorized by 1835. In addition to the London-Birmingham line, there came the Grand Junction Railway linking Birmingham and Warrington, the London and Southampton Railway, the Newcastle and Carlisle and the Great Western linking London and Bristol. In 1846 alone Parliament passed 246 railway acts – not surprisingly since the Stephensons had shown that passengers could be carried comfortably and safely at a speed greater than that of coaches, and in vehicles which ran in all weathers. Before mid-century more than 6,600 miles (10,620 kilometres) of railway line had been laid in Britain.

Although the basic design of the railway steam engine remained much as the Stephensons had conceived it – and was to remain so almost until the end of the century – railway engineers steadily incorporated one new improvement after another. Steam pressure was raised to as much as 300 pounds per square inch (20.68 bars) and pre-heating and super-heating of the steam water was introduced to increase efficiency, although even then it was estimated that only about 6 per cent of the heat produced was actually utilized effectively. More important was adoption of the 'compounding' principle. Here the steam was first used in one cylinder or pair of cylinders and then made available through its expansive power in another and larger cylinder. A saving of between 10 per cent and 15 per cent of fuel was claimed, this more than offsetting the increased complication of the engine and the greater weight.

The two coupled wheels were increased to four, and then to six on some models, and in due course to as many as fourteen. Another innovation was the introduction of wheels which did not form part of the traction system. These smaller wheels were carried on a bogie or undercarriage, first in front of the driven wheels where they helped to lead the main wheels round bends; then, and in addition, behind them, a position from which they helped support the weight of the firebox.

As the efficiency of the railway engine improved so did enthusiasm for new lines grow. But before expansion could be started it was still necessary for the promoters to win Parliamentary approval. This involved the passing of a Bill describing the course that the line was to take and which would enable the railway promoters to deal with objections by landowners across whose property the line was to pass.

A typical example of the Parliamentary process which could hamper the railway engineers of the mid-nineteenth century was illustrated by the battle fought by Isambard Kingdom Brunel for the Great Western line between London and Bristol. Indeed, if it had not been for the extraordinary character of Brunel the Bill would never have been pushed through in the time it was.

Brunel, who was so responsible for turning Victorian engineering ideas into hardware, was the exceptional son of an almost equally exceptional father, the Frenchman Marc Isambard Brunel, who had emigrated to the United States at the beginning of the French Revolution. In America he had helped plan the defences of New York before recrossing the Atlantic to Britain and selling the Admiralty his device for the mechanical manufacture of wooden blocks, some 100,000 of which were required every year by the Royal Navy. The blocks, by means of which ships' rigging was manipulated, were all made by hand but Marc Brunel devised a series of machines which were, in effect, prototypes for the machinery later used by mass production methods. 'A mechanical contrivance', the engineer James Nasmyth has written, 'was made to take the place of the human hand for holding, applying, and directing the motions of a cutting tool to the surface of the work to be cut, by which the tool is constrained to move along or across the surface of the object with such *absolute precision*, that with scarce any expenditure of force on the part of the workman, any figure, bounded by a line, a plane, a circle, a cylinder, a cone or a sphere, may be executed with a degree of accuracy, ease and rapidity, which, as compared with the old imperfect *hand system*, may well be considered a mighty triumph over matter.' The cutting machines, made by Maudslay, represented a triumph of economics. The drilling, the slotting and shaping was all done mechanically and whereas fifty men had previously been needed to complete the shells of the blocks and another sixty to prepare the sheaves, the work was now done by four and six men respectively who between them made 160,000 blocks a year.

Isambard Kingdom Brunel inherited from his father not only mechanical inventiveness but also something just as important, an instinctive 'feel' for what was possible in engineering. The father, seeing for the first time a section drawing of the first suspension bridge over the River Seine at Paris, had commented: 'You would not venture, I think, on that bridge unless you would wish to have a dive.' A few days later the bridge collapsed. In much the same way the son correctly predicted the fall of a building he had watched being built.

The brilliant son, who managed to find time for a variety of miscellaneous works large in number even by the standards of the Victorian workaholic engineers, and who was to win over Parliament in the cause of his beloved Great Western Railway, had started work in his father's office at the age of eighteen. Two years later he was appointed Resident Engineer to the tunnel which Marc Brunel was building beneath the Thames, and here he played a heroic part in mitigating the disasters to which that project was subjected. Recovering at Clifton, outside Bristol, from injuries suffered in the tunnel, he was intrigued by plans for bridging the River Avon at Clifton Gorge. He was eventually given the commission – beating, in the process, the plans of Thomas Telford, the greatest bridging engineer of his day.

Although the building of Brunel's suspension bridge across the 1,000 foot (305 metre) gorge was begun, money ran out and it was not completed until after his death many years later. Lack of money also hamstrung the Bristol to Birmingham railway whose preliminary survey Brunel had been commissioned to make. However, he was also appointed to improve the Bristol Docks a connection with the city which was to bring him to the turning-point of his career.

When in 1832 a company was formed in Bristol to build a railway line to London, Brunel applied for the post of engineer. Although it was, as he knew, his great opportunity, he nearly lost the appointment on learning that the company intended hiring the candidate who produced the lowest estimate for the line. 'You are holding out a premium to the man who will make you the most flattering promises,' he told the company directors, 'and it is quite obvious that he who has the least reputation to lose by the consequences of a disappointment, must be the winner in such a race.' He threatened to withdraw his candidature, was dissuaded from doing so, and was eventually appointed.

Thus it was to Brunel, engineer-designate, that there fell the task of answering the questions when Parliament considered the Bill to approve the line's construction. The first Bill was passed by the House of Commons but rejected by the House of Lords. The second was finally approved, in August 1835, but only after a forty-day debate before a Committee of the House of Lords. Brunel himself had been cross-examined for eleven days in succession by a Commons Committee. And the cost to the promoters before approval was given totalled £88,710 in legal and Parliamentary expenses alone.

The Great Western line from London to Bristol, later continued into Devon before crossing the River Tamar over Brunel's unique Saltash Bridge into Cornwall, was more than any other railway line the creation of one man. Within two months of his appointment as engineer, Brunel had personally surveyed the proposed route, and its alternatives, and once the line had been approved he personally began designing the cuttings, tunnels and bridges which had to be engineered. Travelling in 'The Flying Hearse', a specially designed carriage that could be fixed on a railway truck, and used along the line as it was extended westwards, he became a familiar sight in Berkshire, Wiltshire and Somerset, relentlessly supervising every detail of every detail. He would arrive at his hotel in the evening after a long day surveying or talking with landowners, take a quick dinner, and then work at his correspondence late into the night. 'Between ourselves,' he wrote to his assistant, Mr Hammond, 'it is harder work than I like. I am rarely much under twenty hours a day at it.'

Brunel himself supervised the construction of bridges, tunnel fronts and stations, determined as he put it, to make the whole line 'the finest work in England'. West of Chippenham he was able to incorporate the attractive local Bath stone to give the buildings the 'romantic' outlines that in those times were appreciated. Thus when the ground behind a tunnel front outside Bristol slipped away and made it unnecessary to complete one of the side walls, Brunel had it left unfinished and planted with ivy to present the appearance of a ruined gateway.

At many places he designed 'flying bridges', each end of the single-arch bridge resting on the side of a cutting. The bridges could be built before the cutting was completed and eliminated the need for the expensive wooden centring which would otherwise have been required. He did not limit his interests strictly to railway engineering, and when the line was eventually extended to the Swansea and Briton Ferry docks, it was Brunel who solved the problem of the exceptionally friable coal which broke up badly if loaded in the usual way. He introduced on a large scale, his son has reported, 'the use of trucks carrying four iron boxes, each box about 4 ft. 8 in. in cube, and containing two and a half tons of coal. At the docks, machinery is provided by which each box is lowered down in to the hold of the ship, and the under side being allowed to open, the coal is deposited at once on the bottom of the vessel.'

The ability to tackle successfully the most disparate tasks was a mark of the Victorian engineer, the last of the species before the growth of specialized knowledge began to divide the profession into more nearly watertight compartments. Brunel the younger certainly typified this generation of engineering polymaths. He designed the two 285 foot (87 metre) water towers erected at either end of the Crystal Palace when that building which had housed the Great Exhibition of 1851 was moved from Hyde Park to Sydenham on the outskirts of London. He also submitted plans for the hospital built at Renkiol during the Crimean War, and offered the authorities plans for a revolutionary gunboat. The vessel would have had similarities with the American *Monitor* which in its battle with the *Merrimac* during the Civil War helped change the strategy of naval fighting. Hardly to the surprise of Brunel, a man frequently at war with bureaucracy, the Admiralty was not interested.

In 1835 the protests which Brunel had to face when defending his plans for the Great Western were typical of the times, however curious they look a century and a half later. Dr Hawtree, an Eton master, believed that if a railway passed near the College, pupils would visit London surreptitiously, be attracted by Continental tastes, and would forsake what he called the entire classical tradition in English education. 'Homer, Virgil, Horace, and other classical authors – with the exception of Ovid's "Art of Love" – would be forced into the background in favour of Rousseau, Voltaire, and equally destructive writers,' he maintained.

However, the railway engineers encountered more substantial objections than that, and sometimes from unexpected quarters. Thus Richard Cort, son of the iron founder who had revolutionized the craft, issued in 1834 *Railway Imposition Detected*, a propaganda booklet in which he claimed, 'Veins of water will be cut, springs dried up, and many of the sloping fields along the line so deprived of water that they will become sterile and unfit for pasturage, or the stock driven to a distance for a supply, greatly to its own injury, and with much additional cost. Besides, whole fields and estates will be torn asunder by immense gashes and mounds, over and under which expensive bridges and long and wide tunnels must be constructed.'

The Great Western, among the most notable engineering feats of its time, was to be famous for Box Tunnel, the longest in the world when it was built; for the bridge by which the railway crossed the Thames at Maidenhead on two of the flattest arches ever built in brick; and for its 7 foot (2.1 metre) wide track, the 'broad gauge' which split the railway world into two opposing factions.

About 1⅞ of a mile (3 kilometres) in length, Box Tunnel was described by its critics as 'monstrous and extraordinary, most dangerous and impracticable.' It was feared that passengers might suffocate in passing through it and that since it was built on an incline – of only 1 in 100 – a train would shoot from the tunnel mouth at more than 100 mph (160 kph) if its brakes failed. There might have been further criticism had it been known that the line of the tunnel was so planned that

Brunel's Paddington Station, London, painted by William Frith.

the rising sun shone through its entire length on 9 April, Brunel's birthday.

Work on the tunnel started in September 1836 and was finished in June 1841. The western three-quarters was blasted through clay, blue marl and the inferior colite, and was then brick-lined. The eastern quarter, cut through Bath stone, was left unlined. For months on end a ton of gunpowder and a ton of candles was used weekly. No less than 30 million bricks were produced locally and brought to the site by a hundred horses and carts.

As with everything else connected with the line, Brunel involved himself personally. 'If ever I go mad', he wrote to Charles Saunders, the Secretary and General Superintendent of the G.W.R. Company, 'I shall have the ghost of the opening of the Railway walking before me, or rather standing in front of me, holding out its hand, and when it steps forward, a little swarm of devils in the shape of leaky pickle-tanks, uncut timber, half-finished station-houses, sinking embankments, broken screws, absent guard-plates, unfinished drawings and sketches, will, quietly and quite as a matter of course and as if I ought to have expected it, lift up my ghost and put him a little further off than before.'

The grand opening of the first part of the line finally took place on 4 June 1838 when four trains carried a total of 1,479 passengers from London to Maidenhead. By March 1840 the track was open to Reading and by the summer of 1841 to Bristol. The journey from London to Bristol took only four hours, but some nervous passengers still insisted on leaving the train before it entered Box Tunnel and boarding it again at the western end.

Meanwhile the Great Western had been extending its line into south Wales and into Devon. The effect was dramatically shown in London on 1 May 1844 when Sir Thomas Acland left Exeter at 5.20 pm, arrived in London at 10.00 pm and half an hour later was telling the House of Commons that he had been in Exeter, 194 miles (312 kilometres) away, little more than four and a half hours earlier.

The temporary terminus at which Sir Thomas arrived in London was to be replaced, a few years later, by a building which epitomized Brunel and his approach to everything connected with the Great Western. 'I am going to design, in a great hurry, and I believe to build,' he wrote on 13 January 1851 to Digby Wyatt, 'a Station after my own fancy; this is, with engineering roofs, etc., etc. It is at Paddington, in a cutting, and admitting of no exterior, all interior and all roofed in.' The result with its great roof, still standing today, was one of the minor marvels of the London scene.

It was on the line that Brunel drove west from Exeter that he experienced one of his more costly engineering failures. The first part of the 52 mile (84 kilometre) route to Plymouth follows the course of the Exe Estuary. It then cuts across a number of deep valleys which run down to the Devonshire coast from the high plateau of Dartmoor. Whichever route was taken trains would therefore have to make four steepish ascents totalling about 10 miles (16 kilometres), and Brunel decided to employ the Atmospheric Railway System.

This system had been operated on a short experimental section of line in London in 1840 and had been used with apparent success in Ireland. It involved the laying of a 15 inch (38 centimetre) diameter tube between the two rails. Along the top of the tube ran a slit about 2½ inches (6 centimetres) wide, closed by a long flap of leather which was strengthened with iron plates and secured to the tube at one side of the slit. At the other side, a composition of grease sealed the flap to the tube. Inside the tube there lay a piston bearing an attachment which protruded through the flap and could be attached to a train. At intervals of about 3 miles (5 kilometres) there were steam engines which exhausted the air from the tube in front of the piston, creating a vacuum and thus allowing the pressure of the air behind the piston to push it forward together with the train to which it was attached.

Everything rested on the efficiency with which the flap maintained the vacuum in the tube, and after extensive operations and continual adaptations and improvements, the efficiency was often far from perfect. Freezing weather, percolation of water into the tube, and the attraction to rats of the greased leather, combined to make the system unworkable. On good days it could achieve a running speed of 40 mph (64 kph), but at others it broke down completely, and in September 1848 Brunel reluctantly advised its abandonment.

There were few railway ideas as revolutionary as the failed Atmospheric System, but railway track itself changed regularly if slowly over the years. It continued to be supported by cast-iron chairs which were spiked on to transverse wooden sleepers but in 1857 steel instead of iron was used for the track in Derby. The results were spectacular; whereas iron tracks had to be renewed every three months, steel was to last for sixteen years.

An issue which created considerable argument as railways multiplied during the 1840s was the route that a track could most efficiently follow. The early railway engineers, like the canal builders, had usually taken the flattest route open to them. But as the efficiency of engines improved, and as it was found that they could pull loads up steeper gradients than had been foreseen, a problem arose that became evident in the choice of a western route from London to the Scottish border.

The difficulty presented itself between Lancaster and Carlisle. Joseph Locke, who with the Stephensons and, later, the Brunels, was among the foremost railway engineers of his day, proposed that the line should take the direct route from Lancaster to Carlisle, involving a 4 mile (6.4 kilometre) climb at a gradient of 1 in 75 up the slopes of Shap Fell. The Stephensons ruled this out and proposed, instead, a route following the coast which was 30 miles (48 kilometres) longer but which avoided the hills and mountains of the Lake District. The financial equation was a complicated one since the cost of building an additional 30 miles of track had to be balanced against the cost of the fuel consumed on the climb to Shap summit. Locke won the argument and between July 1844 and the end of 1846, some 10,000 navvies built the line to Shap. Less than twenty years later a survey showed that a potentially successful compromise had been possible: a mile-long tunnel would have eliminated the worst of the climb at little increase in length of line.

Two years after the Shap line was built, the 'battle of the gauges' was finally ended – to Brunel's discomfiture. Whether or not adoption by the early colliery railways, and later by George Stephenson, of a gauge of 4 feet 8½ inches (1.43 metres) arose from the gauge of the Roman chariot, it was employed by most of the early railway engineers in Britain. Chance appears to have played an important part in the decision and Robert Stephenson, asked by a member of the official commission investigating gauges whether his father had deliberately advocated it, replied 'No. It was not proposed by my father. It was the original gauge of the railways about Newcastle-upon-Tyne, and therefore he adopted that gauge.'

Brunel, who had persuaded the directors of the Great Western to use a 7 foot (2.1 metre) gauge, later explained his reasons to the Gauge Commissioners, set up to investigate the question in 1845. 'Looking at the speeds which I contemplated would be adopted in railways, and the masses to be moved,' he said, 'it seemed to me that the whole machine was too small for the work to be done, and that it required that the parts should be on a scale more commensurate with the mass and velocity to be obtained. I think the impression grew upon me gradually, so that it is difficult to fix the time when I first thought a wide gauge desirable; but I daresay there were stages between wishing that it could be so and determining to try and do it.'

The standard gauge lines from the Midlands and the north met the Great Western's broad gauge at Gloucester and many ingenious attempts were made to minimize the difficulties of passengers making through journeys. A competition showed that higher speeds could be obtained on the broad gauge, but official opinion came down in favour of 4 feet 8½ inches (1.43 metres) and the Gauge Act of 1846 decreed that in future all lines should be built to this width unless specifically exempted.

Once traffic on the railways began to grow, two safety meaures became increasingly necessary. One was an improvement in braking methods, the other the development of a more sophisticated signalling system. Braking by steam pressure was at first used, but to be effective it had to be applied not only to the engine but to each of the following carriages or wagons. Before the start of a descent the train was halted for the brakesman to screw down the brake on each carriage. This eventually became impracticable, but it was not until 1867 that the problem was finally solved by George Westinghouse, an American who had learned that the cutting of the Mont Cenis Tunnel through the Alps had been carried out with the use of compressed air, delivered to the working face through tubes more than half a mile (800 metres) long. Why, he asked, should it not be possible for a compressed air tube to be carried the length of a train, from engine to last wagon, and each wagon braked simultaneously by a compressed air cylinder operated from the driving cab? The system, successfully introduced in the United States, soon spread to Europe.

The first signals were erected on the Liverpool and Manchester line about 1834. Each consisted of a rectangular frame on which a red flag was stretched, the frame being fixed to a vertical pole turned by a handle near the ground. When the red flag was facing the oncoming engine it indicated danger; when turned parallel to the line, all was clear. At night, oil lights showing either red or white did the same job. This simple method was superseded in 1841 by the semaphore arm, introduced by Sir C.H. Gregory, which gave the all clear when dipped and indicated 'Stop' when raised to the horizontal position.

The semaphore quickly spread but was eventually overtaken by a variety of more modern, but more complicated, methods. These included the block system which ensured that a minimum distance was maintained by trains on the same line; a code of signalling by beats on an electric bell, and the lock and block system which automatically prevented more than one train being on any particular section of a line.

After the mid-nineteenth century the possibilities of the steam engine were to have a steadily growing impact in one perhaps unexpected sphere, that of agriculture. Although it had been clear, soon after the Stephensons' success, that horses would progressively give way to steam and that the stagecoach would soon be replaced by the railway coach, it was another two decades before it was considered feasible to replace the farm horse as a main source of power. In 1849 a Leeds foundry built the 'Farmer's Steam Engine'. The contrivance, looking like a small railway locomotive, could be driven to the place on the farm where it was needed. Its rear was then jacked up and an axle of the road wheels was used to operate a threshing machine by means of a belt drive.

Attempts were made to adapt such Portable Engines for other agricultural purposes. The most important was ploughing but for long it was ruled out by the fact that any steam engine powerful enough to do the job was so heavy that it got bogged down. Boydell wheels, invented by James Boydell and carrying a succession of broad plates on which the wheel rim ran – almost a prototype caterpillar tractor – were of only limited success, and various engineers began experimenting with systems in which stationary steam engines dragged ploughs across fields by winding them in on long ropes. In practice these schemes brought their own problems and it was not until the later 1850s that John Fowler, an engineer who turned to agriculture problems after seeing the tragedy of the Irish famine, produced a partial answer. It involved the use of two powered winches which drew a plough backwards and forwards between them. Improved over the years not only for ploughing but for land clearance, Fowler's equipment was said to have lowered the cost of cultivation in some areas by nearly a third.

In America, where the areas involved were larger than in Britain, steam agricultural machines were developed more quickly, and Mark Twain had described a typical operation near New Orleans. 'The traction engine travels about on its own wheels, till it reaches the required spot,' he wrote; 'then it stands still and by means of a wire rope pulls the huge plough toward itself two or three hundred yards across the field, between the rows of cane. The thing cuts down into the black mould a foot and a half deep. The plough looks like the fore and aft brace of a Hudson river steamer, inverted. When the negro steersman sits on one end of it, that end tilts down near the ground, while the other sticks up high in the air. This great see-saw goes rolling and pitching like a ship at sea, and it is not every circus rider that could stay on it.'

While agriculture was benefiting from this spin-off from the railway engine, railway transport in Europe continued to spread. Here, where conditions were often similar to those in Britain, and where British influence was considerable, the first line, initially operated by horses, appears to have been built in the 1820s from Budwies to Trojanov in the Austro-Hungarian Empire. In 1831 Belgium became independent from Holland and with her new sovereignty found herself occupying that strategic area in western Europe which centuries earlier had been known as 'the Cockpit of Europe'. Now her rulers saw her as a potential focus for railway transport, and in 1834 a line was opened from Antwerp to Cologne with great plans for further expansion. At the start three British engines were used, the 'Arrow', the 'Stephenson' and the 'Elephant'. George Stephenson remained for some years the leading railway engineer on the Continent, although eventually equalled and then overshadowed by Brunel, and travelled thousands of miles as consultant to the scores of companies which by 1850 had built 3,777 miles (6,078 kilometres) of line in Germany and 1,927 miles (3,101 kilometres) in France.

There were, however, two differences between the task of the railway engineer in Europe and his task in Britain. On the Continent the distances normally travelled were longer. Although the Tsar built in October 1836 a 17 mile (27 kilometre) line from St Petersburg (now Leningrad) to the Imperial pleasure palace at Tsarskoe Selos – a line of 6 feet (1.8 metre) gauge built on an embankment to avoid the effects of drifting winter snow – most lines tended to be longer than those which radiated from London. Thus that to Tsarskoe Selos was followed in 1851 by one from St Petersburg to Moscow, built by the American engineer George Washington Whistler who brought US machinery and workers to Russia and railway lines imported from Britain. The greater European distances soon led to a demand for sleeping cars and for dining facilities. Somewhat primitive sleeping quarters were provided in Germany in 1870 and two years later George M. Pullman introduced into Germany his eponymous coach with compartments for sleeping or for resting in armchairs. Another eight years saw the first dining cars. They were followed, in 1883, by another railway luxury, the Orient Express which carried its passengers without a change, first from Paris to the Danube at Giorgio and later as far east as Constantinople.

The second difference between Britain and the Continent was that the latter faced the railway engineer with the Alps. However ingeniously a line across them was constructed, stiff gradients could not be avoided and it was still considered questionable whether sufficient adhesion could be obtained on the 1 in 40 slopes which would have to be traversed. One man believed that it could be. He was Karl Ritter von Ghega, an Austrian who fervently believed that it would be possible to cross the Alps from Gloggnitz, where the railway ended in 1842, by way of thr 3,215 foot (980 metre) Semmering Pass. It was a project with great possibilities since a line acrosss the Semmering would link Hamburg with Trieste, the North Sea with the Adriatic.

Von Ghega began work in 1848. In places avalanche sheds were erected to protect the track. Between Gloggnitz and Murzzuschlag, 34 miles (55 kilometres) to the south, fifteen tunnels had to be built as well as sixteen viaducts. Thousands of workers from all over Europe were employed in building the line and the 1,564 yard (1430 metre) summit tunnel. Then, in 1851, the authorities had renewed doubts about the traction problem and like the proprietors of the Liverpool-Manchester railway in 1829 decided that only a competition would settle the matter satisfactorily.

They chose a stretch of line near Eichberg with a gradient of 1 in 40, and here four locomotives were set the task of hauling 140 tons at a minimum of 7 mph (11 kph). The 'Bavaria' came from Munich, the 'Seraig' from Belgium and the 'Wiener-Neustadt' and 'Vindobona' from Austrian engine yards. All four vehicles passed the test, the 'Bavaria' being judged the winner. But none of the performances was considered perfect, the results of the competition were embodied in another locomotive, and it was this which pulled the first carriages across the Semmering on 12 April 1854.

This was the year which proved that the railway had a military potential, one which was to be exploited for the rest of the century and up until the outbreak of World War I. In 1854 the Crimean War was being fought and the Russian forces in the apparently impregnable fortress of Sebastopol were under siege by Anglo-French troops. Within six weeks, a 20 mile (32 kilometre) double track line was laid from the port of Balaclava. Along it, heavy guns and ammunition were carried to the front-line troops outside Sebastopol. The bombardment, causing the fall of the key position, would have been impossible without the railway.

The lesson was not lost. The unification of Italy under Victor Emmanuel II two decades later was helped by the railway lines built by the British contractor Thomas Brassey on the North Italian plain. Across the Atlantic, the North won the American Civil War partly because of its superior railway network, and half a century later the Germans' Schlieffen Plan for attacking France through Belgium at the start of World War I rested on the quick concentration of the German forces by railway.

All that lay in the future as the conquest of the Semmering showed that even the Alps could be crossed by rail. Where one man has succeeded others will follow. This was so with the railway engineers, and the Semmering railway was the first of several. Next came the Fell railway from Lanslebourg to Susa across the Mont Cenis, using a central rail grasped by horizontal wheels to brake the carriages. The track rose in sweeping zigzags and for nearly 16 miles (25 kilometres) had a gradient steeper than 1 in 15. 'In some places,' wrote the British mountaineer, Edward Whymper who attended its opening, 'it rose one foot in twelve and a half! An incline of this angle, starting from the base of the Nelson Column in Trafalgar Square, would reach the top of St. Paul's cathedral if it were placed at Temple Bar!'

The Fell railway was followed by the conventional nearby Mont Cenis route; by a railway across the Brenner; and by the St Gotthard, built by Louis Favre, which had 324 bridges, a main 10 mile (16 kilometre) tunnel and eighty minor tunnels. The latter included the famous helicoidal tunnels which enter the mountainside at one height, make one or more complete circles inside it and emerge at a level higher than that of the entrance. Between Lucerne and Chiasso there are no less than 28 miles (45 kilometres) of such spirals, the Pfaffensprung rising 120 feet (36 metres) in a multiple loop.

While the railways continued to tunnel through the Alps, Niklaus Riggenbach took his rack railway to the top of the Rigi beside the Lake of Lucerne, the forerunner of those numerous mountain railways which culminated in 1912 in the Jungfrau railway which tunnels through the Eiger in the Bernese Oberland to reach the Jungfraujoch station, at 11,000 feet (3,353 metres) still the highest in Europe.

Nevertheless, Europe did not lead the way in mountain railways since it was in 1869, two years before the Rigi line was opened, that a railway was engineered to the 6,300 foot (1,920 metre) summit of Mount Washington in the White Mountains of the United States. This was perhaps to be expected since the Americans had been quick to exploit the potential of the steam locomotive. 'The father of American railroads' was Colonel John Stevens who had initially interested himself in steamboats and had built, in 1806-8 with Nicholas J. Roosevelt, a vessel which plied between Trenton and Philadelphia. As early as 1812 Stevens was petitioning Congress to support a national railroad but his efforts were frustrated, and it was not until 1826 that he built, and operated on his own estate in Hoboken, New Jersey, the first American steam engine to run on rails. It was purely a private experimental enterprise, however, and it was left to Horatio Allen, of the Delaware & Hudson Canal Company to run publicly America's first steam locomotive, the 'Stourbridge Lion', built at Stourbridge in England and brought to America by Allen in 1829. The 7-ton 'Lion', equipped with four driving wheels made of oak but rimmed with iron, was taken by canal to Honesdale, Pennsylvania, but was too heavy for the Honesdale-Carbondale track.

The 'Lion' was followed the 1-ton 'Tom Thumb', built by the wealthy Peter Cooper of New York. But although successful, 'Tom Thumb' was used only for demonstration purposes and it was left to 'The Best Friend of Charleston', designed by a Charleston engineer but built at the West Point Foundry in New York, to start the first regular railway service in America. Operating on the South Carolina Railroad, it pulled up to fifty passengers at speeds of more than 20 mph (32 kph) for a year and a half before being destroyed after a safety valve had been mistakenly tied down.

In 1831, the year in which 'The Best Friend' was wrecked, Matthais Baldwin, to become the leading pioneer of the early American railroad industry, built a model locomotive. It was followed by 'Old Ironsides', which when put into service on the Philadelphia, Germantown & Norristown Railroad, averaged 28 mph (45 kph). 'From this experiment', wrote the Philadelphia *Chronicle* on 24 November 1832, 'there is every reason to believe that this engine will draw 30 tons gross, at an average speed of 40 miles an hour. The principle superiority of the engine over any of the English ones consists in the light weight – which is between four and five tons – her small bulk, and the simplicity of her working machinery. We rejoice in the result of this experiment, as it conclusively shows that Philadelphia always produces steam engines for railroads combining so many superior qualities as to warrant the belief that her mechanics will supply nearly all the public works of this description in the country.'

The claim was perhaps the result of over-optimistic local pride, but 'Old Ironsides' was nevertheless to run without trouble for twenty years and was the first of the scores of successful engines built by what was to become the internationally known Baldwin Locomotive Works. Other works sprang up in the Eastern States as throughout the 1830s and 1840s numerous lines were opened. No more British engines were to be imported, mainly because of their weight. But it was not only the locomotives that developed independently in America. So did the railways themselves, and in more than one way. In Britain, densely populated compared with the United States, the shorter distances and the greater concentration of industry almost ensured a good return on capital, so that the railway promoters could afford major engineering works such as tunnels and bridges. In the United States there was no wealthy aristocracy and as yet only few wealthy families to fund the railways, many of which were built to open up the West to immigrants who had only limited resources. Thus the tracks that spread across America in the 1830s and 1840s, and the engines that ran on them, developed in a context very different from that of Britain.

The tracks themselves tended to be built with sharper curves than in Europe which produced troubles that sometimes ended in derailment. The situation was dealt with in 1832 when John B. Jarvis, chief engineer of the Mohawk and Hudson Railroad, invented and designed the bogie independently of what had been produced in Britain. This was a four-wheeled carrier attached beneath the forward part of the engine by a centre-pin which allowed it to swivel as the engine went round a curve. The same device was used when the short passenger coaches were superseded by the longer Pullman coaches with seats arranged facing each other for the full length of the carriage. A bogie at either end enabled the coach to take sharper curves.

In the United States two other innovations followed from lack of coke and from the steeper gradients which often had to be dealt with. Wood, rather than the coke used in Britain, was usually the fuel, and deeper and narrower fireboxes had to be built as containers. The steeper gradients demanded more power, with the result that boiler pressures were increased – to an extent that would have alarmed George Stephenson and possibly even Richard Trevithick. The cold American winters and the fact that railway tracks traversed open country rather than the cultivated landscape of Europe also affected design. The rigours of temperatures rarely known except in the far north of Britain, where railways were few, quickly led to engine cabs for drivers and firemen. The railway across the open prairie demanded bells and whistles to warn pedestrians by day and headlights to warn them at night. Sanding apparatus was required to deal with insect swarms which would otherwise have reduced traction, while the cowcatcher became for many in Europe a symbol of the American locomotive.

Materials and manufacturing processes were similar to those in Britain, although the Americans appear to have been more cost-conscious and soon began to use wrought iron for many parts of the locomotive. At first America tended to follow the British practice of a 4 foot 8½ inch (1.43 metre) gauge. But in the South a gauge of 5 feet (1.5 metres) was laid and for the Lake Erie railway one of 6 feet (1.8 metres). By mid-nineteenth century there were more than a dozen different gauges in existence and the difficulties that this created when amalgamations were proposed forced Congress to insist on all future lines being built to a single gauge. The British width of 4 feet 8½ inches (1.43 metres) was chosen.

The railway in Britain not only made the Industrial Revolution possible but had dramatic social results as populations which had previously been tied to local

An American 2.8.0. Damens Class S.160 railway engine
built during World War II for transporting US troops in Europe.

areas were able to visit the large cities, and even to have seaside holidays. In the United States of the mid-nineteenth century, its population spread out more thinly across vaster distances, the social factor was even more important. As Carlton J. Corliss has said of the early settlers in *Main Line of Mid-America: The Story of the Illinois Central*, 'The railway, in whose shadow they built their shingled homes, remained their guardian angel. It gave the signal for an advance of civilisation without parallel. When, for example, the great north to south line was built across the prairies, no settler could remain a hermit. Behind the fence, farms, factories, shops, inns and warehouses grew like mushrooms.' This no doubt does much to explain the fact that as early as 1850 American track mileage was 9,072 miles (14,600 kilometres) compared with Britain's 6,658 miles (10,715 kilometres), Germany's 3,777 miles (6,078 kilometres) and the French 1,927 miles (3,101 kilometres). By 1900 the contrast had become still greater with America having 192,000 miles (309,000 kilometres) of track and Britain 20,000 miles (32,000 kilometres). However, much of the American mileage was single-track whereas in Britain lines were predominantly double-track.

However, the greatest social impact of the railways in North America was that of the two transcontinental lines, one in the United States, the other in Canada, which linked the older east coast communities with those of California and British Columbia. Both were epic enterprises overcoming the engineering difficulties of crossing the Rockies as well as the opposition of the native Indians.

The Pacific Railroad Act was passed by the US Congress on 1 July 1862, while the Civil War was still raging fiercely. Under it, the Central Pacific Railroad was to build eastwards from Sacramento, the Union Pacific to build westwards from Omaha, Nebraska. Each company was given a 400 foot (122 metre) right of way for the track, support from government funds for construction and alternate square-mile sections of public lands on both sides of the track.

From the start of the work in June 1863, construction soon developed into a race between the two companies, each intent on laying more track than its rival and thereby getting most of the land grant. The Union Pacific had an advantage in the early days since west of Omaha there lay 500 miles (800 kilometres) of easy grassland; the Central Pacific, on the other hand, was soon faced with blasting nine tunnels to get through the mountains. Thus when the summit tunnel was finished in November 1867, only 131 miles (211 kilometres) of the western stretch had been built compared with many hundreds of miles from the Omaha base.

The end of the Civil War in 1865 had eased the labour problems but a disastrous thaw the following spring swept away many of the bridges, culverts and embankments which had been built at such cost of labour. At the same time the Indians launched major attacks on the line, and an account by Edwin L. Sabin in *Building the Pacific Railway*, not only illustrates the size of the railway engineers' task also but the constant danger of slaughter by American Indians.

'The snaky, undulating double row of glistening rails stretching on, on, on into the west; the sidetracks filled with supply trains bearing hundreds of tons of iron and thousands of ties; the last terminal base, brimming with a riotous life, grotesque with makeshift shacks and portable buildings, amidst which Casement

Tracks of the elevated railway at Chicago, Illinois.

Brothers' huge takedown warehouse with dining-room loomed portentous; then out to end o' track, a route impeded every few miles by more construction trains awaiting their turn, and at the end o' track the boarding train for the track gangs, of dining-cars, bunk-cars, the combined kitchen, stores-car and office-car, each 80 ft. long, with beds made up atop and hammocks slung to braces and trucks underneath; the dusty line of wagons toiling still on and on, bearing ties, hay, what-not, up the interminable grade; on the grade, ant-like figures, delving, ploughing, scraping, cursing; beside the grade, the 'grader's forts' – of dug-outs half beneath the ground, roofed maybe with sheet-iron, sheltering two or four or six men apiece in time of Indian attack.'

In addition to the Indians there were the natural hazards, avalanches in the mountains sweeping away track and bridges, and buffaloes on the plains removing the essential wooden markers placed by the surveyors. Despite the recruitment of demobilized soldiers and of thousands of Irish labourers, the Central Pacific continued to bring in Chinese coolies of whom there were eventually some 10,000 at work.

Towards the end of 1868, as only a few hundred miles separated the two lines, competition grew. The Central Pacific maintained it would lay a mile of track a day. The Union Pacific responded by laying 6 miles (9.7 kilometres), and the Central Pacific answered by laying 10 miles (16 kilometres). It had been part of the companies' contracts that they should lay a telegraph wire alongside the track and news of the day's 'score' was thus available each evening.

Eventually Congress ordered the two companies to meet at a designated place, Promontory Point to the north of the Great Salt Lake, Utah. By the end of December 1868, only 312 miles (502 kilometres) separated the approaching lines, and it was calculated that the job would be completed by May – after six years instead of the estimated fourteen. The calculation was correct and on 10 May 1869, Leland Stanford, one of the four Sacramento merchants who had financed the Central Pacific, and D.C. Durant, vice-president of the Union Pacific, used silver sledge-hammers to drive golden spikes holding down the final rails.

It was more than a decade before the second transcontinental line was completed. The link between the east and west coast of Canada would not have been achieved even then had it not been for fear that British Columbia, almost isolated from the rest of the country, might secede from the Canadian Confederation, which she had joined only in 1871, unless she were given a rail connection with the east.

On 21 October 1880 the Confederate Government signed a contract with a syndicate, giving the Canadian Pacific Railway 25 million dollars and 25 million acres. In return a railway was to be built across the Rockies via the Yellowhead Pass. The engineering and other problems were similar to those which had been faced by the Americans a decade earlier, although the subjection of the Indians had lessened that threat and the experience of the Central and Union Pacific companies obviously helped. Eventually it was decided to cross the Rockies not by the Yellowhead but by a shorter although more difficult route over Kicking Horse Pass. The change of plan was justified. The undertaking was completed in four and a half years and on 7 November 1885, the last iron spike was driven at Craigellachie. Canada now had her 4,325 mile (6,960 kilometre) railway line from the Atlantic to the Pacific.

While transcontinental railway lines continued to spread throughout both

Europe and the Americas, steam remained the only motive power until the end of the 1870s. Even the details of the railway engine remained remarkably similar to those of Brunel's Great Western engines a quarter of a century earlier. Not until 1879 did Ernst Werner von Siemens use a small electric motor to power a locomotive at the Berlin Exhibition. It was little more than a demonstration unit, and four years were to pass before Siemens used hydro-electric power in Ireland to provide current for an electric tramway at Portrush, the water turbines driving a Siemens dynamo which powered a motor on the tramcar. The lack of fumes gave electricity one main advantage over steam as a source of power. and when an underground railway was opened in London in 1890 between King William Street and Stockwell, electricity was chosen.

While it was to become a popular source of locomotive power in the twentieth century, one of electricity's early applications was in conjunction with the diesel engine. In 1893, almost two decades after Niklaus August Otto had built the first internal combustion engines, Rudolf Diesel published a paper 'On the Theory and Construction of a Rational Heat Engine'. It was to dispense with the artificial ignition of Otto's engine, the highly compressed air in the cylinder itself igniting the crude oil fuel. The system was adopted without much success in locomotives but there was developed from it the diesel – electric system in which the power developed from a diesel engine was transformed into electric power by a generator, and the electric power then used to turn the driving wheels.

In the twentieth century, the steam locomotive has steadily disappeared, being replaced by improved diesel engines and also by electric power, much of it distributed by overhead wires above the tracks. Rocket power, used by Fritz von

Union Pacific diesel locomotives at Green River, Wyoming.

The Wuppertal overhead railway in Western Germany. Built before World War II,
it has been renovated recently. The cars have been updated but the cabinets are modern.

Opel in 1928, proved impracticable, although one rocket-powered train reached 254 mph (409 kph) before exploding. Speeds of the same order were reached in Germany by the propeller-driven aluminium railcar designed by Frank Kruckenberg in 1931.

In the later years of the twentieth century, roughly a century and a half after George Stephenson had started to revolutionize the communication patterns of the world, railways have reached a peak of speed, safety and comfort far beyond the imaginings of the pioneers. Speeds well in excess of 100 mph (160 kph) have become regular in France and Japan, while in the early 1980s Britain had ninety-five High Speed Trains, most of them in daily operation at up to 125 mph (200 kph). Accidents, when weighed against the huge number of passenger-miles involved, are low. And in many countries super trains provide luxurious dining, wining and sleeping in addition to fast travel. The factor limiting contemporary railways is no longer engineering technology but finance, and it is significant that in most developed countries the railways have to be subsidized by the State.

Bullet train at Tokyo Station, Japan.

NEW METALS AND BETTER BRIDGES

The half-century which followed the opening of George Stephenson's Stockton to Darlington railway was the heyday not only of the railway promoter but also of the bridge builder. The heavily loaded waggons carrying coal to the ports, the trains which took hundreds of passengers at a time from one part of Britain to another, made demands on the constructional engineer far greater than had ever been made. This same half-century – roughly from 1830 to 1880 – also witnessed two other developments which in both Britain and the United States raised impressive examples of engineering enterprise, many of which are still in existence. One was the extended exploitation of cast iron for more purposes. The other was the invention of new steel-making processes which enabled this remarkable material to be produced more cheaply and in greater quantity than before. Both cast iron and steel along with wrought iron were to be employed by the bridge builders but their growing availability brought them into other fields at a time when industry was developing with almost explosive force.

Cast iron had, it is true, been used as far back as 1664 for the pipes supplying water to the fountains in the Gardens of the Palace of Versailles, and throughout the eighteenth century a succession of experiments and investigations began to reveal both the possibilities and the limitations of the material. Pieter van Musschenbroek of Leyden found that the buckling resistance of slender wrought-iron struts decreases with the inverse ratio of the square of their length, and published a table showing the strengths of various materials when subjected to compression, tension and bending. Comte Georges Louis Leclerc de Buffon, best known for his 44-volume *Natural History*, 1749-1804, measured the extent of sagging in iron beams before they broke, while Claude Louis Marie Henri Navier took comparable measurements on the loads that wrought-iron plates would bear before rupturing.

While information on the safe use of iron was thus being accumulated, the scientific approach to building in general was receiving encouragement from a move within the Vatican. One of the greatest structural achievements of the Renaissance had been the completion of St Peter's in Rome. The architect commissioned by Pope Julius II, Donato di Angelo Bramante, died only eight years after the foundation stone had been laid in 1506. Michelangelo worked on it from 1546 until his death in 1564. Then, after an interval of over twenty years, the architects Giacomo della Porta and Domenico Fontana who, in 1586, had successfully moved the great obelisk to a new position before St Peter's, completed the dome between 1586 and 1590, greatly modifying Michelangelo's original design, and used three great iron girders, embedded in the base of the dome to counteract lateral thrust.

By 1742, however, cracks were appearing in the structure and Pope Benedict XIV ordered a commission of three men, Le P. Le Seur, François Jacquier and

Roger Boscowich, to calculate the dimensions of the iron reinforcement required. They made their report largely on the basis of Van Musschenbroek's tables.

From this time onwards it became more usual to build on the basis of scientific knowledge, so that when iron came into its own during the first years of the Industrial Revolution there was a growing knowledge of what could safely be done with it. As early as 1784 James Rennie had used cast iron for most of the machinery in the Albion Flour Mills on the banks of the Thames near London's Blackfriars Bridge. The machinery here was driven by steam engines designed by Boulton & Watt who employed Rennie to design and supervise the erection of the mills. By the start of the nineteenth century cast-iron beams were well established as structural units in mills and factory buildings, while arched beams held by horizontal iron ties were being used for road bridges.

Meanwhile, scientific investigation continued. In 1803, Charles Bage carried out the first full-scale load tests of iron roof frames. Thomas Tredgold published a new edition of his work on the strength of cast iron. William Fairbairn, a leading engineer, and Eaton Hodgkinson, mathematician and specialist on the strength of metals, carried out extensive experiments that in 1831 were published in Manchester as 'Theoretical and Experimental Researches to Ascertain the Strength and Best Forms for Iron Beams', and republished the following year in the *Journal of the Franklin Institute* in Philadelphia. The work quantified for the first time the differences between wrought and cast iron, and showed that while cast iron (with high compressive but relatively low tensile strength) was best for columns (compression members), wrought iron (higher in tensile strength) was best for beams – those members subject to tension.

On the Continent iron was becoming popular for a number of ambitious structures. Among them was the Paris Corn Market whose cupola was supported on cast-iron ribs spanning 131 feet (40 metres) when they were erected in 1809. In 1827 iron was used for the cupola roofing of the east choir of Mayence Cathedral. The early ventures continued to influence architects and later in the century Henri Labrouste employed iron to dramatic effect in two of his buildings, the Bibliothèque Sainte-Geneviève (1850) and the Bibliothèque Nationale (1861), both in Paris. In the first, Labrouste made no attempt to conceal the iron of the columns holding the metal roof-vaulting.

An important development which greatly reduced the cost of making iron had been initiated in 1828. Before that date it had been generally accepted that the efficiency of a blast furnace was directly linked to the coldness of the air current injected into it. Indeed, some ironmasters passed the air over cold water before use or even through pipes packed with ice. However, in 1828 James Nielson, foreman of the Glasgow gas works, patented a method involving a hot air blast. At the Clyde Ironworks it was found that with a blast temperature of 150°C., fuel consumption per ton of iron produced dropped from 8 tons to 5 tons. Later Alfred Cowper, an associate of Friedrich von Siemens, brother of Werner, found that tapping the hot furnace gases was even more efficient – a step along the road to contemporary practice in which a ton of iron can at temperatures of up to 1,350°C. be smelted with less than a ton of fuel.

With cast iron becoming cheaper it could be used for a growing number of non-industrial purposes. The first decades of the nineteenth century saw a great increase of terraced houses in Britain's expanding cities. It became fashionable for the balconies of such rows to be bordered by railings, and at the same time it was necessary for those with basements to have the basements protected with railings. Cast iron was the ideal material and within a few years the streets of London, Bristol, Oxford, Cambridge, and many other towns and cities became show places for the ironfounder's art. Cast-iron railings had been raised round St Paul's as far back as 1714, but a century and more later they were to be seen not by the yard but by the hundred yard. Cast iron was little affected by salt-laden air, and around Britain's coasts it became the popular material of piers, shelters and bandstands.

The most spectacular use of the much desired metal was pioneered by James Paxton, engineer and horticulturalist. Paxton was put in charge of the gardens at the Duke of Devonshire's Chatsworth estate and here he designed the Great Conservatory, 277 feet (84 metres) long, 132 feet (40 metres) broad and 67 feet (20 metres) high, built between 1836 and 1840. Containing 70,000 square feet (6,503 square metres) of glass held in cast-iron channelling, the Conservatory was heated by eight boilers beneath the building which fed hot air into 7 miles (11 kilometres) of 4-inch (10-centimetre) pipes. Everything about the building was larger than lifesize and when Queen Victoria came to see it in 1843, the huge glass house was lit by 12,000 lamps.

Decimus Burton subsequently built a comparable but smaller glass and iron Palm House at Kew Gardens on the outskirts of London, and between 1849 and 1850 Paxton conceived and built for Chatsworth the glass Victoria Regia House to accommodate the extraordinary lily whose seed had been sent from British Guiana. Within a year it had produced 112 flowers and 140 leaves, and had twice required replanting from the original tank from which it had been successfully grown under the protection of Paxton's creation. The combination of glass and iron was already popular and was now to be used to house the Great Exhibition of 1851 – in a building which was one of the most extraordinary creations of an extraordinary age.

None of the plans first proposed for the Exhibition building was accepted by the Building Committee – which included Isambard Kingdom Brunel and Robert Stephenson – and for a while it looked as if an 'official' design was inevitable. But Paxton heard that a 'late entry' would be considered and at a meeting of a railway board in Derby on 11 June 1850 jotted down on a piece of blotting paper a proposed cross-section and end elevation, based on his work at Chatsworth. Within eight days, he and his staff had calculated the required strength of the columns and girders and had completed the plans. Despite the expected support of some members of the Building Committee, Paxton decided to take no chances and appealed to the public by having his plan published in the *Illustrated London News* together with his own description of the building.

It was to be erected in Hyde Park, and a serious problem was presented by the tall trees which stood on the site. 'I went direct with Mr. Fox to his office,' Paxton later wrote in the *Illustrated London News* 'and while he arranged the ground so as to bring the trees into the centre of the Building, I was contriving how they were to be covered. At length I hit upon the plan of covering the Transept with a circular roof similar to that on the great conservatory at Chatsworth, and made a sketch of it, which was copied that night by one of the draughtsmen, in order that I might have it to show Mr. Brunel, whom I had agreed to meet on the ground the next day. Before nine the next morning Mr. Brunel called at Devonshire House, and brought me the height of all the great trees.'

Paxton successfully incorporated the group of elms in the building which he described in this article, a description which exemplifies the confidence which Victorian engineers were now taking for granted.

'[It] is a vast structure,' he wrote, 'covering a space upwards of twenty-one acres; and, by the addition of longitudinal and cross-galleries, twenty-five per cent more space may be obtained. The whole is supported by cast-iron columns, resting on patent screw piles: externally it shows a base of six feet in height. At each end there is a large portico or entrance veranda; and at each side there are three similar entrances, covered in for the purpose of setting down and taking up company. The longitudinal galleries running the whole length of the building, together with the transverse galleries, will afford ample means for the display of lighter articles of manufacture, and will also give a complete view of the whole of the articles exhibited. The whole being covered in with glass, renders the building light, airy, and suitable. Every facility will be afforded for the transmission on rails from the entrances to the different departments; and proper means will be employed for hoisting the lighter sorts of goods to the galleries; in which, and on the columns, there will be suspension-rods, chains, &c., on which to hang woollen, cotton, and linen manufactures, and all other articles requiring to be suspended. Magnifying glasses, worked on swivels, and placed at short distances apart on the galleries, would give additional facilities for commanding a more perfect general view of the entire Exhibition.

'The extreme simplicity of this structure in all its details will, Mr. Paxton considers,' he continued, writing in the third person, 'make this a far more economical building than that proposed in the *Illustrated London News* of the 22nd of June. One great feature in its erection is that not a vestige of stone, brick or mortar is necessary. All the roofing and upright sashes would be made by machinery, and fitted together and glazed with great rapidity, most of them being

A lithograph by Joseph Nash
showing machinery exhibited at the Great Exhibition, Hyde Park, London, 1851.

finished previous to being brought to the place, so that little else would be required on the spot than to fit the finished materials together. The whole of the structure is supported on cast-iron columns, and the extensive roof is sustained without the necessity of interior walls: hence the saving internally of interior division-walls for this purpose. If removed after the Exhibition, the materials might be sold far more advantageously than a structure filled in with bricks and mortar, and some of the materials would bring in full half the original outlay.

'Complete ventilation has been provided by filling in every third upright compartment with luffer boarding, which would be made to open and shut by machinery: the whole of the basement will be filled in after the same manner. The current of air may be modified by the use of coarse open canvas, which by being kept wet in hot weather, would render the interior of the building much cooler than the external atmosphere.

'In order to subdue the intense light in a building covered with glass, it was proposed to cover all the south side of the upright parts, together with the whole of the roofs outside, with calico or canvas, tacked on the ridge rafters of the latter. This would allow a current of air to pass in the valleys under the calico, which would, with the ventilators, keep the air of the house cooler than the external atmosphere.

'To give the roof a light and graceful appearance,' Paxton continued, 'it should be on the ridge and furrow principle, and glazed with sheet glass. The ridge and valley rafters will be continued in uninterrupted lines the whole length of the structure, and be supported by cast-iron beams. These beams will have a hollow gutter formed in them to receive the rain water from the wooden valley rafters, which will be thence conveyed through the hollow columns to the drains. These drains will be formed of ample dimensions under the whole of the pathways throughout.'

The 'Crystal Palace', as it was christened by *Punch*, was to be 1,848 feet (563 metres) long and 408 feet (124 metres) wide, with an extension 939 feet (286 metres) by 48 feet (15 metres) on the north side. The side aisles were 24 feet (7 metres) and 48 feet (15 metres) high, the nave 63 feet (19 metres) and the transept 108 feet (33 metres). There were 1¾ miles (2.8 kilometres) of galleries to which access was provided by ten double staircases 8 feet (2.44 metres) wide. Nearly 4,000 tons of cast iron and 700 tons of wrought iron were used; other figures were equally impressive – 3,300 columns; 2,150 girders; 372 roof trusses; 24 miles (39 kilometres) of Paxton's patent guttering with its three grooves, the central groove to hold the glass and the other two to carry away condensation from inside the building and rain from outside it; 205 miles (330 kilometres) of sash bar; 900,000 square feet (83,600 square metres) of glass and 600,000 cubic feet (17,000 cubic metres) of timber.

All this arrived on schedule and was erected without a hitch by 2,260 workmen on time for the State Opening by the Queen on May 1. Of the successes which can be claimed for the Great Exhibition not the least was the construction of the extraordinary building itself.

Paxton's use of cast iron soon followed elsewhere.

In the United States, where the use of iron had been slow to start, the first iron bridge was not built until 1836-9 when a traditional arch bridge was built at Brownsville, Pennsylvania. But following the Hyde Park Great Exhibition a scaled-down version of the Crystal Palace was built in New York to house the

New York Exhibition, America's response to London. In Washington, moreover, there were to be two important developments. In 1851 Thomas U. Walter was appointed architect for a greatly extended Capitol. After its destruction by the British in 1814 a new building had been completed by Charles Bulfinch in 1830. Now it was decided that wings should be added and that the building should be crowned by what was to become its famous dome. At that time there was only one completed iron dome in existence, that of the Cathedral of St Isaac in St Petersburg. But Walter studied the Russian plans and prepared, probably in 1856, plans for a Washington building based on cast-iron ribs. There were numerous delays, partly due to the Civil War, and it was only in 1864 that the building was completed. It was altered in 1870 when further cast iron work was added. There was also the Washington Monument, ostensibly a granite skyscraper, 555 feet (169 metres) high, topped by an aluminium-capped pyramid. Inside this there is, in fact, an iron framework for the elevator and the iron stairway leading to the top – the highest iron frame in America when the monument was completed in 1884.

In Paris, the vaulted ceiling of the Reading Room in the Bibliothèque Nationale was supported on sixteen slender cast iron columns, and simple cast-iron grids were used to floor the stack rooms in order to let in as much light as possible. In London, Captain Fowke, who was to help design the Royal Albert Hall, produced in 1860 a Winter Garden for the Royal Horticultural Society designed along the lines of the Crystal Palace.

Throughout the nineteenth century the use of cast iron, available at prices which would have been considered unbelievably low even a decade or so earlier, continued to increase. In 1784 Henry Cort's patented puddling process began to produce wrought iron and even the conservative Royal Navy began to use it. By 1811 wrought-iron chain link had taken over from hempen anchor chains and was in common use. In 1845 the steamship *Great Britain* became the first ocean vessel to be made of wrought iron. York Railway Station, completed in 1877, was then not only the largest station in the world but a structure consisting of wrought-iron ribs and girders supported on cast-iron columns. And in 1889 Gustave Eiffel, fresh from erecting the huge iron framework for Auguste Bertholdi's Statue of Liberty at the entrance to New York Harbour, used 7,300 tons of iron in his 331 yard (303 metre) high tower on the Champs de Mars, Paris.

Although the Crystal Palace was both an engineering and an aesthetic success, the coming of iron brought for a while a divorce between two approaches. This has rarely been better described than by Hans Straub who has written in *A History of Civil Engineering*, of the coming of iron as a construction material: 'The scientific treatment and solution of the structural problems of statics and strength permitted a more rational design of the structure, and thus made it possible to cope with extensive and difficult structural tasks in an economic way, without prejudice to safety requirements. At the same time, however, it was now possible to design structures according to two points of view, different in principle: the one emphasizing the engineering aspect, i.e. structural analysis and calculation, and the other stressing the architectural aspect, i.e. the aesthetic appearance. All according to the nature of the task, the one or the other of them prevailed: the engineering aspect in the case of utility buildings, and the architectural aspect in the case of monumental buildings. With certain engineering works, such as bridges in towns and the like, both aspects must be reconciled as far as possible.'

The difference in attitude – basically between the use of stone and brick and the use of iron – might not have been so great had not building in traditional materials been greatly aided in the early years of the nineteenth century by the discovery of how to make the first artificial cement. This was Portland cement, produced by the English mason and building contractor, Joseph Aspdin, and so called by him because it had the hardness of Portland stone. 'My method', he said in his patent application of 21 October 1824, 'OF MAKING A CEMENT OF ARTIFICIAL STONE FOR stuccoing buildings, waterworks, cisterns, or any other purpose to which it may be applicable . . . is as follows: I take a specific quantity of limestone, such as that generally used for making or repairing roads, and I take it from the roads after it is reduced to a puddle or powder; but if I cannot secure a sufficient quantity of the above from the roads, I obtain the limestone itself, and I cause the puddle or powder, or the limestone, as the case may be, to be calcined. I then take a specific quantity of argillacious earth, or clay, and mix them with water to a state approaching impalpability, either by manual labour or machinery. After this proceeding I put the above mixture into a slip pan for evaporation either by the heat of the sun or by submitting it to the action of fire or steam conveyed in flues or pipes under or near the pan till the water is entirely evaporated. Then I break the mixture into suitable lumps and calcine them in a furnace similar to a lime kiln till the carbonic acid is entirely expelled. The mixture so calcined is to be ground, beat or rolled to a fine powder, and is then in a fit state for making cement or artificial stone. This powder is to be mixed with a sufficient quantity of water to bring it into the consistency of mortar, and thus applied to the purposes wanted.'

The first extensive use of the new material came in the 1840s when Sir Joseph Bazalgette ordered that all the brickwork of his London sewage scheme should be laid with Portland cement; nearly a million cubic yards (760,000 cubic metres) of it were used in the work. Manufacture was concentrated in the Lower Medway Valley of Kent, in south-east England, but during the latter part of the century European manufacturers, working in France and Germany, improved the strength of the material.

While the effect of Portland cement on building practice can hardly be over-estimated, it was the revolutionary new availability of steel which was to transform purely engineering possibilities during the later decades of the nineteenth century. An iron-carbon alloy, steel had until the 1850s been made only in small quantities and only at considerable expense. Cast iron contains from between 3.5 and 4.5 per cent carbon and to produce steel the percentage must be reduced to below 1.5 per cent. However, the characteristics of different steels vary considerably, being dependent on the percentage of carbon which is allowed to remain in them. The position is further complicated by the fact that other impurities in the iron also affect the steel's characteristics, as does the treatment which it receives during manufacture.

In earlier times, when little was known about the reasons for the qualities of steel, small amounts were made by using a very pure iron ore, such as the magnetic oxide of iron, heated with a nearly pure carbon such as wood charcoal. But the quantity of carbon in the steel produced could not be controlled and the articles beaten from the resulting metal were frequently hard on the inside although soft outside.

Since the beginning of the seventeenth century steel had also been made by the cementation process in which carbon was added to good quality bar iron, frequently imported from Sweden. It was then covered with charcoal and subjected to great heat in a furnace for up to a week. The result was 'blister steel' which was broken into small pieces, bound into bundles and covered with sand, heated again, and then forged. This Newcastle or shear steel, as it was also called, was customarily used for articles of small size such as springs, for weapons and for cutlery; even though it had considerable engineering advantages it was ruled out for major engineering projects by cost and by the impossibility of producing it in bulk.

During the mid-eighteenth century Benjamin Huntsman managed to achieve higher temperatures and to melt wrought-iron bars which, with a little carbon, gave him more homogeneous steel. Quite incidentally, the wrought iron rose to the top and the resulting steel could be poured off. This crucible steel, made in small quantities and still of variable quantity, was the only steel available until the Englishman Henry Bessemer (later Sir Henry) arrived on the scene in the 1850s.

An inveterate inventor, Bessemer turned his mind in 1854, at the outbreak of the Crimean War, to the rifling of shells in order to give them greater accuracy. 'In thinking over this subject, it occurred to me that it would be possible to give rotation to a projectile, when fired from a smooth-bore gun, by allowing a portion of the powder gas to escape throughout longitudinal passages formed in the interior, or on the outer surfaces of, the projectile', he wrote. 'If such passages terminated in the direction of a tangent to the circumference of the projectile, the tangential emission of powder gas (under enormous pressure) would act as in a turbine, and produce a rapid rotary action of the projectile.'

The British War Office turned down the idea. The French Emperor backed it and at Vincennes Bessemer showed that it worked. But Bessemer was asked by Commandant Minié – the inventor of the Minié ball – whether any artillery could stand the strain. 'This simple observation', he wrote, 'was the spark which has kindled one of the greatest industrial revolutions that the present century has to record.'

With a man of Bessemer's energy, events moved quickly and within three weeks of the experiments at Vincennes, he had found a first solution to the problem and applied for a patent for his 'Improvements in the Manufacture of Iron and Steel', granted in 1856. His basic idea for producing an iron which could be shaped as readily as wrought iron, but could also be melted and cast in a mould, was to burn out the carbon from pig iron by blowing air through its molten mass.

The initial 'Bessemer converter' consisted of a vertical, but tippable, furnace, around the base of which were six horizontal tuyères or nozzles through which air could be blown. 'All being thus arranged, and a blast of 10 or 15 lb. pressure turned on, about 7 cwt. of molten pig iron was run into the hopper provided on one side of the converter for that purpose,' Bessemer wrote of his first experiment. 'All went on quietly for about ten minutes; sparks such as are commonly seen when tapping a cupola, accompanied by hot gases, ascended through the opening on the top of the converter, just as I supposed would be the case. But soon after a rapid change took place; in fact, the silicon had been quietly consumed, and the oxygen, next uniting with the carbon, sent up an ever-increasing stream of sparks and a voluminous white flame. Then followed a series of mild explosions, throwing molten slags and splashes of metal high up into the air, the apparatus becoming a veritable volcano in a state of active eruption. No one could approach the converter to turn off the blast, and some low, flat, zinc-covered rocks, close at hand were in danger of being set on fire by the shower of red-hot matter falling on them. All this was a revelation to me, as I had in no way anticipated such violent results. However, in ten minutes more the eruption had ceased, the flame died down, and the process was complete. On tapping the converter into a shallow pan or ladle, and forming the metal into an ingot, it was found to be wholly decarbonised malleable iron.

'Such were the conditions under which the first charge of pig iron was converted in a vessel neither internally nor externally heated by fire.'

It was not to be quite as simple as Bessemer suggests. He was lucky during his early experiments since he had used pig iron which was almost free from phosphorus. This was unusual, as most iron ores used in Britain contained significant amounts of the impurity which produced only brittle steel if it were made by the Bessemer process. The difficulty was removed by a number of developments. First Robert Forester Mushet proposed the use of an alloy of manganese and iron at the end of the blow; this removed surplus oxygen and left the necessary level of carbon required in the finished steel. However, there was still a shortage of suitable ores and it was only in 1879 that Sidney Gilchrist Thomas and Percy Gilchrist solved the problem by producing a stable basic lining for the Bessemer converter which allowed the large available supplies of phosphoric ores to be used.

There were other problems, but all were overcome and the process finally justified the words of the engineer George Rennie, brother of James, – 'an unquestioned success; no fuel, no manipulation, no puddle-balls, no piling and welding; huge masses of any shape made in a few minutes'. The age of steel was about to begin.

As often happens in engineering, science and technology, other men in other countries were moving forward independently along similar lines. In the case of steel, the most important of them was William Kelly, a small-time iron founder of Pittsburgh, Pennsylvania. Kelly noticed that his pig iron appeared to become hotter if air was blown over it and appealed to the iron founders to support his experiments to discover why this was so. When they refused he carried them out himself and in 1851 began to build his own steel converters. Five years later he attempted to lodge a patent for his process, only to find that Bessemer had already been granted one. However, Kelly was granted a patent the following year and it was under his, rather than Bessemer's that much early steel was made in the United States.

In Europe, Mushet and Thomas were not the only men who turned their minds to improving the steel-making process once Bessemer had shown that it was possible to make the metal in comparatively large quantities at comparatively little cost. From the mid-1850s onwards a succession of improvements in methods and techniques changed the steel-making industry year by year so that Bessemer's importance lay not only in the process which he had pioneered but in the impetus which his example gave to others.

Among the earliest, as well as the most important developments was the open hearth method, the groundwork for which was laid in 1856 when Friedrich von Siemens patented a process using what came to be known as the regenerative furnace. Friedrich had with his brother Karl Wilhelm (later Sir William) been experimenting with such furnaces in the hope of improving the efficiency of steam-engines. They had little success but realized in the mid-1850s that their investigations might be applied in other fields. The idea was comparatively simple. The products of combustion were not led direct from the furnace to the chimney but were conducted through a chamber lined with refractory brickwork to which they gave up their heat. As soon as the chamber had been made sufficiently hot, the flow of combustion products was cut off and replaced by the air supply to the furnace which thus reached the burning fuel hot instead of cold. Two such chambers were used, alternately receiving and giving out heat, and the process was thus made continuous. A further improvement, patented in 1861, allowed the use of gas instead of solid fuel.

A further step forward was taken in 1863 when Pierre and Emile Martin used in France a regenerative furnace built by Siemens engineers to produce steel of a predetermined carbon content. By using these various methods in an open-hearth furnace, a high-working temperature of about 1650°C. could be attained; the system was economic since low-grade scrap iron and low-grade coal could be used; and while the processes were slow they had the advantage that strict control of the final product was possible.

Throughout both Europe and the United States the last decades of the nineteenth century continued to be marked by new patents regulating the burning of specific fuels in various ways to produce particular kinds of steel. It was soon found that at least four basic grades of steel could be produced merely by varying the amount of carbon that remained in them. Low carbon steel containing about 0.15 per cent of carbon could be cold-pressed. Mild steels containing 0.25 per cent were useful for making beams and girders. Medium carbon steels contain up to 0.5 per cent carbon, while high carbon steels with up to 0.7–1.0 per cent carbon were best for springs, hammers and axes.

These humdrum applications of steel had been preceded by its use for railway lines, laid at Derby in 1857, made from crucible steel by Mushet. Within a few years the much cheaper Bessemer steel was available; costing only a little more than iron rails and lasting four or five times as long, it quickly took over. Shortly afterwards it was being incorporated in locomotives and rolling stock. In 1859 a number of steel paddlesteamers were built on the Thames for service by Russia on the Black Sea, and in the United States steel ships were a few years later built to run the blockade in the Civil War. But the British Admiralty was suspicious of the newly available material and did not come round to its use until 1877 when the steel HMS *Iris* was built at Pembroke Dockyard. There also remained the same conservative suspicion about the use of steel in the railway bridges without which the engineers would have been unable to criss-cross Britain with their tracks.

New bridges had, of course, been required in growing numbers as the canal age developed. However, the coming of the railway presented the bridge builders with problems of an entirely different order. The weights of the new steam locomotives, and of the loads they pulled, were greater than those which any road bridge had so far had to sustain; and railway lines, taking the straightest possible course across hill and dale, could be diverted from difficult river crossings only at much cost and inconvenience. Thus the bridge builders, not only in Britain but in Europe and in the United States, where railways spread more quickly than elsewhere, were encouraged to exploit new ideas and new materials. They had plenty of opportunity and they turned first to iron, then to steel.

The first iron railway bridge was built in the early 1820s for the Stockton to Darlington line, a 50 foot (15 metre) bridge in four 12 foot 6 inch (3.81 metre) spans across the River Gavorless. The 30 mile (48 kilometre) line between Liverpool and Manchester demanded no less than sixty-three bridges, and for the 100 mile (160 kilometre) route between London and Birmingham 100 over- and 110 under-bridges were built. Towards the end of the railway age, it was estimated that there were nearly 600 bridges on 287 miles (462 kilometres) of main line, while by the end of the nineteenth century nearly 4,000 had been built on the iron girder and jack system which obviated the use of the arch. There were others similar to George Stephenson's High Level Bridge at Newcastle where the problem was to carry both road and railway across the valley that separated Newcastle and Gateshead. Stephenson dealt with the problem by building huge masonry piers which supported a double bridge carrying railway above and road beneath. The railway level was supported by cast-iron arched members, and the road level on the horizontal ties which held the arched beams together and resisted their thrust.

Until the end of the eighteenth century, the technique of bridge construction had changed but slowly. The French, who set up their Corps des Ponts et Chaussées, put bridge building on a basis more scientific and more businesslike than their predecessors. In Britain, Robert Myles, one of the last men to be notable as both architect and civil engineer, devised a revolutionary method of centring in which arch bridges were supported by a wooden structure while in process of construction. This method was first used in the brick and masonry Blackfriars Bridge across the Thames at London which stood for a century until it was replaced by the present wrought-iron bridge in 1869.

Geography, which presents the problem, and geology, which governs the material available to solve it, have, of course, imposed differences on the ways in which bridge builders have tackled their work. Yet the force of gravity is universal, the stresses and strains imposed on stone, brick, wood or iron are the same whether the raw material is in China, France, Britain or the United States, and the result is that the bridges of the world usually fall into one of the same four categories.

There is first the beam or girder bridge, consisting of a horizontal member supported at each end and originating, almost certainly, in the tree which had fallen accidentally across a stream. Next there probably came the primitive cantilever bridge with fixed spans jutting out from opposite sides of a river and linked by a third member which joined them both, as Benjamin Baker was to explain in the later 1860s when describing the Forth Railway Bridge. At some unknown date, probably more than 6,000 years ago, it was also realized that an arch, built up from individual voussoirs – wedge-shaped stones – would be kept stable by the force of gravity pressing down on each of the stones. While primitive stone arches have been found in many parts of the Near East, it was the Etruscans who in central Italy devised the semicircular arch and the idea of supporting it during construction by a temporary wooden framework. The Romans perfected the idea, extending throughout their Empire a form of bridge construction that dominated the western world for centuries.

There were also truss bridges, reputedly the invention of Leonardo da Vinci and Andrea Palladio during the Renaissance, but little used until the spread of railways in the nineteenth century. Where the distance to be spanned was greater than could conveniently be dealt with by an arch, the truss bridge had obvious advantages, consisting as it did of a number of small units connected together so that they behaved like a single rigid unit. Since the triangle is the only straight-sided figure that is itself rigid, individual units of the truss were frequently triangular, and this was true of the early wooden truss bridges, of the later ones which incorporated iron, and of those made from the late eighteenth century onwards which were built entirely of iron, and subsequently of steel. They sometimes incorporated features from other kinds of bridge; and as well as such hybrids there were also a number of different truss designs, braced in a variety of different ways. 'When a truss supported at its ends is subject to a load,' Carl J. Condit, the American architectural historian, has explained, in *American Building*, 'it is deflected downwards, like a simple beam. The result ordinarily is that the top chord is placed under compression, the bottom chord under tension, the vertical members or posts under compression, and the diagonals under tension. But many early trusses were designed in such a way that the verticals were in tension and the diagonals in compression. Moreover it is often impossible to predict how the diagonals will be stressed, for any one may be stressed alternately in tension and compression as the load moves over the structure.'

A fifth type of bridge, the suspension bridge, is a development of the primitive ropes or cables slung from one side of a river to the other. The first suspension bridge of modern times is often claimed to have been the chain footbridge spanning the River Tees in north-east England, which was built in 1741. This, however, like others in the next few decades, had a flexible curving roadway which while suitable for pedestrians or packhorses, could not be crossed by vehicles of any sort.

The modern suspension bridge with its stiff level floor was the invention of James Finney of Pennsylvania who in 1801 built in the United States a 70 foot (21 metre) bridge 12½ feet (3.81 metres) wide, warranted 'for fifty years (all but the flooring) for the consideration of six hundred dollars'. Seven years later he patented his method.

In Britain, the first significant suspension bridge was the Union Bridge over the Tweed at Berwick, which had a span of 449 feet (137 metres) and was finished in 1820. It was followed, six years later by Thomas Telford's suspension bridge across the Menai Straits which, when Robert Stephenson's Britannia Railway Bridge was built nearby in 1850, gave the Straits a double wonder of the engineering world. These two extraordinary constructions, in fact, epitomize the energy and enterprise with which the early nineteenth century engineers tackled their problems.

In the first part of the nineteenth century Telford was given the task of rebuilding the road from Shrewsbury to Holyhead, the port on the Isle of Anglesey, North Wales, which had been chosen as the starting point for a new service of ships to Dublin. Between the mainland of North Wales and the Isle of Anglesey there lay the Menai Straits, from 200 yards (183 metres) wide at some places to 2 miles (3 kilometres) at others. 'In common with other straits which separate any island from the continent, or from a larger island, this area of the sea exhibits peculiarities in its tide, which, twice in every twelve hours, runs in different directions, and frequently with great velocity,' John Rickman has written in the *Life of Thomas Telford*. 'The rise at ordinary spring-tides is about twenty-two feet, sometimes as much as thirty feet; and being in the vicinity of the Snowdon range of mountains, it is subject to violent gusts of wind, from which liability, and from the ferry passage being made in the night, this part of the journey was rendered a disagreeable object of anticipation, and was sometimes really dangerous.'

Telford's first plans were for a conventional cast-iron bridge. He offered two alternatives, one consisting of three arches, each of 260 feet (79 metres), and the other of a single 500 foot (152 metre) span and approaches. Both were ruled out by the British Admiralty's insistence that naval vessels must be able to pass through the Straits with their masts erect. Greatly daring, Telford then submitted plans for a suspension bridge with a single span of 579 feet (176 metres). The carriageway was to be 100 feet (30.5 metres) above high water and supported by sixteen chains which would be taken over the tops of 153 feet (46.6 metres) high stone piers to be built on either side of the Straits. From the chains there were to hang one-inch rods supporting two carriageways and a pedestrian way.

In the initial plan, each chain would have been made up of wrought-iron bars, welded together at the ends. At an early stage, however, it was decided to substitute for them the wrought-iron links designed, and soon to be patented, by Captain Samuel Brown, R.N. Telford was well aware that he was working on the edge of accepted knowledge – Brown's own bridges were to have a disastrous record – and took every possible step to test his raw materials. 'A very accurate and powerful proving machine having been constructed, every piece of iron was submitted to the same proportional strain,' he later wrote; 'and after being proved, every separate piece of iron was well cleansed, put into a stove, and when brought to a gentle heat, was immersed in a trough containing linseed oil. After remaining a short time, it was again put into the stove, and, when dried, appeared covered with a varnish. When taken out of the stove the second time, each piece was finished with a coat of linseed oil paint; and this dry, the iron bar was considered fit to be sent to the bridge.'

The two huge masonry towers were completed by the spring of 1825 and before the end of April the first of the sixteen chains had been anchored behind the Anglesey tower and taken to its top. Telford himself now arrived to watch the dramatic next step. 'On Tuesday, the 26th April, 1825,' runs an account of the event, 'the first Chain of this stupendous work was thrown over the Straits of Menai; the day was calm, and highly propitious for the purpose. An immense concourse of persons, of all ranks, began to assemble on the Anglesey and Caernarvonshire shores, about 12 o'clock at noon, to witness a scene which our ancestors had never contemplated. Precisely at half past 2 o'clock, it being then about half-flood tide, the raft, prepared for the occasion, stationed on the Caernarvonshire side, near Treborth Mill, which supported a part of the Chain intended to be drawn over, began to move gradually from its moorings, towed by four boats, with the assistance of the tide, to the centre of the river, between the two grand Piers; when the raft was properly adjusted and brought to its ultimate situation, it was made fast to several buoys, anchored in the Channel for that specific purpose. The whole of this arduous process was accomplished in twenty-five minutes.

'A part of the Chain, pending from the apex of the suspending Pier, on the Caernarvonshire side, down nearly to highwater mark, was then made fast by a bolt, to that part of the Chain lying on the Raft, which operation was completed in ten minutes.

'The next process was fastening the other extremity of the Chain (on the raft) to two immense powerful blocks, for the purpose of hoisting the entire line of Chain to its intended station, the apex of the Suspending Pier on the Anglesey side; the tension of the Chain then being 40 tons. When the blocks were made secure to the Chain, (comprising 25 ton weight of iron) two Capstans, and also two preventive Capstans, commenced working, each Capstan being propelled by 32 men.

'To preserve the equanimity in the rotatory evolutions of the two principal Capstans, two fifers played several enlivening tunes, to keep the more regular in their steps, for which purpose they had been previously trained.

'At this critical and interesting juncture, the attention of the numberless spectators, assembled on the occasion, seemed rivetted to the novel spectacle, now presented to their anxious view; the Chain rose majestically, and the gratifying sight was enthusiastically enjoyed by each individual present.

'At fifty minutes after four o'clock, the final bolt was fixed, which completed the whole line of Chain, and the happy event was hailed by the heavy acclamation of the numerous spectators, joined by the vociferations of the workmen, which had a beautiful effect from the reiteration of sound, caused by the heights of the opposite banks of the river. Not the least accident, delay, or failure in any department took place during the whole operation, which does infinite credit to every individual employed in this grand Work.'

So great was the excitement at this first success that three workmen on the Anglesey side – Hugh Davies, stonemason; John Williams, carpenter; and William Williams, labourer, – walked from shore to shore along the 9 inch (23 centimetre) wide chain, more than 100 feet (30.5 metres) above the water.

On July 9 the sixteenth chain was hauled into position. Once this crucial operation had been completed, the bridge was finished without further problems, the Straits were formally reopened and the suspension bridge began its service life which continued until 1940 when it was almost completely reconstructed and the original wrought-iron links replaced by links of high-tensile steel.

The fame of Telford's Menai Bridge started a fashion for suspension bridges, although there was continuing argument about the length to which the span could safely be extended. The first to be built for railway traffic, Samuel Brown's bridge carrying the Stockton and Darlington line over the Tees, became unusable a few years after it had been opened in 1830. 'The undulations into which the roadway was thrown, by the inevitable unequal distribution of the weight of the train upon it,' said a report, 'were such as to threaten the instant downfall of the whole structure.'

Most early suspension bridges were held up by chains, but in 1822 a Swiss engineer, and the Frenchman Marc Séguin, began to build at Geneva Europe's first major cable suspension bridge. This had two 131 foot (40 metre) spans and was followed by a number of similar bridges in Switzerland and France. It was only in 1834, however, that the Frenchman Joseph Chaley, built his Grand Pont across the valley of the Fraine at Fribourg. With a span of more than 870 feet (264 metres) the bridge, described by Saint-Venant, the French engineer-author, as 'the boldest structure ever built', carried a 22 foot (6.7 metre) wide roadway suspended by four cables each of which consisted of 1,056⅛ inch (0.3 centimetre) wires, later strengthened by the addition of two more cables. It was not demolished until 1923.

Construction of the Britannia Bridge
across the Menai Straits for the Chester and Holyhead Railway, 1849.

It was the apparent failure of the suspension bridge to deal with the strains and stresses of railway traffic which presented Robert Stephenson with his great problem when, in the 1840s he was faced, as Telford had been, with building a bridge across the Menai Straits. This one, however, was to carry the Chester-Holyhead railway line. As in Telford's case, the Admiralty demanded that navigation should not be impeded, thus ruling out Stephenson's first plan for two cast iron arches of 350 foot (107 metre) span, to be built either side of Britannia Rock which lay in mid-channel and could support a central pier.

Only one option appeared to be left: some form of flat girder bridge which would rest on a pier rising from Britannia Rock, and on two other piers to be built either side of the Straits. Yet this would involve two 460 foot (140 metre) spans – and the longest girder bridge so far built had a span of only 60 foot (18 metre). Stephenson's proposed solution was to build a tubular bridge made of wrought-iron plates, supported by chains, and big enough to allow a train to be driven through its interior. There was some argument about whether the chains were necessary. Eaton Hodgkinson, one of the two men whose advice Stephenson sought, thought they were. William Fairbairn, the engineer who was to superintend construction of the new bridge, thought not and carried out experiments which finally supported his view. Cylindrical, elliptical and rectangular tubes from 17 to 30 feet (5 to 9 metres) long were tested. Finally a tube 75 feet (23 metres) long 'was tested to destruction', says Fairbairn's biographer, William Pole, 'by hanging weights on it till it gave way, the object being to find out the weak places. After each trial the injured and defective parts were cut out and the tube restored to its original form, with plates of altered strength, as indicated by the nature and appearance of the fracture, and as circumstances might require. This was done seven different times, until proportions were arrived at which appeared to be satisfactory, as giving all the strength of which such a tube was capable.'

Each of the large tubes which were now built contained 327,000 rivets, and the whole bridge about 2,000,000. 'The rivet-holes in the plates are formed by a machine which punches out a piece of iron the exact size of the required hole,' explained an official report of the operation. 'The plates are fastened upon a sliding-table, which advances at every stroke double the required space between each rivet. Two punches then descend through the plate, and form the holes. In this way about forty holes are punched per minute.' The rivets were put into the holes red hot and heads formed by beating the projecting end of the rivets with a heavy hammer. 'In cooling, they contract strongly, and draw the plates together so powerfully, that it requires a force of from four to six tons to each rivet to cause the plates to slide over each other.'

Before it was decided that suspension chains were unnecessary, work had begun on the towers from which they were to be slung, and it was decided to continue with their construction, apparently for aesthetic reasons as much as anything else. So while the iron plates were being bolted together into two independent continuous rectangular beams, each 1,511 feet (460 metres) long and each weighing 4,680 tons, the work of building the towers continued. The highest was the Great Britannia Tower, 230 feet (70 metres) high on the Britannia Rock in the centre of the Straits; in addition, there were the two side-towers, 212 feet (65 metres) high, and two abutment towers 177 feet (54 metres) high. Much of the interior masonry was of soft red sandstone from Runcorn in Cheshire but the exterior was of Anglesey marble, a hard carboniferous limestone which could be given a high polish. 'The design of the masonry,' wrote one of Stephenson's biographers, 'is such as to accord with the form of the tubes, being somewhat of an Egyptian character, massive and gigantic rather than beautiful, but bearing the unmistakable impress of power.' The impression was increased by the massive masonry lions resting on the abutments of the bridge.

An army of men had been hired for building the bridge, a village complete with school was constructed by the site, and a surgeon and chaplain installed. A distinct form of communal life developed, and on at least one occasion a concert for the workers was held in one of the giant tubes. 'The floor was boarded, and the interior being illuminated by several thousand of candles, had a most imposing and gorgeous appearance. The cells of the top and bottom form excellent speaking tubes and conversation may be carried on through them even in a faint whisper. By elevating the voice persons may converse through the entire length of the bridge, a distance of more than 500 yards.' The building of the bridge had become a spectacle for all who could reach the site and Sir George Biddell Airy, the Astronomer Royal, wrote to his wife from the Straits, 'The Tube Works are evidently the grand promenade of the idlers about Bangor. I saw many scores of ladies and gentlemen walking that way with their baskets of provisions evidently going to gypsy in the fields nearby.'

By mid-June 1849 the great tubes were ready and were floated on pontoons to a pre-arranged position in the Straits. At 7.30 on the evening of 20 June 1849, the first of the pontoons swung out into the current. 'The success of this operation', says a contemporary report on it, quoted by Edwin Clark in Volume II of his *The Britannia and Conway Tubular Bridges*, 'depended mainly on properly striking the "butt" beneath the Anglesey tower, on which, as upon a centre, the tube was to be veered round into its position across the opening. This position was determined by a 12-inch line, which was to be paid out to a fixed mark from the Llanfair capstan. The coils of the rope unfortunately over-rode each other upon this capstan, so that it could not be paid out. In resisting the motion of the tube, the capstan was bodily dragged out of the platform by the action of the pawls, and the tube was in imminent danger of being carried away by the stream or the pontoons crushed upon the rocks. The men at the capstan were all knocked down and some of them thrown into the water, though they made every exertion to arrest the motion of the capstan-bars. In this dilemma Mr. Rolfe, who had charge of the capstan, with great presence of mind, called the visitors on shore to his assistance; and handing out the spare coil of the 12-inch line into the field at the back of the capstan, it was carried with great rapidity up the field and a crowd of people, men, women, and children, holding on to this huge cable, arresting the progress of the tube, which was at length brought safely against the "butt" and veered round. The Britannia end was then drawn into the recess of the masonry by a chain passing through the tower to a crab on the far side. The violence of the tide abated though the wind increased, and the Anglesey end was drawn into its place beneath the corbelling in the masonry; and as the tide went down, the pontoons deposited their valuable cargo on the welcome shelf at each end. The successful issue was greeted by cannon from the shore and the hearty cheers of many thousands of spectators, whose sympathy and anxiety were too clearly indicated by the unbroken silence with which the whole operation had been accompanied.'

The job had been completed as planned, exactly at the start of the fifteen minutes during which, at the turn of the tide, there was virtually no current passing through the Straits. The valves in the pontoons were opened and the tube was sunk on to the timber beds placed to receive it. The tube now had to be raised into position, a task performed by hydraulic presses already installed on the stone towers; the huge mass of metal was raised a few inches at a time, with wooden stacking being placed under it at every move. The remaining tubes were floated in and raised later in the year. On 5 March 1850 Stephenson himself put the last rivet in the last tube before being driven across the bridge in one of three trains carrying about a thousand passengers.

The Britannia Bridge was not only the first of what were to be scores of similar, if shorter and less spectacular bridges. It combined to a greater degree than most such projects the theoretical expertise of the scientist and the practical expertise of the engineer. As Stephenson himself was to say, 'The true and accurate calculation of all the conditions and elements essential to the safety of the bridge had been a source not only of mental but of bodily toil; including, as it did, a combination of abstract thought and well-considered experiment adequate to the magnitude of the project.'

Two years after Stephenson's triumph, Isambard Kingdom Brunel, the creator of the Great Western Railway who had watched the Britannia Bridge being erected, was faced with an even more intractable problem, and one whose solution was a landmark in civil engineering. As Chief Engineer to the Great Western Railway, he was confronted, when driving a railway line from London to Penzance, with the construction of a bridge across the River Tamar which outside Plymouth forms the boundary between Devon and Cornwall. At Saltash, the most suitable place for a bridge, the river was 1,100 feet (335 metres) wide. The Admiralty, as with the Menai Straits, vetoed any interference with navigation. But the Tamar, unlike the Straits, had no rock in mid-stream on which to build a supporting pier. Indeed, soundings quickly showed that the rock forming the river bed lay below no less than 80 feet (24 metres) of water and mud.

To solve the problem Brunel used a caisson, only the third to be seen in Britain and by far the largest. From the earliest times engineers encountering the problems of building the foundations of a bridge in mid-stream had to decide whether to lay them actually in the water or within an enclosure from which the water had been drained. If they chose the first alternative, which might be the most practicable if the river was not deep, it was necessary to raise the river bed at least to low water level by dumping loads of rubble, confined if possible within a palisade of vertical piles driven into the river bed. These artificial islands or 'starlings' not only acted as foundations for the bridge's mid-stream masonry piers but were usually extended upsteam to break the force of the water. The disadvantage of starlings was that they obstructed the flow of the river, not only making navigation difficult but increasing the speed of the water between them. This could be important as in the case of the first London Bridge, begun in 1176, where the difference in water level upstream and downstream from the bridge at mid-tide could be many feet.

If the engineer decided to build his foundations from dry ground he had to construct a coffer dam which involved driving two rows of piles to form an square or rectangle, fixing horizontal beams between them and then filling the intervening space with puddled clay. The square or rectangle forming the coffer dam then had to be pumped or baled free of water. The driving of the piles was a laborious task, the least inefficient kind of pile-driver raising a hammer by a treadmill and releasing it when it had reached its maximum height. Taking out the water from the coffer dam was an equally time-consuming job, performed either with an endless chain of buckets, a pump that could only occasionally be operated by a river-turned waterwheel, or sometimes with an Archimedean screw.

But the use of coffer dams was in any case restricted by the depth and speed of a river, and by the inefficiency of wood as a material for making them. The answer was the caisson. Introduced into Britain by Charles Dangeau de Labelye when he began to bridge the River Thames at Westminster in 1738, it consisted of a flat base-platform to which waterproof sides were attached by wedges. Sheet-piling was driven in to surround the area where the caisson was to be sunk, and the river bed was dredged level. The first courses of masonry which were to provide foundations for the pier were then built on the caisson which was sunk on the required spot, and its sides removed.

A caisson had been used in 1851 when a bridge was built across the Medway at Rochester, and Brunel himself had used one the following year when bridging the Wye at Chepstow. His plans for the Saltash Bridge were sophisticated. First he constructed a trial cylinder 6 feet (1.8 metres) across and 85 feet (26 metres) long. This was attached to tackle that could raise and lower it from two gun-brig hulks. Once it had been lowered, five borings were taken within it, reaching through the mud to the rock below. The positions of the borings were carefully recorded, the cylinder was moved to a different position, and five further borings were made. The process was repeated until the positions of 175 borings had been recorded and there had thus been built up a detailed survey of the rock surface on which the central pier would have to be built.

The caisson within which the pier would be constructed was 35 feet (10.7 metres) wide, says Isambard Kingdom Brunel's son. 'It was determined to provide for the possibility of having to employ the pneumatic process,' he has written in *The Life of Isambard Kingdom Brunel, Civil Engineer*. 'The cylinder had a diameter of 35 feet at the bottom, and about 20 feet above the lower end of it a dome was made to form the roof of the diving-bell; from the centre of the dome rose a tube 10 feet in diameter to the level of the top of the great cylinder. As a diving-bell of this size, under 80 feet of water, might have proved unmanageable, an annular space, forming a gallery or jacket of four feet in width and 20 feet high was formed round the inner circumference of the bottom of the cylinder below the dam. This annular space was divided by radial vertical partitions into 11 compartments, and was connected at the top by an air-passage with a six-foot cylinder, which was placed eccentrically inside the 10-feet cylinder already mentioned, and served as a communication between the outside and the annulus. On the top of the six-foot cylinder were placed the air locks of the pneumatic apparatus which had been used at Chepstow. Thus air might be pumped into the annular space, the water expelled, and the work carried on without having to use air pressure under the whole of the dome. In that part of the 10-foot cylinder which was not occupied by the six-foot cylinder a powerful set of pumps was fixed to keep down the water in the central space, and diminish the pressure under which the men worked, thus utilising whatever advantage could be gained from the great cylinder acting as a coffer-dam. As it had been ascertained that the surface of the rock dipped to the south-west to the extent of about six feet in the width of the pier the bottom of the cylinder was made oblique so as to fit the surface of the rock.'

While the central pier was rising from the caisson, and the piers on the Devon and Cornish banks were under construction, the two main spans of the bridge were being assembled. With an arched bridge ruled out by the requirements of the Admiralty and a suspension bridge still thought to be inadequate, Brunel was attracted to the enclosed tube which had solved Stephenson's problem at the Menai Straits. But in his case he used, for each of the two spans, two arched tubes of elliptical cross-section, joined at their ends. The deck on which the line was laid was supported by these bow-shaped units, the upper section of which was in compression while the lower was in tension. The units were towed into position on pontoons and then raised 100 feet (30.5 metres) on hydraulic jacks. The bridge was opened in 1859.

By now the effect of Bessemer's invention, which made steel available on a new scale, was increasingly being felt throughout the engineering industry. However, steel was so new, and so comparatively untried, that its use in bridges was still forbidden in Britain, and it fell to the Americans to make spectacularly successful use of it for the first time in a major bridge, which was to cross the Mississippi at St Louis. The building of the bridge, which began in 1867, was an epic enterprise which had to overcome multiple difficulties.

James Buchanan Eads, an engineer who had built ships but had no experience of bridges, was the man chosen to carry out the daunting task of bridging the 1,500 foot (457 metre) wide river whose volume and flow presented special problems. Eads' design was for a construction of three spans, 502, 520 and 502 feet (153, 158 and 153 metres) long. Each arched span consisted of eight hollow steel tubes, 18 inches (46 centimetres) in diameter and braced together with metal ties. The design was considered so revolutionary that the chief engineer of the Pennsylvania Central Railroad whose trains were to cross the bridge reported: 'I cannot consent to imperil my reputation by appearing to encourage or approve of its adoption. I deem it entirely unsafe and impractical, as well as in fault in the questions of durability.'

Despite this devastating statement, the St Louis and Illinois Bridge Company continued to support Eads and in 1869 work was begun on the two caissons from which the central supporting piers were to be built. Eads insisted that the piers should rise from bed-rock, but in places this lay below a 103 foot (31 metre) covering of water and sand, the use of caissons at such a depth had never before been attempted. In spite of all precautions – and not surprisingly since so little was known about the cause of *the bends*, or caisson disease – fourteen men died before work on the pier foundations was completed. In December 1869 the men in one caisson were isolated by ice floes for fifteen days and survived only because supplies had previously been laid in.

Once construction of the spans began, Eads found it necessary to keep a tight control on the suppliers. Specification for much of the metal was that its strength should be 60,000 pounds per square inch (4,134 bars) and when it was found to be only 55,000 it was sent back – not once but three times on one occasion.

The spans were successfully built out above the river, held in place by cantilevers, and by September 1873 the first two half-arches were approaching each other. Now there came the climax of the operation, not only engineering but also financial. Eads had negotiated in London a loan of half a million dollars, but on condition that the first arch was closed by 19 September.

The summer weather was hotter than usual and on 14 September it was found that the heat had brought the two parts of the arch so close together that the tubes prepared for joining them could not be inserted. With the gap 2½ inches (6 centimetres) shorter than planned, cloth was wrapped round the tubes and packed with fifteen tons of ice. The two parts of the arch contracted slightly, and the gap grew larger; but not large enough, and the addition of more tons of ice ended with the same result. But Eads' knowledge of the weather at St Louis had led him to prepare for such potential disasters, and shorter, adjustable joiners had been made. Late on the night of 17 September, with no sign of the weather changing, the emergency joiners were used to close the arch. After that, all went well and the St Louis bridge was officially opened on 4 July 1874.

Steel also played an important part in what was to be one of the most famous of all American bridges. This was the Brooklyn suspension bridge, spanning the East River and the work of John Augustus Roebling. In the United States the lack of facilities for producing large iron castings made the suspension bridge particularly attractive. So did the number of swift rivers which, with the American advance across the continent, had to be bridged with the least possible disruption of navigation. Charles Ellet, whose 1,010 foot (308 metre) bridge across the Ohio River at Wheeling, West Virginia, claimed in 1848 the world's longest span, was one of the pioneers.

Another was Roebling, an immigrant from Germany who set up a wire rope factory, repaired the Wheeling Bridge after it had been damaged by a tornado and went on to design and build the Monongahela Bridge at Pittsburgh. In 1852, there came Roebling's Clifton suspension bridge at the Niagara Falls. Its span of 821 feet (250 metres) was far longer than that of the Britannia Bridge across the Menai Straits and its iron cables held up not only one deck for the railway track but another for coaches and carriages. Its success did much to spread the fame of suspension bridges in North America, and it was a suspension bridge that Roebling proposed in a letter to the *New York Journal of Commerce* in 1857. Nothing was done before the outbreak of the Civil War, and it was not until 1867 that the New York Bridge Company was formed and Roebling was appointed chief engineer.

He had no doubt about the importance of the project, writing: 'The contemplated work when constructed in accordance with my designs, will not only be the greatest bridge in existence, but it will be the greatest engineering work of this continent, and of the Age. Its most conspicuous features, the great towers, will serve as landmarks to the adjoining cities, and they will be entitled to be ranked as national monuments.' Grandiose as the claims sound, they were to be justified by the massive construction that was soon rising. The two towers, some 1,596 feet (486 metres) apart, were 140 feet by 60 feet (43 by 18 metres) at the water line and rose for 272 feet (83 metres). The 168 feet by 103 feet (51 by 31 metre) caisson constructed for the building of the foundations was the largest in the world. And the four main cables, each of 5,296 galvanized steel wires, oil-coated, weighed nearly 1,000 tons and contained 14,357 miles (23,105 kilometres) of wire which was spun into cable by a device designed by Roebling. It is still being used, in improved versions, to spin cable for American suspension bridges.

Roebling was not to see the result of his work since he was injured during the building of the towers and shortly afterwards died of tetanus. However, his son, Washington R.A. Roebling, took over as architect and was present at the opening

ceremonies on 24 May 1883. The Brooklyn Bridge not only introduced the revolutionary form of cable-spinning, but gave a great boost to the popularity of the suspension bridge, two more of which were to be thrown across the East River early in the twentieth century, the Williamsburgh Bridge in 1903 and the Manhattan Bridge in 1909.

It was steel which also made possible what has been called 'the crowning victory of engineering science in the nineteenth century', the spanning of the Firth of Forth with the great cantilever bridge, a mile and a half long, which almost a hundred years later is still one of the most impressive sights in the world. The firth, cutting deep into the heartland of Scotland, had until the closing decade of the nineteenth century been a serious impediment to travel. From Dunbar to Anstruther, for instance, a straight line across the firth is a mere 18 miles (29 kilometres); but until the huge cantilevers were completed the land journey involved a circuit of 150 miles (240 kilometres).

As early as 1805 it was proposed that a double tunnel should be built under the firth from South Queensferry, a few miles from Edinburgh. The plan was abandoned, and the same fate befell a scheme for throwing a chain bridge across the water eleven years later. This bridge was to use only some 2,500 tons of iron and would have had, according to the proposals, a very light and slender appearance – 'so light indeed', it was later remarked, 'that on a dull day it would hardly have been visible, and after a heavy gale probably no longer to be seen on a clear day either.'

Not until 1860 was the next proposal for a railway bridge put forward. This also fell through, mainly due to the restructuring of the railway companies involved, and it was 1873 before the Forth Bridge Company was formed. Design was entrusted to Sir Thomas Bouch, manager and engineer of the Edinburgh and Northern Railway. Bouch produced plans for a great suspension bridge which would have made use of the island of Inchgarvie, lying almost in mid-stream on the line of the proposed Queensferry crossing and thus being a suitable base for a pier. A foundation stone was ceremoniously laid for the Inchgarvie pier in 1879 and the first courses were built.

Then, in December of the same year, the Tay Bridge, spanning the Firth of Tay which bites into the Scottish mainland some 20 miles (32 kilometres) north of the Firth of Forth, was swept away as a train was crossing during a heavy gale. More than seventy persons were drowned. It was Thomas Bouch who had designed the Tay Bridge, and in the words of a semi-official account of the Forth Bridge, 'the investigations into the causes of [the Tay Bridge] disaster, and the disclosures made, shook the public confidence in Sir Thomas Bouch's design, and rendered a thorough reconsideration of the whole subject necessary.'

However, it was an age when disasters did not deter. The Forth would have to be bridged, whatever the effort, and from the past of almost a century ago there comes what may sound a curious reason. 'In these days of high pressure, of living and working and eating and drinking at top speed,' went W. Westhoven's account of the bridge in *Engineering*, 'the saving of an hour or two for thousands of struggling men every day is a point of the greatest importance, and every delay,

---

The George Washington Bridge, New York, is slung with over 100,000 miles of cable. It was opened in 1931, sixty-three years after Roebling said the Hudson could be spanned.

Clifton suspension bridge spanning the Avon Gorge outside Bristol.
Initially planned by Brunel, the bridge was finally built some years after his death.

however excusable and unavoidable, is fatal to enterprise.'

The railways' consulting engineers now commissioned plans from the civil engineers Sir John Fowler and Benjamin Baker. Their choice was of a cantilever bridge, previously called a continuous girder bridge and a type which had been known from earliest times. Baker later explained its principles in simple terms. 'Cantilever is a two-hundred-year-old term for a bracket, and the Forth Bridge spans are made up of two brackets and a connecting girder,' he said. 'Imagine two men trying to shake hands across a stream a little too wide for their hands to meet. One man extends his walking-stick, and the other grasps it and so the stream is bridged. There we have the two arms or brackets and the connecting girder. In the Forth Bridge the arms are supported by great struts, as in a living model, where raking struts extended from the man's wrists to the points of support.'

Baker, as his predecessors had planned, used the island of Inchgarvie as the foundation for the main central pier. On either side, to north and south there were to be built the Fife and the Queensferry piers, each nearly a third of a mile away. From all three, huge cantilevers reached out, joined by 350 foot long girders. The bridge thus consisted of two main spans of 1,710 feet (521 metres); two side spans of 689 feet 9 inches (210 metres) whose cantilevers met approach viaducts; and the two viaducts, each consisting of fifteen 168 feet (51 metre) spans.

This immense structure, which used 54,160 tons of steel and was to present 135 acres (59 hectares) of metal to be painted, was of a revolutionary size, and even before work was started it attracted criticism, although from an unusual quarter. On 19 October 1882, *Nature* published a letter 'On the Proposed Forth Bridge' written by Sir George Biddell Airy, until recently Astronomer Royal.

After repeating the engineers' description of the bridge, he continued, 'This statement is enough, I think, to justify great alarm. No specimen, I believe, exists of any cantilever protruding to a length comparable, even in a low degree, to the enormous brackets proposed here. The only structures of this class, in ordinary mechanics, known to me, are the swing-bridges for crossing dock-entrances and the like, and these are absolutely petty in the present comparison.' He then dealt with the forces he assumed would be involved and added: 'This leverage is considerably greater in the instance of the proposed Forth Bridge than it was in that of the unfortunate Tay Bridge, and we may reasonably expect the destruction of the Forth Bridge in a lighter gale than that which destroyed the Tay Bridge.'

Now Sir George was not only innocent of any engineering experience. It was he, it had been revealed in the Tay Bridge enquiry, who had advised Bouch on the wind pressures the bridge might have to withstand – advice which almost certainly played its part in the disaster.

Airy's warning was, therefore, rightly ignored and in January 1883, survey work was begun to ensure that the huge masonry piers were correctly sited. Workshops were set up on both sides of the Firth, and construction was begun on the southern shore of a 2,200 foot (671 metre) jetty along which much of the building material was to be taken. A considerable amount of the steel work had to be assembled twice, first at the onshore works where the important components were put together. The individual girders were then numbered and taken apart before being rebuilt in their permanent positions.

The first of the two giant caissons needed for assembly of the main piers was launched in May 1884, the second towards the end of the year. The second caisson was sunk with just sufficient concrete to allow it to rise and fall with the tide but on 1 January 1885, it failed to rise, then tilted. Nine months' work was needed for repairs – the only major interruption to construction in the whole enterprise. From April 1886 until 7 July 1887, the 2,000-ton girders were slowly raised into position by hydraulic jacks, the piers on which they rested being erected at the same time. Throughout 1888 and the first half of 1889, the massive steel cantilevers were built out towards each other.

At one point in the operations roughly 5,000 men were employed on the work and there was much rivalry to be the first to cross the bridge. On the afternoon of 26 September 1889, when the gap was only a matter of feet, 'two of them proposed to make an effort to out-distance their neighbours,' it has been written, 'and the knowing one secured the prize. He suggested the use of a ladder which was at hand, provided a rope could be got to fasten it. But when his neighbour was away finding a rope, he placed the ladder across between the jibs of the cranes working on the ends of the two arms, and made the perilous passage at a height of about 200 feet above the water. Two days later a party of ladies and gentlemen crossed the gap, but on a safer connection, and by the 10th of October it was completely filled up.'

All the work was carried out with an accuracy demanded not only by the stresses and strains caused by the wind pressure of typical Scottish weather, but by the effects of the sun. When the sun shone, the plates on which it fell expanded, producing a bend away from the sun that could only be corrected later in the day. These complications, the result of expansion and contraction due to temperature changes present in all metal constructions, were excessive in a bridge the length of the Forth.

The dangers were also considerable, with ample opportunities for men dropping tools or material from a lethal height. More than 6,500,000 rivets were used, some of them nearly a foot long, and remarkable escapes were numerous. 'I saw a hole one inch in diameter made through the four-inch timbering of the staging by a spanner which fell about 300 foot and took off a man's cap in its course,' Benjamin Baker once said. 'On another occasion, a dropped spanner entered a man's waistcoat and came out at his ankle, tearing open the whole of his clothes but not injuring the man himself in any way.'

Before the bridge was even completed it had become one of the sights of Europe, and after its formal opening by the Prince of Wales on 4 March 1890, Don Pedro of Brazil, the kings of Saxony and of Belgium, as well as the Shah of Shahs were among those who visited it. There were also hundreds of ordinary visitors. 'As in most other matters,' it was written, 'ladies were to the fore, pluckily climbing into every nook and corner where anything interesting might be seen or learned, up the hoists and down the stairs and ladders, and frequently leaving behind the members of the so-called stronger sex.'

Once the bridge was in use, with trains thundering across it day and night, a key question was how much the cantilevers would 'give', under the regular passage of the trains and the changes in temperature. The question was answered with the help of an extraordinarily simple piece of equipment which Sir John Fowler had fitted at the expansion and contraction joint at the end of the south main span. It consisted of a marker attached to the girder and pressing against a strip of paper which unwound at a constant speed. When the cantilever

The Forth Bridge, built between 1883 and 1888,
spanning the Firth of Forth, and one of the wonders of Victorian engineering.

contracted, the marker was pushed horizontally in one direction: when it expanded, the pencil was pulled in the opposite direction and throughout the day thus made a curving trace on the paper. When a train came on to the far end of the cantilever, its presence was noted by a vertical movement, and the single strip of paper thus automatically recorded all the information needed.

The Forth Bridge, to be joined almost three-quarters of a century later by the Forth Road Bridge – a suspension bridge using a total of 29,000 tons of steel, including 7,400 tons in the cables – was not only an enormous engineering success, but to many a beautiful achievement. There were, however, to be exceptions. 'There never would be an architecture in iron,' declared William Morris, poet, artist and designer who helped transform late Victorian taste, 'every improvement in machinery being uglier and uglier until they reached the supremest specimen of all ugliness, the Forth bridge.'

While Morris protested, plans were already being drawn up for a project even more ambitious than the Forth Bridge. This was to be a bridge spanning the English Channel and turning into reality that persistent dream – or nightmare – of a land-link between Britain and the Continent. The bridge was to be 24 miles (39 kilometres) long, and its 120 piers were to support a track 180 feet (55 metres) above high water. Mid-channel cantilevers would provide a 1,640 feet (500 metre) wide opening for ships and the structure was expected to use a million tons of steel. However, nothing came of the project – presumably this was because of the enormous cost, estimated at £34,000,000 which is a vastly greater sum in today's terms.

The Europe Bridge, near Innsbruck, Austria. The bridge's six spans, the longest 600 feet (183 metres), carry a road 624 feet (190 metres) above the valley floor.

ENGINES ON THE OCEANS

Before the end of the eighteenth century steam was already being investigated for a use quite as revolutionary as that of transforming coal mining and factory life or spreading the railway network which soon followed. Since earliest times, movement on rivers, or on the high seas, had been largely dependent either on human muscle power working on the long oars of galleys, or on the fitful and uncertain winds which filled – or failed to fill – the sails of ships. The limitations of such vessels are emphasized in the descriptions and paintings of great fleets of merchantmen or warships lying becalmed in harbour, their crews immobilized and their captains impatient. Surely, it began to be asked towards the end of the eighteenth century, the new-found wonder of steam could alter all that?

The idea of using steam, first for inland navigation, then on the high seas, came at the end of a story which had begun in ancient times when men realized that a hollowed-out log could carry them downstream and that help from hand-wielded paddles could often take them upstream. Such craft were eventually hand-hewn to carry an upright curved prow which made it easier to land on beaches, and were supplied with a long stern paddle to increase the vessel's manoeuvrability. There followed the use of a simple sail, sometimes the triangular-shaped lateen sail which could be handled easily in a sea such as the Mediterranean where unexpected stormy winds were more usual than the steady trade winds of other areas.

Although the ancient people of Mesopotamia used banks of rowers to help move their river craft in still weather, it was only with the confrontation between Greece and Persia about 500 BC that galleys with their two, three or four rows of crew were developed into fighting vessels. These early warships, like those, very similar, which were used for trade, were carvel-built of wood, the beams forming their sides being butted up end to end. Greater strength was given only some centuries later when Viking galleys, built in Scandinavia with the aim of navigating rough open seas, began to be clinker-built, a method of construction in which the beams were overlapped.

During the Middle Ages, sturdier, larger, and more complicated vessels having superstructures such as forecastle, aftercastle and topcastle began to explore the world's oceans at the same time as the introduction of gunpowder made it necessary to pierce a vessel's side with embrasures through which cannon could be aimed. Together, these developments brought an end to the manned galley, and the last battle between rower-powered ships was fought off Lepanto in 1571 when the Holy League's fleet, commanded by Don Jon of Austria, virtually destroyed a Turkish fleet – and liberated 10,000 Christian galley slaves.

For the next two centuries the war fleets of the world, as well as commercial navigators, relied on sail alone. Moreover, the engineering advances which

170

followed the development of the scientific approach, and the reliance on experiment rather than intuition, were less evident at sea than they were on land. What changes there were came about mainly from the need to make ships larger and this itself sprang from the use of cannon and the change of emphasis which made the result of sea battle more dependent on a broadside of heavy fire than on a successful boarding after two ships had grappled with each other.

But the number of vessels continued to increase, particularly after the Portuguese, Spanish and British had intensified their exploration of the seas and the countries that lay beyond the Atlantic and Indian Oceans. With this increase there came a growing demand for lights which would warn vessels of the more dangerous rocks and reefs.

The Pharos of Alexandria was the most famous lighthouse of ancient times, but not the only one; for hundreds of years the shores of the Mediterranean were dotted with round, or sometimes octagonal, stone towers on top of which in bad weather a fire would be burned or a light lit. Wood was the usual fuel, its thick column of black smoke as useful a sign by day as the blaze of a fire during the night. Olive oil, easily obtainable around the Mediterranean, was the usual fuel for the lamps which were sometimes kept burning. Cranelike erections were often sited by the towers to help raise the wood or oil to the level of the fire-basket on top of the tower.

Lighthouses were built in considerable numbers by the Romans and so usual did they become that the outline of a lighthouse became the sign of a port. These constructions showed little engineering advance and it was not until 1611 that the 186 feet (57 metres) Tour de Condonan at the mouth of the Garonne became the first lighthouse with a warning oil flare that revolved, sending out the beams from its light first over one section of the waves then over another. It was also the only lighthouse to incorporate a Royal Chapel.

The greatest problems facing the engineers who built the masonry lighthouse towers around and near Europe's coasts were presented by the red rocks of Eddystone, three lines of red gneiss 14 miles (22 kilometres) south-west of Plymouth. As trade with the American colonies grew during the seventeenth century, so did the number of ships endangered by these half-submerged razor-sharp ridges. The task of building here was taken up in 1696 by Henry Winstanley who spent the summer of that year cutting twelve holes in the rock to take iron bars. The following year a masonry base was built, keyed to the bars, and carried to a height of 15 feet (4.6 metres). On this there was constructed a hollow tower containing living quarters and, above this, a domed roof housing a glazed lantern in which twenty-six candles could be burned simultaneously.

Winstanley spent part of 1697 strengthening the tower during which time he was captured by a French privateer but returned on the orders of Louis XIV who said he was 'at war with England, not humanity' – and on 14 November 1698 the first light blazed out above Eddystone.

In November 1703 the lighthouse was entirely swept away in a storm. Winstanley was on Eddystone at the time and it was not until 1706 that work on a new lighthouse was started by John Rudyerd who began by drilling thirty-six holes in the rock and sealing wrought-iron rods into them with the help of melted pewter. To these metal ties, which John Smeaton found in position half a century later, Rudyerd bolted a layer of oak beams. He built courses of stone, then more timber and on this more stone until the foundation was some 30 feet (9 metres)

Detail from painting by an unknown artist of the Battle of Lepanto fought off Greece on
7 October 1571, between the Turks and a fleet of the Holy Roman League commanded by Don John of Austria.

from sea level. Only then came rooms for lighthousemen and, nearly 70 feet (21 metres) above the waves, the lantern floor and lantern. The entire tapering tower was sheathed in wood which had to be regularly replaced on account of attack by marine worms. Nevertheless Rudyerd's lighthouse might have lasted indefinitely if a fire in the lantern had not broken out in 1755 and led to the destruction of the building.

The following year John Smeaton was called in. Already known for his improvements to mill machinery and soon to become the prototype of the civil engineer able to build bridges, harbours or canals, Smeaton designed a masonry lighthouse reputed to have been based on the form of an oak tree trunk whose individual granite blocks were ingeniously dovetailed together. Each course was cunningly tied in to the ones above and below, while double oak wedges and marble dowels gave the structure immense strength. This interlocking system was reinforced by a binding agent that Smeaton chose only after making extensive experiments with different 'mixes'. Eventually he settled for pozzuolana, the cement which the Romans had used seventeen centuries earlier. Smeaton's system of interlocking, which was used in a number of subsequent rock lighthouses, was so good that the Eddystone lighthouse light shone without trouble for 123 years. Only in the late 1870s was it found that the rocks on which the lighthouse was built were being undermined. The upper part of Smeaton's masterpiece was dismantled and reconstructed on Plymouth Hoe, the foundation was left on the reef, and a new tower was built nearby, designed by James N. Douglass, Engineer-in-Chief of Trinity House, the organization controlling lighthouses around the coasts of England, Wales and Ireland.

Already, as seaborne trade continued to grow, men were asking whether the steam engine could not be used to overcome the uncertainties that reliance on sail imposed. As early as 1736 a patent was granted to Jonathan Hulls who, according to his application, had 'with much labour and with great expense, invented and formed a machine for towing ships and vessels out of, or into any harbour or river, against wind or tide, or in a calm, which the petitioner apprehends may be of great service to our Royal Party and merchant ships, and to boats and other vessels, of which the petitioner has made oath that he is the sole inventor.' It was to have been a stern-wheel tugboat worked by one of Savery's atmospheric engines, and Hull's patent suggested ambitious possibilities. But Savery's engines, like their Newcomen successors, proved impracticable for water-borne craft – largely due to the small horsepower per pound weight that they could produce – and it was only after Watt had patented his parallel linkage almost half a century later that it was possible to apply steam to ships.

In France the Marquis de Jouffroy designed in 1781 a boat which was to be driven on the Seine by a paddle wheel working by chains from a steam engine amidships, and two years later he ascended the River Sadne near Lyons in his 'Pyroschape'. In another two years Jacques-Constantin Perier was operating his own steamboat on the Seine. But these experiments were not successful and the same was true of Lord Stanhope's plans in 1790 for a double-bowed vessel, powered by a steam engine driving a propeller, which would operate like the foot of an aquatic bird. Stanhope was supported by the British Admiralty, and a 200-ton vessel named the *Ambi-Navigator, Kent,* was built in the Greenland Dock, Rotherhithe. Though fitted with a 12 horsepower Boulton and Watt engine, it could travel at a maximum of only 3 mph (4.8 kph) and no more was heard of it.

By this time there had been rapid developments in the United States, the country which was to witness the real birth of the steamboat. Here, however, the successful end-product was to be different from the vessels soon being constructed in British yards. In America the aim was to build comparatively small boats which could navigate the great rivers of the west, rivers which flowed through only partially explored territory traversed by no railways and only a few roads. In Britain, and to a lesser extent in Europe, the long-term aim was to be for larger, ocean-going vessels which would no longer be dependent on the weather for their transatlantic crossings or for their even more hazardous journeys round the Cape of Good Hope to the Far East.

The American view was well expressed by Robert Fulton, the successful pioneer of steam navigation. 'It will give a cheap and quick conveyance to the merchants on the Mississippi, Missouri and other great rivers which are now laying open their treasures to the enterprise of our countrymen', he wrote to his financial supporter, Joel Barlow; 'and although the prospect of personal emolument has been some inducement to me, yet I feel infinitely more pleasure in reflecting on the immense advantages that my country will draw from the invention.'

Fulton had a number of predecessors, among them James Rumsey of Virginia who led the way with a vessel intended to use hydraulic jet propulsion by means of a steam pump which drew in water at the bow and forced it out at the stern. In 1785, 1787 and 1788 Rumsey tested his invention on the Potomac and succeeded in driving it at 4 mph (6.4 kilometres) against the current. But the engines were inefficient, used far more fuel than expected, and encouraged Rumsey to travel to England in search of money, ideas or both. However, in Britain he died, nothing further accomplished.

Rumsey's more successful rival was John Fitch of Connecticut who on 27 September 1785 laid before the American Philosophical Society drawings, and a model, of a vessel in which a steam engine would operate an endless chain of baskets, dipping into and out of the water and thus propelling it forward. The device was adapted into another vessel, 34 feet (10.4 metres) long, of 8 foot (2.4 metres) beam and with a draught of 3 feet (0.914 metres) in which the steam engine 'by sprocket gearing actuated six oars placed vertically in a frame on each side of the boat'. It was successfully tested on the Delaware in July 1786, and Fitch then began to build a number of improved successors.

Success came in 1790 when, on 15 May, the *United States Gazette* reported from Burlington: 'The friends of sciences and the liberal arts will be gratified on hearing that we were favoured, on Sunday last, with a visit from the ingenious Mr. Fitch accompanied by several gentlemen of taste and knowledge in mechanics in a steamboat constructed on an improved plan. From these gentlemen we learn that they came from Philadelphia in 3 hours and a quarter, with a head wind with the tide in their favour. On their return by accurate observation, they proceeded down the river at the rate of upwards of seven miles an hour.' Fitch's vessel was an engineering success but a commercial failure. It ran for some weeks during the summer of 1790 but was then withdrawn from service, and Fitch died eight years later, his success unexploited.

The way was now left open for Fulton, the Pennsylvanian who can fairly be claimed to have shown the steamboat to be a practical proposition and thus led the way towards the giant vessels of the nineteenth century.

Trained as an artist, Fulton emigrated to Britain in 1786 at the age of twenty-three. But he had already shown the stirrings of mechanical ingenuity and had devised a boat driven by manually operated paddle wheels. In Britain he was caught up in the canal mania then sweeping the country, and turned his mind to a range of inventions that included an inclined plane to obviate the need for canal locks, a canal dredger, and plans for the cast iron aqueducts and bridges that the canal builders were demanding in numbers. As a sideline he devised a mill for sawing marble and a flax-spinning machine.

Then he turned to steam navigation. His first ideas, as revealed in a letter to Lord Stanhope, were almost bizarre. 'In June 1793', he wrote, 'I began the experiments on the steamship; my first design was to imitate the spring in the tail of a Salmon; for this purpose I supposed a large bow to be wound up by the steam engine and the collected force attached to the end of a paddle as in No. 1 to be let off which would urge the vessel forward. This model I have had made of which No. 1 is the exact representation and I found it to spring forward in proportion to the strength of the bow, about 20 yards, but by the return of the paddle the continuity of the motion would be stopped. I then endeavoured to give it a circular motion which I affected by applying two paddles on an axis, then the boat moved by jerks. There was too great a space between the strokes. I then applied three paddles forming an equilateral triangle on which I gave a circular motion by winding up the bow. I then found it to move in a gradual and even motion 100 yards with the same bow which before drove it but 20 yards.'

Nothing came of Fulton's schemes for imitating the spring in the tail of the salmon, and so he devoted his energies largely to canals. Then he travelled to France, where he conceived the idea of an underwater vessel and built a prototype christened the *Nautilus*, a name to be immortalized first by Jules Verne and then by the US Navy's first nuclear submarine. But neither the French nor the British, to whom Fulton later offered his plans, finally supported either the submarine or the 'torpedoes' with which he claimed to be able to destroy enemy shipping.

Before the collapse of negotiations about underwater craft and weapons, Fulton's interest in the steamboat was revived by the appearance in Paris of Robert Livingston, the newly appointed US Minister to France. A former Chancellor of the State of New York, Livingston owned the rights to operate ships throughout the State and in 1802 signed an agreement with Fulton to build a vessel 120 feet (36.6 metres) long and 8 feet (2.4 metres) wide to carry 'sixty passengers from New York to Albany at eight miles an hour in still water.'

A prototype in which the steam engine, mounted amidships, turned two primitive paddle wheels, was built on the Seine. While this first vessel sank under the weight of the engine, its successor performed successfully before a select committee of the French National Academy by moving at 4½ mph (7 kph) up-river against the current. The vessel was, said the *Journal des Débats*, 'a boat of curious appearance, equipped with two large wheels mounted on an axle like a cart, while behind these wheels was a kind of large stove with a pipe, as if there was some kind of small fire engine intended to operate the wheels of the boat . . . At six o'clock in the evening, assisted by three persons, [Fulton] put his boat in motion with two other boats in tow behind it, and for an hour and a half he afforded the curious spectacle of a boat moved by wheels like a cart, these wheels being provided with paddles or flat plates and being moved by a fire engine.'

So pleased was Fulton with his success that he promptly ordered an engine from Boulton & Watt to follow him across the Atlantic. There it was to power the vessel which marks a turning point in the history of steam navigation. This was *The Steamship*, later renamed *Clermont* after Livingston's country residence. Built during the spring of 1807 in Charles Browne's yard at Corlear's Hook on the East River, the *Clermont* was described by one onlooker as 'an ungainly craft looking precisely like a back-woods' sawmill mounted on a scow and set on fire.'

On 9 August 1807 Fulton ran the vessel a mile up the East River, then turned her round and steamed her back into the berth she had just left. 'She answers the helm equal to anything that ever was built, and I turned her twice into three times her own length,' he wrote proudly to Livingston.

The first official run, from New York to Albany, came a week later. Already the vessel had been nicknamed *Fulton's Folly*, but those who crowded the banks of the East River were to be surprised. 'It was in the early autumn of the year 1807' says one account, 'that a knot of villagers was gathered on a high bluff, just opposite Poughkeepsie, on the west bank of the Hudson, attracted by the appearance of a strange-looking craft, which was slowly making its way up the river. Some imagined it to be a sea-monster, while others did not hesitate to express their belief that it was a sign of the approaching judgment. What seemed strange in the vessel was the substitution of a lofty and strange black smoke-pipe rising from the deck, instead of the gracefully tapered masts that commonly stood on the vessels navigating the stream, and in place of the spars and rigging, the curious play of the working beam and piston, and the slow turning and splashing of the huge and naked paddle-wheels, met his astonished gaze. The dense clouds of smoke, as they rose wave upon wave, added still more to the wonder of the rustics.'

With forty passengers on board the vessel covered the 150 miles (240 kilometres) upstream to Albany in thirty-two hours, and the return voyage equally successfully. 'As we passed the farms on the borders of the river' an English passenger wrote in the *Naval Chronicle*, 'every eye was intent, and from village to village, the heights and conspicuous places were occupied by the sentinels of curiosity, not viewing a thing they could possibly anticipate any idea of, but conjecturing about the possibility of the motion.'

The journey up the Hudson was to be the start of a financial as well as an engineering success. 'Perpetual motion authorizes you to calculate on a certain time to land', wrote the author of the *Naval Chronicle* report; '[the ship's] works move with all the facility of a clock; and the noise when on board is not greater than that of a vessel sailing in a good breeze.'

During the next few years, Fulton was to build no less than another sixteen steamboats, the largest of them the 526-ton *Chancellor Livingston*, which served the Hudson and the Rariton, the Potomac and the Mississippi.

When the war with England broke out in 1812 Fulton designed his *Demologo* or *Fulton*, the first steam-powered warship ever to be built. It was more than a manageable vessel which did not have to depend on the weather since it included a number of ingenious devices. There were special furnaces for heating shot to a red-heat; there were submarine guns which could fire 100 pound (45 kilogram) balls 12 feet (3.7 metres) below the water line and an engine for discharging an immense column of water upon decks and through portholes. But the war with Britain ended before the *Fulton* was completed, and she was stationed at Brooklyn Naval Yard until destroyed by an accidental explosion in 1829.

Fulton's achievements had not been lost on the Canadians to the north and as early as 1809 the *Accommodation*, with berths for twenty, was running on the St Lawrence between Quebec and Montreal. 'The great advantage attending a vessel so constructed', the *Quebec Mercury* recorded of her maiden voyage, 'is that a passage may be calculated to a degree of certainty, in point of time, which cannot be the case with any vessel propelled by sails only. The steamboat received her impulse from an open-spoked, perpendicular wheel, on each side, without any circular band or rim. To the end of each double spoke is fixed a square board, which enters the water, and by the rotary motion of the wheel, acts like a paddle. The wheels are put and kept in motion by steam, operating within the vessel. A mast is to be fixed in her for the purpose of using a sail when the wind is favourable, which will occasionally accelerate her headway.' Within the next few years, three other vessels were put into service on the route.

While Fulton had been pioneering his ideas on the rivers of North America, progress was being made in Britain. William Symington, who in the late 1780s had failed to finance a twin-hulled vessel powered by a Boulton & Watt engine, obtained the support of Lord Dundas, a governor of the Forth and Clyde Canal, and in the first year of the nineteenth century was successfully operating the *Charlotte Dundas* on the Clyde.

The *Charlotte Dundas* was followed in 1812 by the *Comet*, a 25-ton ship only 40 feet (12 metres) long and the first successful European passenger vessel. Though carrying only a handful of passengers she raised as much opposition as larger steamships elsewhere and one Clyde skipper used to pipe hands on deck when the *Comet* passed and order them: 'Kneel down and thank God, that ye sail

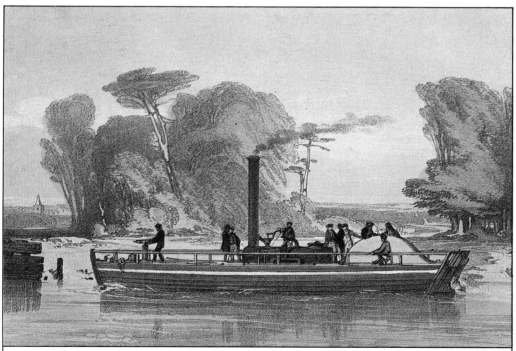

The *Charlotte Dundas* on her trials in 1802. The engine with its single horizontal cylinder turned a paddle wheel in the stern of the 56-foot (17-metre) long wooden vessel.

wi' the Almighty's ain win', an' no' wi' the deevil's sunfire an' brimstane like that spluttering thing there.'

As the possibilities of steam power for ships began to be appreciated in both the United States and Britain, another change started to make itself felt. This was the transition from wood construction to the use of iron. Just as the success of the steam engine on land and the growth of the coal industry were to be almost inextricably linked, so were the change from sail to steam and from 'wooden walls' to the 'iron-clad'.

In 1787 John Wilkinson's barge *Trial*, 70 feet (21 metres) long, 6 feet 8½ inches (2 metres) wide and carrying 32 tons, was launched on the River Severn. She had a shell of ⁵⁄₁₆ inches (0.79 centimetres) iron plates, although wood was used for bracing inside the hull. 'Yesterday week my iron boat was launched', he wrote to a friend. 'It answers all my expectations and has convinced the unbelievers who were 999 in every 1,000.' Many spectators had apparently expected the vessel to sink immediately.

The first steamship in Britain to be built entirely of iron was the *Aaron Manby*, whose hull was built in sections at the Horsley Ironworks outside Birmingham and then assembled on the Thames at Rotherhithe. On 9 May 1822, a party of naval officers boarded the boat at Parliament Stairs, says a report on steam navigation laid before the House of Commons. The ship 'immediately got under weigh and proceeded to Battersea Bridge; she then descended to Blackfriars and manoeuvred for several hours between the two bridges in a very superior style. . . . She is the most complete specimen of workmanship in the iron way that has ever been witnessed, and draws one foot less water than any steamboat that has ever been built. She is 106 feet long and 17 feet broad, and is propelled by a 30 h.p. engine and Oldham's revolving bars.' Shortly afterwards she crossed the Channel to Le Havre and steamed up the Seine to Paris.

Two years later the *Garry Owen*, built by William Laird who had started building iron ships at Birkenhead, dramatically illustrated the advantages of iron over wood when she was driven ashore with a number of wooden ships in a storm. She survived with little damage while most of the wooden ships broke up. One reason was no doubt the ability to give, with metal, the firm end-to-end longitudinal connection that was impossible with wooden ships built from planks which were limited in length. Metal, by contrast with wood, dealt adequately with the stresses and strains, put on a vessel when, for instance, her bows and stern were supported by wavecrests and her midships' sections were left unsupported by a wave trough.

When metal took over from wood as the main raw material of shipbuilding it brought with it engineering problems as well as benefits. New skills and new trades were called for as the practices of centuries became obsolete, and beneath the clanging of metal on metal new crafts evolved. The extent of the change was well described by Sir Westcott Abell in *The Shipwright's Trade*. 'The kind of skill needed was new as well as rare', he has written; 'there was little knowledge of iron structure or of ocean ships. The shipwrights had to give guidance as to the parts, based upon wooden ships; they hated ironwork and their help was grudging. The smiths alone knew how to handle iron in simple fashion, but they had little in common with the shipwright. Iron was in most use for engines and boilers, and the class of workers had grown up who could cut, bend, shape and join sheets of iron with angles.

'Thus it came about that the firms who built engines also began to build ships because the machines and plant needed to fashion iron served the common purpose. All the tools needed to deal with iron are quite unlike those which were used for wood. Treatment is quite changed. Heating, hammering, rolling, punching, shearing and welding were part of the craft for workers – the smiths, the fitters and the boilermakers. Again, the plater and the riveter took over the work of the shipwrights in regard to the skin of the ship. Later, as machines were more used to work the iron, there came about the closer contact with the engine-builder.'

The transition from sail to steam, and later from wooden ships to iron, came about with some speed. Ten years after the first steam vessels had appeared on the Mersey and the Thames in 1815 there were no less than forty-four ships of 250 to 500 tons each on the stocks at Liverpool, while in London forty-five companies had been established to operate steam packets. Almost inevitably, innovation brought protests from those with vested interests in old ways and on 14 December 1826 a meeting was held in Swansea by die-hards whose business was with sail. A resolution was passed calling for a petition to be sent to the House of Commons which would ask for the intervention of Parliament to protect sailing vessels against the further increase of steamers.

Parliament refused to intervene and the move from sail to steam continued, as did the replacement of wood by metal. A pioneer in both transformations was Isambard Kingdom Brunel, the engineering all-rounder whose dedicated enthusiasms had built 'the finest thing in England', the railway from London to the West. Indeed, it was as a direct result of the London-Bristol line that Brunel embarked on the projects which were to open up the Atlantic to steam.

At a London meeting of the Great Western Company in 1835, one of its directors had expressed doubts about the length of the line being planned between London and Bristol, the 100 miles (160 kilometres) which made up an extraordinarily ambitious distance for the times. 'Why not make it longer?' Brunel replied, 'and have a steamboat go from Bristol to New York and call it the *Great Western*?' At first the suggestion was hardly taken seriously, although everyone, apparently including Brunel, appears to have overlooked the fact that the Atlantic had already been crossed by steam.

The first crossing by a steam-powered ship had been by the paddle steamer *Savannah* which had left Savannah en route to St Petersburg in 1819 and after coaling at Kinsale had reached Liverpool in twenty-seven days, eleven hours. However, the vessel was a fully rigged sailing ship and used her auxiliary paddle power for only eighty-five hours on her transatlantic voyage.

But by the time of Brunel's proposal there had also appeared the *Royal William*, built by the Quebec and Halifax Steam Navigation Company at Quebec in the winter of 1832-3. Powered by 180 horsepower engines built by Boulton & Watt in Birmingham, she was a 1,370-ton vessel of 176 feet (54 metres) which for several months after her launching had plied between Montreal and Pictou in Nova Scotia. Then it was announced she would cross the Atlantic to Britain. This she did in August 1833, leaving Quebec on the 5th, picking up coal at Pictou and arriving at Cowes, Isle of Wight, 2,500 miles (4,000 kilometres) away, seventeen days later. Despite such an auspicious start to her career, the *Royal William* was sold to the Spanish Government, which converted her into a six-gun man-of-war. Eventually she was wrecked off the harbour of Santander.

Under Brunel's persuasion the Great Western Steamship Company was founded, and on 28 July, 1838 William Patterson of Bristol set up the sternpost of a vessel which was to be 236 feet (72 metres) long and with a displacement of 2,300 tons. Controversy rested not so much on the size of the vessel as on its ability to carry enough fuel to cross the Atlantic. Brunel had been reassuring in his report to the company, having seen the significance of two facts missed by others. 'It is well known', he wrote, 'that the *proportional* consumption of fuel decreases as the dimension and power of engines are increased and consequently a large engine can be much more economical than a small one. The resistance of vessels on the water does not increase in direct proportion to their tonnage. The tonnage increases with the cubes of their dimensions while the resistance increases at about their squares, so that a vessel of double the tonnage of another capable of containing an engine of twice the power does not really meet with twice the resistance. Speed would therefore be greater with the larger vessel or the proportion of power in the engine and consumption of fuel may be reduced.'

This had no effect on Dr Dionysius Lardner, who had been critical of the plans for Box Tunnel. 'Take a vessel of 1,600 tons provided with 400 horse-power engines', he told the British Association which was meeting in Bristol in 1836. 'You must take 2⅓ tons for each horse-power, so the vessel must have 1,348 tons of coal. To that add 400 tons, and the vessel must carry a burden of 1,748 tons.' Once these figures were considered, he went on, it must be seen that a voyage direct from Liverpool to New York was chimerical and his audience 'might as well talk of making a voyage from New York or Liverpool to the moon.'

Completely unconvinced, Brunel continued to maintain that a steamship would be able to cross the Atlantic to New York in twenty days and to return in thirteen, compared with the current times of thirty-six and thirty-four under sail. Construction of the *Great Western* went ahead as scheduled.

Although Brunel did not take advantage in the *Great Western* of the great longitudinal strength that metal was already making possible, he gave the vessel a rigidity that foreshadowed his plans when he turned to iron. 'The ribs were of oak', his son has written in the life of his father, 'of scantling equal to that of line-of-battle ships. They were placed close together, and caulked within and without before the planking was put on. They were dowelled and bolted in pairs; and there were also four rows of 1½ in. iron bolts, 24 feet long, and scarfing about 4 feet, which ran longitudinally through the whole length of the bottom frames of the ship. She was deeply trussed with iron and wooden diagonals and shelf pieces, which, with the whole of her upper works, were fastened with bolts and nuts to a much greater extent than had hitherto been the practice.' And on 19 July 1837 she was floated out of Patterson's dock and prepared for her voyage under sail to the Thames where her 750 ihp engines, made by Maudslay's, were to be fitted. Installation and testing of the engines took seven months despite Brunel's insistence on speed.

Speed was necessary since neither London nor Liverpool had taken easily to the idea that a Bristol-built vessel might make the first of what were planned to be regular steam-powered crossings of the Atlantic. While Bristol had once been the most flourishing of Britain's ports, Liverpool and London were both displacing her and both had prepared vessels to challenge the *Great Western*. In March 1830, therefore, there started what was to be a desperate race.

On March 28, while Brunel's men were still installing the *Great Western*'s engines, the 703-ton *Sirius*, chartered by the Transatlantic Steamship Company of Liverpool, steamed down the Thames with forty passengers and a crew of thirty-five. The *Great Western* followed her early on the morning of the 31st. The race was still open since it was known that the *Sirius* was to take on fuel at Cork.

The *Great Western* had been steaming for little more than two hours when flames and smoke began pouring from the for'ard boiler room. George Pearne, the ship's first engineer, gallantly climbed down to investigate, discovered that the boiler lagging had been fixed too close to the boiler flues, had caught fire and had then set some of the timbering alight. Pearne dealt with the crisis; the captain, not knowing what would happen next, headed inshore and grounded the ship temporarily on Chapman Sound. Brunel, descending into the boiler room to see the situation for himself, fell from a burned-through ladder rung and was saved from death by the captain.

That evening the *Great Western* was floated off. On April 2 she arrived in Bristol. Repairs were hurried through but it was not until the morning of April 8 that she steamed down the Severn estuary into the Atlantic, four days after the *Sirius* had left Cork. The smaller vessel pressed on, but as her log reached the 3,000 mile (4,800 kilometre) mark the coal was seen to be dangerously low. Four barrels of resin from the cargo helped to eke out the supplies and when she docked in New York on the 23rd, she had only 13 tons of coal left. The *Great Western* arrived a few hours later, having taken fifteen days five hours from Bristol compared with the *Sirius*'s nineteen days at sea. Moreover, she had nearly 200 tons of coal left in her bunkers.

A small but significant factor in the *Great Western*'s crossing time may have been a navigational aid put in by Brunel. 'I had a chart drawn and engraved of the sea (that is, the lines of latitude and longitude, and the bearings of the compass, and the coast and soundings) on a cylindrical projection of a great circle from Bristol to New York' he later wrote, 'and we found it very useful for the captain to *see* his great circle sailing, and to see how much he was deviating from it.'

The *Sirius* had won the race but it was the *Great Western* with her faster crossing time and her reserve of fuel which had shown that on the oceans of the world sail was giving way to steam. She was also the more impressive ship, as was made clear in the *Morning Herald* by James Gordon Bennett, the journalist who more than forty years later was to send H.M. Stanley in search of Dr Livingstone in darkest Africa. 'The sky was clear – the crowds immense' he wrote of the *Great Western*'s arrival. 'The Battery was filled with the human multitude, one half of whom were females, their faces covered with smiles, and their delicate persons with the gayest attire. Below, on the broad blue water, appeared this huge thing of life, the four masts and emitting volumes of smoke. She looked black and blackguard – rakish, cool, reckless, fierce and forbidding in sombre colours to an extreme. As she neared the *Sirius*, she slackened her movements, and took a sweep round, forming a sort of half circle. At this moment the whole battery sent forth a tumultuous shout of delight, at the revelation of her magnificent proportions. After making another turn towards Staten Island, she made another sweep, and shot towards East River with extraordinary speed.'

The performance of the *Great Western* so captured American imagination that she carried sixty-eight passengers on her return voyage which took her to Bristol in fifteen days, and during the next eight years she was to make sixty-seven crossings of the Atlantic, thus ensuring that the steamship had come to stay.

By the autumn Brunel was planning a successor. This was originally planned as another wooden paddle steamer, but instead turned out to be a vessel revolutionary in two different ways. The *Great Britain*, Brunel's second transatlantic vessel, was not only to be an iron ship but was to be driven by screw propellers rather than paddles.

Brunel was already considering the merits of iron instead of wood when, in October 1838, the small iron paddle steamer *Rainbow* docked in Bristol to take on a load for Antwerp. It was an opportunity too good to miss and he persuaded two of the Great Western Steamship Company's directors to join her for the voyage to Belgium and report on how the vessel behaved in the open sea. Their report was so favourable that work was begun in July 1839 on what was for those days a huge iron vessel of 3,443 tons.

One of the first problems that arose concerned not the metal structure of the ship but its size. 'I find' wrote Francis Humphries, Brunel's engineer in charge of the work, 'that there is not a forge hammer in England or Scotland powerful enough . . . .' What was he to do? Brunel asked James Nasmyth, a former pupil of Maudslay and now one of Britain's leading engineers.

From this simple enquiry there was to come one of the most revolutionary machines of the industrial age, Nasmyth's Steam Hammer, which, first as hammer and later when converted into pile-driver, was to become familiar in British workshops and on civil engineering sites.

'. . . A few minutes' rapid thought satisfied me that it was by our rigidly adhering to the old traditional form of the smith's hand hammer – of which the forge and tilt hammer, although driven by water or steam power, were mere enlarged modifications – that the difficulty had arisen;' Nasmyth later wrote, 'as, whenever the largest forge hammer was tilted up to its full height, its range was so small that when a piece of work of considerable size was placed on the anvil, the hammer became "gagged"; so that, when the forging required the most powerful blow, it received next to no blow at all, as the clear space for the fall of the hammer was almost entirely occupied by the work on the anvil.'

Within little more than half an hour, he had devised a solution and drawn its details in his Scheme Book.

'My Steam Hammer, as thus first sketched, consisted of, first a massive anvil on which to rest the work; second, a block of iron constituting the hammer or blow-giving portion; and, third, an inverted steam cylinder to whose piston rod the hammer block was attached. All that was then required to produce a most effective hammer was simply to admit steam of sufficient pressure into the cylinder so as to act on the underside of the piston rod. By a very simple arrangement of a slide-valve, under the control of an attendant, the steam was allowed to escape and thus permit the massive block of iron rapidly to descend by its own gravity upon the work then upon the anvil.

'Thus, by the more or less rapid manner in which attendant [*sic*] allowed the steam to enter or escape from the cylinder, any required number or intensity of blows could be delivered. Their succession might be modified in an instant. The hammer might be arrested and suspended according to the requirements of the work. The workman might thus, as it were, *think in blows*. He might deal them out to the ponderous glowing mass, and mould or knead it into the desired form as if it were a lump of clay; or pat it with gentle taps according to his will, or at the desire of the forgeman. . . . The hammer could be made to give so gentle a blow as to crack the end of an egg placed in a wine glass on the anvil; whilst the next blow would shake the parish, or be instantly arrested in its descent mid-way.'

Only a few years after the steam hammer was firmly established, it was converted by Nasmyth into a pile-driver and used at Devonport where he was enlarging the harbour for the Admiralty. Here the new machine drove a 70 foot by 18 inch (21 metre by 46 centimetre) square pile into position in four and a half minutes, compared with the twelve hours necessary to do the same thing by hand. This was in 1845 and within a few years Nasmyth's pile-driver was being employed in the building of the Border Bridge at Berwick-upon-Tweed, new docks at Tynemouth, Birkenhead and Grimsby, the new Westminster Bridge, and a number of engineering projects in Russia, including fortifications at Cronstadt.

However, the *Great Britain* whose demands had given birth to Nasmyth's machine, was not to need the steam hammer after all since in 1840 there arrived in Bristol the 237-ton *Archimedes*, driven not by paddle wheels but by a screw propeller. The innovation, which was to transform steamship practice, was a practical development of the Archimedean screw which had been used for water-lifting more than 2,000 years previously. There had been various attempts to incorporate the screw movement and as far back as 1785 Joseph Bramah had obtained a patent claiming that instead of the paddle wheel there might be introduced 'a wheel with inclined fans or wings similar to the fly of a smoke-jack or the vertical sails of a windmill. This wheel or fly may be fixed on the spindle, and may be wholly under water, when it would, being turned round either way, cause the ship to be forced backwards or forwards, and its power will be in proportion to the size and velocity of the wheel.' There were other attempts at adapting the screw principle, but none was successful until the 1830s. Then the problem was successfully solved quite independently by two men.

One was the Englishman, Francis Pettit Smith; the other was the Swede, John Ericsson, who had in July 1836 patented a propeller consisting of 'two thin broad hoops or short cylinders made to revolve in contrary direction round a common cylinder, each cylinder or hoop being kept moving with a different velocity from the other, each hoop or cylinder being also situated entirely under the water at the stern of the boat and furnished each with a series of short spiral planes or plates – the plates of each series standing at an angle the exact converse of the angle given to those of the other series and kept revolving by the power of the steam engine.'

Shortly after the patent had been granted, the *Francis B. Ogden*, 45 feet (14 metres) long and of 8 feet (2.4 metres) beam was launched on the Thames to test the efficiency of its two 5 foot 3 inch (1.6 metres) propellers, so fitted in the stern of the vessel that either could be used. 'So successful was the experiment that when steam was turned on for the first time, the boat at once moved at a speed of upward of ten miles an hour, without a single alteration being required in her machinery' wrote John Bourne in his *Treatise on the Screw Propeller*. 'This miniature steamer had such power, too, that she towed a schooner of one hundred and forty tons burden at the rate of seven miles an hour, and the American packet ship *Toronto* at the rate of more than four and a half knots . . . against the tide.' 'The demonstration', said John O. Sargent in a lecture before the Boston Lyceum in December 1843 'excited no little interest among the boatmen of the Thames, who were astounded at the sight of this novel craft moving against wind and tide without any visible agency of propulsion, and, ascribing it some supernatural origin, they united in giving it the name of the "Flying Devil". But the engineers of London regarded the experiment with silent neglect.'

Similar neglect was shown by the British Admiralty who after a successful demonstration of the *Francis B. Ogden* rejected the idea of screw propellers on the grounds that it would be impossible to steer a ship which was powered from the stern. Despite the lack of official interest, Ericsson in 1837 designed the successful 10 horsepower *Novelty* which plied on the Manchester-London canal, and the propeller-powered steam schooner, the *Robert F. Stockton* which successfully crossed the Atlantic in 1839 and was then used as a tug on the Delaware and Raritan Canal for nearly thirty years.

Believing there were no opportunities for him in Britain, Ericsson now emigrated to the United States, a journey whose significance could not have been known at the time. But more than two decades later, after the outbreak of the Civil War, he was to build the ironclad *Monitor* for the North. In her famous battle with the South's *Merrimac* on 8 March 1862 the *Monitor* was hit twenty-two times but not critically damaged. The *Merrimac*, moreover, was forced to retreat. Had it not been for the *Monitor* the North's blockade of the South might have been broken with all that that would have meant for the outcome of the Civil War.

Ericsson's move to America in 1839 left the field open for Pettit Smith who in May 1836 had patented a screw propeller to be fitted 'in a recess or open space formed in that part of the after part of the vessel commonly called the deadrising or deadwood of the run.' He had subsequently helped form the Screw Propeller Company and equipped with propellers the *Archimedes* which had arrived in Bristol while Brunel was planning the *Great Britain*.

Brunel was so impressed with the *Archimedes* that he sent one of the company's directors up to Liverpool in it and then hired the ship for six months. The outcome was that his plans for a wooden paddle steamer were scrapped and work went ahead for a vessel incorporating two major innovations. Thus the *Great Britain* was not only to be built of metal but driven by a screw propeller.

There was, in fact, to be yet a third innovation, since the screw propeller was to be powered by the first V type engine, which Brunel called the 'Triangle' engine. This had two cylinders arranged at 90 degrees to each other and driving upwards to a crankshaft above them. The engine's governor had its weights rotating in a vertical plane, with the force of gravity countered by a spring, and centrifugal force being the only one operative.

On 19 July 1843, the *Great Britain* was named and launched by Prince Albert. But the launch was only into the waters of the Bristol docks. All went well until mid-December 1844, when she was due to be taken through the narrow lock gate at Bristol in preparation for her journey down the Severn and round to the Thames for her final fitting-out. 'We have had an unexpected difficulty with the *Great Britain* this morning', wrote Brunel, according to his son, 'She stuck in the lock; we *did* get her back. I have been hard at work all day altering the masonry of the lock. Tonight, our last tide, we have succeeded in getting her through; but, being dark, we have been obliged to ground her outside, and I confess I cannot leave her till I see her afloat again, and all clear of her difficulties. I have, as you will admit, much at stake here, and I am too anxious about it to leave her.'

The *Great Britain* eventually left for the Thames on 23 January 1845, was successfully completed, and steamed from Liverpool on her maiden transatlantic voyage on August 26.

By this time, the screw propeller had come through a decisive test when in 1845 a tug-of-war was organized between two Admiralty steam sloops, the 888-ton propeller-driven *Rattler*, commissioned after the early success of the *Archimedes*, and the 800-ton paddle-driven *Alecto*. Both vessels were equipped with 200 horsepower steam engines and on a calm April morning were linked up, stern to stern, by iron cables. The engines were started and the order was given 'full steam ahead'. Before long the *Rattler* began to pull away, towing the *Alecto* against the pull of the latter ship's paddles. Soon the *Rattler* was pulling its rival, stern first, at a speed of 2.8 knots, a demonstration which eventually convinced the authorities that the propeller's virtues outweighed its disadvantages. The most important of these was, as Brunel found, that the propeller required a higher speed of shaft revolution than the paddle wheel so geared engines operating pitch-chains or toothed wheels were necessary to produce the extra speed.

The *Great Britain*'s first voyages were a great success. The number of passengers continued to increase and when she sailed from Liverpool in September 1846 she carried 185.

Then disaster struck. On the evening of the 22nd, when the vessel was believed to be rounding the Isle of Man, she ran ashore in Dundrum Bay, Ireland. The cause of the accident later appeared to have been the absence, on the chart being used, of an essential lighthouse. Dramatic attempts to float her off were made during the next few days but after her passengers had been taken off it was to be more than eleven months before she could finally be moved. The catastrophe had proved at least one thing: that an iron ship could withstand a hard winter's battering that would have destroyed any wooden vessel. After repairs and

The *Rattler* and the *Alecto* struggling
against each other in 1845 – screw against paddle-power.

a trial run to the United States, the *Great Britain* was transferred to the Australia run and continued to steam on it successfully for more than thirty years.

To Brunel, the experience with the *Great Britain* confirmed his belief that the advantages of size were more important than the disadvantages. When, in 1851, he was consulted by the Australian Mail Company, he saw an opportunity to implement his ideas. The result was *Leviathan*, operated under the name the *Great Eastern*, the largest steamship to be built in the nineteenth century and an exemplar of marine engineering.

The design was to be for a ship big enough to carry all the fuel to take it to Australia, a feat made possible by the use of iron, Brunel told the company's directors. Compared with the 236 foot (72 metre) length of the *Great Western* and the 322 feet (98 metres) of the *Great Britain*, the *Great Eastern* was to be 692 feet (211 metres) long. Its hull would be 82.7 feet (25 metres) broad compared with its predecessors' 35.3 feet (10.8 metres) and 51 feet (15 metres) while her displacement, instead of 2,300 tons and 3,618 tons in the earlier ships, would be 32,000 tons. She would have paddles and screw propeller, would be able to carry 4,000 passengers or 10,000 troops, and capable of stowing 12,000 tons of coal.

In the *Great Eastern*, moreover, Brunel carried to its conclusion his theory of cellular construction. 'The whole of the vessel', he told the directors in a report in February 1855, 'is divided transversely into ten separate perfectly water-tight compartments by bulkheads carried up to the upper deck, and consequently far above the deepest water lines, even if the ship were water-logged, so far as such a ship could be; and these are not nominal divisions, but complete substantial bulkheads, water-tight, and of strength sufficient to bear the pressure of the water, should a compartment be even filled with water; so that if the ship were supposed to be cut in two, the two separate portions would float; and no damage, however great, to the ship's bottom, in one or even two of these compartments would endanger the floating of the whole, or even damage the cargo in the rest of the ship, or above the main-decks of the compartment in question, and all damageable cargo would be stowed above that deck. . .'

The *Great Eastern* was also one of the first ships if not the very first, in which the principle of standardization of components was followed. 'In [the vessel]', wrote John Scott-Russell who built the ship for Brunel, 'there is one thickness of plates (¾ inch) for skin, outer and inner, one thickness of internal work (½ inch), one size of rivet (⅞ inch), one pitch (3 inch), and one size of angle-iron (4 inch × 4 inch × ⅝ inch). The exceptions that happen to have got in . . . were accidental, and formed no part of the design.'

It was not only in the design of the hull that Brunel intended the *Great Eastern* to be the wonder of the world. He sought advice from Sir George Biddell Airy on the appointment of observers on the ship who would keep a constant record of her position, and planned to instal a gyroscope to give them a suitable foundation for their work. As with his building of the Great Western Railway, he concerned himself with detail. He intended that a stream of surface water should be constantly pumped up into the astronomical observers' cabin, so that the presence of icebergs might be indicated by its change of temperature, and even designed a device enabling the look-out men to keep their eyes open in a gale. 'This consisted of two sets of vertical plates of tin placed one behind another, diverging from the direction of the wind, with a clear wide passage between the two sets of plates. The wind, entering at the end of the apparatus, became separated by the first two

The launching of the *Great Eastern* in 1857 showing (left to right) the engineers J. Scott-Russell and Henry Wakefield, Isambard Kingdom Brunel and Lord Derby.

inclined plates, and the residue that passed on in the direct line was again subdivided, so that at the end of the last set of plates there was no rush of air between them, and a man looking through the aperture, with his face to the wind, was in a perfect calm. This was a useful arrangement, the look-out man's eyes being as well protected as though behind a glass. A glass would not answer the purpose, as it would become obscured with spray.'

The hull of the *Great Eastern*, consisting of 30,000 iron plates weighing 10,000 tons and joined by 3 million rivets, was completed on the Thames at Millwall by the late summer of 1857. Brunel had in his usual fashion tried to keep his finger firmly on the progress of events and was perturbed to find, according to John Scott-Russell, that foremen were being allowed to tackle many jobs in their own way, considering this to be 'disorganized'. 'He would even', it has been said, 'stipulate which unfinished engine parts were to be "gone on with". As for the stern and bow drawings he had just received, they were not "such drawings as might be given to a manufacturer and contractor to work from without further instructions". "I must have them", he demanded. They were such drawings, however, as foremen platers and erectors could work from with their customary discretion on the site, a discretion which they continued to use – much to Brunel's annoyance – even when the drawings or his instructions were very explicit, as in the case of some cabin deck beams which he wanted to be fixed with wood screws instead of bolts.'

Once the great ship was finished, plans were started for the launching. It was to be a daunting problem since Brunel had decided that the ship must be built parallel to the river; so it would have to be launched sideways on into the water.

Brunel's *Great Eastern* later named *The Leviathan*.

Despite Brunel's wish to keep crowds away from the event, thousands assembled at the Millwall yards on the morning of 3 November 1857. The directors, he discovered to his consternation, had even sent out invitations to distinguished guests. From the first, almost everything that could go wrong did go wrong. The great ship moved a few feet down the guide-rails along which it should have passed into the water, then stuck. There was a hush for some minutes, then a terrific report as the ship began to move again. The winch handles on one of the huge drums being used to let the vessel down spun round unexpectedly killing one man and injuring a number of others. Then the ship came to a halt once again.

A second attempt at getting the ship into the water was made on 19 November, but with no more success. One of the important mooring-chains broke. Work was hampered by dense fog. Then it was found that a number of other chains were deficient. During subsequent attempts, launching equipment broke down or burst, and it was not until 31 January 1858, that the great ship finally floated free in the river. It had cost an additional £120,000 to launch her.

The trials and tribulations of the longest ship the world had seen did not end with the disastrously long drawn-out difficulty of her launching. It was September 1859 before she was ready for making the transatlantic crossing planned as a precursor to her making the first of the Australian runs for which she had been built. But while being taken from the Thames to Weymouth the casing round one of the funnels of the paddle engines exploded. The funnel was thrown up on to the deck, boiling water was driven down into the boiler room and a number of men were injured, some of them fatally.

The accident was given prominence by a new process which in the years ahead was to have a growing effect on engineering. This was photography, the use of light to affect sensitized paper chemically so that permanent images of people, places or events could be produced. The discovery in France by Louis Jacques Mandé Daguerre that it was possible to make a photographic image and the almost simultaneous discovery in England by William Henry Fox Talbot that a negative allowed numerous prints to be made, had come in the 1830s. But although they had been preceded in France by Nicephore Niepce who discovered the principles of what was to become photography, it was only in the following decades that the art or craft became a practical proposition. Before this, the world would have seen the results of the explosion on the *Great Eastern* only through sketches or drawings. But a news photographer provided a graphic photograph of the smashed funnel with members of the crew crowding round it.

Like the impressive photographs of the *Great Eastern*'s launching – and like the equally remarkable photographs of the Great Exhibition in 1851 – photographic prints enabled a newly enlarged public to witness the facts of engineering progress. Even more important was the impact of photography on engineering itself. This was emphasized by no less a person than Richard Owen, the British scientist, in his presidential address to the British Association at Leeds in September 1858. 'The engineer at home', he told the packed hall of members, 'can ascertain by photographs transmitted by successive mails the weekly progress, brick by brick, board by board, nail by nail, of the most complex works on the Indian or other remote railroads.' The same was true of bridge building, shipbuilding, and other engineering enterprises. But a greater impact on engineering was to be made by photography later in the nineteenth century with the development of high-speed photography which enabled engineers to watch and analyze the movements of machinery previously invisible to the human eye.

After the *Great Eastern* had been repaired at Weymouth the ship steamed on a trial voyage to Holyhead. Here an unexpected and violent storm blew up destroying the nearby *Royal Charter*, but revealing the *Great Eastern*'s advantage in having both paddles and screw. 'A portion of the temporary staging erected by the contractor at the breakwater was carried away, and drifted down upon the ship', Brunel's son has written. 'During the gale her engines were kept going, in order to relieve the strain on the cables. The timbers of the staging got foul of both paddle wheels and screw; but, as it was always possible to keep one of the engines at work, the ship was saved from drifting.'

The *Great Eastern* subsequently made a number of successful transatlantic crossings and in 1861 took 2,500 troops and two hundred artillery horses to Quebec. But bad luck continued to dog her and later in 1861 her rudder was seriously damaged in an Atlantic storm. The following year, approaching Long Island Sound, she passed over a reef of sunken rocks not marked on the charts and fractured her hull in ten places, an accident which would have proved fatal to almost any other ship.

But in 1864 she was hired for the historic task of laying the Atlantic cable. Half way across the ocean the following year the cable broke, only to be recovered and satisfactorily relaid in 1866. The vessel, said a memorandum drawn up by the engineers and scientists on board, 'from her size and constant steadiness, and from the control afforded over her by the joint use of paddles and screw, renders it safe to lay an Atlantic cable in any weather.'

Before Brunel's *Great Eastern* had at last found this useful niche for which it had not, in fact, been built, John Ericsson, whose exploitation of the propeller was already transforming steam craft, had helped change the concept of naval warfare, the shape of naval vessels, and the ideas of what they could do. 'An impregnable and partially submerged instrument for destroying ships of war has been one of the hobbies of my life', he wrote to a colleague in the 1850s. 'I had the plan matured long before I left England. As for protecting war engines for naval purposes with iron, the idea is as old as my recollection.' These thoughts were the natural outcome of the fact that the paddle wheels normally propelling steam vessels made them particularly vulnerable to enemy fire, while Ericsson's propeller, operating below the waves, presented a minimum target. But only with the outbreak of the American Civil War in April 1861, and the need for the Union forces to blockade the Confederate South, did there come an opportunity for Ericsson to implement his ideas.

In August the Navy Department in Washington publicly asked for proposals for iron-clad steam vessels, and on the 29th Ericsson wrote to President Lincoln offering 'to construct a vessel for the destruction of the rebel fleet at Norfolk and for scouring the Southern Rivers and inlets of all craft protected by rebel batteries.' The offer was accepted and before the end of October work started on what was to become the *Monitor*. The plan of the vessel, Ericsson later wrote, 'was based on the observations of the behaviour of timber on our great Swedish lakes. I found that while the raftsman in his elevated cabin experienced very little motion, the seas breaking over his nearly submerged craft, these seas at the same time worked the sailing vessels nearly on their beam ends.'

This explained the unusual form of the vessel which was launched the

The battle, 9 March 1862, between
the *Merrimac* (right) and the *Monitor*, a key engagement in the American Civil War.

following January and completed the following month. The deck of the 172 foot
(52 metre) long, 41½ foot (12.6 metre) beam, vessel was only 18 inches above the
water and, like the rest of the vesel, was heavily sheathed in iron. From the deck
there rose a squat turret which carried two 11 inch (28 centimetre) guns and
revolved round a full circle.

The *Monitor* left New York on 6 March, two days before the Confederate
Merrimac, 4,636 tons and carrying ten guns, had successfully attacked Union ships
lying in Hampton Roads. Her arrival there on the night of the 8th led to huge
crowds lining the shore for what it was realized would be a spectacular encounter
the following morning. It was later described in a letter to Ericsson from Chief
Engineer Alban C. Stimers, USN, the naval inspector of ironclads who was on
board the *Monitor*. 'Iron-clad against iron-clad, we manoeuvred about the bay
here, and went at each other with mutual fairness ... I consider that both ships
were well fought. We were struck twenty-two times, pilot-house twice, turret nine
times, deck three times, sides eight times. The only vulnerable point was the pilot
house ... She tried to run us down and sink us as she did the *Cumberland*
yesterday, but she got the worst of it. Her horn passed over our deck, and our
sharp, upper-edged rail cut through the light-iron shoe upon her stem and well
into her oak. She will not try that again.'

At first glance the famous battle looked indecisive. The *Merrimac* was able to
limp away – 'The *Monitor* ought to have sunk her in fifteen minutes', said Ericsson
when he heard of the engagement – while the *Monitor* failed to follow up its
advantage immediately. Yet, as the lessons of the engagement sank in, it became
clear that the death knell had sounded for wooden men-of-war. William Swinton
did not exaggerate when in his *The Twelve Decisive Battles of the War*, he wrote:
'The story of the battle of Hampton Roads created the profoundest sensation in
the court of every maritime nation. For months, not only the scientific, but the

popular journals were filled with the discussion of its merits and its meaning; the professional naval world was profoundly agitated; admiralty boards and ministers of marine conned its details; in fine, Russia and Sweden promptly accepted the *Monitor* as the solution of the naval problem of the age, and followed the lead of America in reconstructing their navies on that system.'

Yet it was not until 1873, nearly three decades after the famous naval tug-of-war, that HMS *Devastation*, a steam-powered iron-clad with two propellers, with armoured hull and armoured gun turrets and without auxiliary sail, ushered in the age of the modern navies. She was followed three years later by the Italian *Duilio*. Britain responded with the more heavily armoured and armed HMS *Inflexible*, which Italy countered with the even more powerful *Dandalo*. The Americans in 1891 launched the first warship of their *Indiana* class and the French, two years later, the first of their *Charles Martel* class. The competition in leap-frogging gun power reached a critical stage in 1906 with the launching of HMS *Dreadnought*, very heavily armoured and carrying ten 12 inch (30 centimetre) guns in five gun turrets.

From then until the end of World War I, naval construction concentrated very largely on gun power and armoured protection combined with what speed was possible. Germany, France, Italy, the United States and Japan each produced what were, for a brief while, ships which could be expected to out-fight their rivals. Only at the Washington Conference in 1922 were limitations put on warship construction and naval engineers enabled to concentrate on passenger liners once more – the French *Normandie* and the British *Queen Mary* and *Queen Elizabeth*, all of more than 80,000 tons.

The *Queen Elizabeth* approaching New York.

MAN MAKES MATERIALS

While sail was giving way to steam, a revolution was taking place in a totally different field. This was the production of man-made materials which could take the place of natural products and which in some cases were even to have advantages over them. There were two different areas in which the revolution took place. One was dyeing where the manufacture of synthetic dyes soon led to the development of the organic chemistry industry; the other was that of raw materials in which, quite independently, chemicals were soon being fashioned into substitutes for wood or for ivory, a development which was to lead, within less than a century, to the enormous plastics industry.

Both advances, which were to produce such varied and exciting man-made creations, were illustrations of chemical discovery and of chemical engineering, two human occupations which had been affecting man's life since earliest times. The utilization of salt, soap, starch, glass and leather, all operations involving chemical reactions, had been carried on almost since prehistoric ages, as had the preparation of wine and beer. But only empirical methods were used and only with the spread of Arab culture from the seventh century onwards did the rudiments of chemistry begin to be understood.

Progress was slow, and it was not until 1661 that Robert Boyle, in *The Sceptical Chymist*, started to put chemistry on a scientific basis. Among the most important of his statements in the book was the first scientific definition of a chemical element, a word used today to describe a substance consisting entirely of atoms which have the same number of protons, or positively-charged particles, in their nuclei. 'To prevent mistakes', wrote Boyle, 'I must advertise to You, what I now mean by Elements, which not being made of any other Bodies or of one another, are the Ingredients of which all those call'd perfectly mixt Bodies are immediately compounded and into which they are ultimately resolved: now whether there be any one such Body to be constantly met with in all, and each, of those that are said to be Elemental bodies, is the thing I now question.'

Boyle was followed by a succession of British, German, French and Italian chemists who, during the eighteenth century, slowly but steadily laid the foundations for an understanding of how the world's elements reacted with each other. While their work helped to explain many of the processes which men had been making use of for years, it was only towards the end of the eighteenth century that conditions in Europe led to the foundation of the first industrial chemical process to be operated on a major scale and involving the use of chemical engineering works. This was the Leblanc method of making soda, a product increasingly in demand by makers of textiles, soap and glass as the Industrial Revolution gathered pace.

A shortage of soda was general throughout Europe and in 1775 the French Academy offered a prize of 2,400 livres for a process which would start with salt as a raw material and finish with soda as an end-product. The prize was won by Nicolas Leblanc, physician to the Duke of Orléans, whose scheme was to treat common salt with sulphuric acid, mixing the resulting sodium sulphate with coal and limestone and then roasting it. Soda and water was extracted from the 'black ash', as it was called, which was then evaporated to leave the soda. Factories for the production of soda by the Leblanc process were opened in France before the end of the eighteenth century, but in Britain its first major use did not come until 1822 when James Muspratt founded a Liverpool chemical works for making it. Three years later came the Glasgow factory of Charles Tennant, soon the largest in Europe, covering 100 acres (40 hectares) and employing more than a thousand workers. The Leblanc process, which had the disadvantage of producing noxious fumes and large quantities of unuseable by-products, was eventually superseded in the 1860s by the Solvay ammonia-soda process. But by this time it had given a hint that the Industrial Revolution would involve not only factories and mills of unprecedented size but works in which chemical engineers would have to develop new methods of handling sometimes noxious materials of which there was, as yet, only limited knowledge.

The second half of the nineteenth century thus saw the growth of numerous industrial complexes aimed at producing specific chemicals. They were not used only in industry, and as early as 1813 publication of Humphry Davy's *Elements of Agricultural Chemistry* had started a process which led to the manuring of the soil with chemical fertilizers. Chemicals were needed in growing quantities for the production of newsprint as the Victorian age advanced; and the continuing spread of steam power led to the chemical vulcanizing of rubber belting, discovered in the United States by Charles Goodyear but soon afterwards being carried out in a Scottish factory. Meanwhile, in an era often considered an age of peace, minor wars continued and Alfred Nobel and George Kynoch, among other men, used chemistry to perfect explosives and cartridges.

Amid this glut of development from which the professional chemical engineer was to emerge there was the almost startling production of man-made dyes and the manufacture of the first plastics – using that word in its contemporary meaning.

At Easter 1856, the year in which Henry Bessemer patented his new and revolutionary method of making steel, an eighteen-year-old English youth accidentally discovered how to make the first synthetic dye, and in doing so paved the way for the organic chemical industry. His name was William H. Perkin and his achievement was not only to make brilliant dyes more available and more reliable than ever before but to show how coal tar, previously an embarrassing waste product of the gas-lighting industry, could be directed towards innumerable useful purposes. Coal tar is a black sticky liquid of complex chemical composition in which substances related to benzene are predominant. It was a waste product, often disposed of by dumping in the nearest river. Only slowly were uses found for the materials which could be produced by its distillation, a process carried on in a series of specially engineered stills, operated at carefully regulated temperatures, and producing the volatile constituents of the coal tar which are condensed in separate receivers. Each of these condensed distillates is then redistilled to produce a further set of chemicals. The wide variety of the products can be judged from the fact that ammonium sulphate, an important artificial fertilizer, is obtained from 'ammonia water' one of the distillates, and that from another,

'crude naphtha', there is produced benzol, from which there are made benzene and toluene, used in the manufacture of dyes, drugs and explosives. Phenol, or 'carbolic acid'; anthracene, the raw material for alizarin dye; creosote and pitch are among other intermediate products which can be turned into a yet further range of chemicals.

In the first decades of the nineteenth century few of the potentials of coal tar were known, although Michael Faraday, the English physicist and chemist, had discovered benzene in 1825, and Charles Macintosh used naphtha, made from one of the distillates, in a rubber solution for waterproofing materials which gave his name – usually with the addition of a 'K' – to the raincoat. And with the spread of railways it was found that creosote made from coal tar could preserve railway sleepers. Nevertheless, most coal tar was still going to waste when William Perkin made his momentous discovery that was to found more than one new industry.

As a boy, he wrote, he was led 'to take an interest in mechanics and engineering, and I used to pore over an old book called *The Artisan*, which referred to these subjects and also described some of the steam engines then in use, and I tried to make an engine myself and got as far as making the patterns for casting, but I was unable to go any farther for want of appliances . . . But when I was between twelve and thirteen years of age, a young friend showed me some chemical experiments and the wonderful power of substances to crystallise in definite forms, and the latter, especially, struck me very much, with the result that I saw there was in chemistry something far beyond the other pursuits with which I had previously been occupied. The possibility also of making new discoveries impressed me very much. My choice was fixed, and I determined if possible to become a chemist, and I immediately commenced to accumulate bottles of chemicals and make experiments.'

In October 1853 Perkin was enrolled at the Royal College of Chemistry under August Wilhelm Hofmann, who had studied coal tar in Germany and had been invited to Britain on the suggestion of Prince Albert, later the Prince Consort. Perkin was so enthusiastic that he worked at home as well as at the college. 'My own first private laboratory' he wrote, 'was half of a small but long-shaped room with a few shelves for bottles and a table. In the fireplace a furnace was also built. No water laid on or gas. I used to work with old Berzelius spirit lamps and in a shed I did combustions with charcoal. It was in this laboratory I worked in the evening and vacation times.'

And it was here, at Easter 1856, that Perkin started on the work which was to make him famous. A few years earlier Hofmann had considered the problem of synthesizing quinine. 'As a young chemist', Perkin later wrote, 'I was ambitious enough to work on this subject of the artificial formation of natural organic compounds. Probably from reading [Hofmann's] remarks on the importance of forming quinine, I began to think how it might be accomplished, and was led by the then popular additive and subtractive method to the idea that it might be formed from toluidine by first adding to its compositon C_2H_4 by substituting allyl for hydrogen, thus forming allyl-toluidine, and then removing 2 hydrogen atoms and adding 2 atoms of oxygen. Allyl-toluidine, prepared by the action of allyl

Perkin's original crystals
of alizarin and mauveine, and yarn coloured with mauveine dye.

iodide on toluidine, was converted into a salt and treated with potassium bichromate; no quinine was formed, but only a dirty reddish-brown precipitate.'

Perkin's main aim had not been to synthesize quinine but to learn more about chemical reactions and he now repeated the experiment using aniline as a starting point. This time the outcome was a black sludge instead of a brown one; but this time the addition of alcohol produced a deep violet solution. This liquid, he soon discovered, could dye silk a brilliant mauve. Moreover, it appeared that the colour was not removed by washing and did not easily fade.

In 1856 the bright Tyrian purple of the ancient world, made from shellfish found near the Mediterranean city of Tyre, was no longer available. During the reign of Charles II a citizen of Cork used to make laundry marks on the linen of the nobility with a red dye from local whelks, and there were other minor uses throughout the world where suitable shellfish were found. But only far duller violet dyes were generally available and the potentialities of the artificial dye were quickly apparent.

Perkin immediately sent a sample of his dyed silk to Pullars of Perth, one of Britain's leading dyers. 'If your discovery does not make the goods too expensive, it is decidedly one of the most valuable that has come out for a very long time', came the reply. 'This colour is one which has been very much wanted in all classes of goods, and could not be obtained fast on silks, and only at great expense on cotton yarns. I enclose a pattern of the *best* lilac we have on cotton – it is dyed only by one house in the United Kingdom, but even this is not quite fast, and does not stand the tests that yours does, and fades by exposure to air. On silk the colour has always been fugitive.'

Sir Robert Pullar later became more expansive: 'Any one who knows anything about silk dyeing is aware that, prior to this invention purple dyes were so fugitive that sometimes a lady would have a new violet ribbon in her hat in the morning and by evening it would be a sort of red colour. That is the kind of thing Sir William Perkin [as he later became] and I tried to overcome at the works at Perth for a considerable time before he took out a patent.'

Before he could lodge his patent, however, Perkin had to dismiss suggestions that at the age of eighteen he was too young to file one. His father and elder brother supported him and after starting production in the back garden of the family home, it was decided to build a factory at Greenford Green, near Harrow.

This early chemical engineering was the work of a youth who was later to say: 'At the time neither I nor my friends had seen the inside of a chemical works, and whatever knowledge I had was obtained from books.' Problems of industrial production were far greater than those of laboratory production. While some of the aniline dyes, it was found later, could be made comparatively simply, they all needed at least four separate steps and some required a large range of raw materials in addition to the coal tar.

Production of the first synthetic dyes was complicated enough, but the process became even more so as chemists learned more about coal tar and the structure of the intermediate materials which had to be made before the dye itself came into being. Thus anthracene had to pass through a dozen distinct chemical stages, each resulting in material of an ever more complicated structure, before the dye itself was produced. In making certain other dyes no less than twenty-two intermediate processes were required. Some could be carried out simply enough in wooden vats, but others demanded special metal vessels able to withstand great pressures and high temperatures. In fact, Perkin's first equipment, which was ready in 1857, included a vessel in which the reactants were heated with high pressure or superheated steam while being continuously stirred by a steam engine. Even when the apparatus was completed, there was difficulty about raw materials. Coal-tar waste was being regularly dumped and Perkin had a long search for the benzene from which aniline could be made. Eventually it was found in Glasgow – but at a price of 5s. a gallon and of such poor quality that it had to be re-distilled. It then had to be 'nitrated' into nitro-benzene but no nitric acid could be found, so Perkin was forced to make it from saltpetre and sulphuric acid. The nitrobenzene was then reduced with hydrogen and special equipment had to be designed and made for the task. 'Everything in connection with the new industry'. Perkin was later to say, 'had to be worked out from the very beginning – the methods for the isolation and preparation of the raw materials, as well as the manufacture of the new dyestuff, and the prejudices of the dyers and printers against innovation had also to be overcome.'

He found that in a dye-bath containing soap, silk showed even better results but that with cotton the dye did not seem to operate properly. The solution lay in finding the right mordant, a chemical which would fix the dye on the fibres of the material. Tannic acid was found to be the answer, and soon after Perkin had shown that his dye was applicable to a wide range of materials, he was told by Pullar: 'I am glad to hear that a rage for your colour has set in among that all-powerful class of the Community – the Ladies. If they once take a mania for it and you can supply the demand, your fame and fortune are secure.'

The ladies obliged, Queen Victoria going so far as to wear a purple dress at the opening of the International Exhibition of 1862 even though she was still in mourning for the Prince Consort. At the exhibition she inspected a glistening block of pure dye 'the quantity obtainable from 2,000 tons of coal tar and sufficient to print over 100 miles of calico.' The new dye was used to colour the penny stamp and according to *Punch* London policemen invented their own pun and advised loiterers to 'Mauve on'.

It was not only the industrialists who appreciated the changes that Perkin was to bring about ... 'Already', said Richard Owen, President of the British Association for the Advancement of Science, at the Association's Leeds meeting in 1858, 'natural processes can be more economically replaced by artificial ones in the formation of a few organic compounds ... It is impossible to foresee the extent to which chemistry may ultimately, in the production of things needful, supersede the present vital agencies of nature.'

Within a few years the popularly named 'mauve mania' fell away, but it was of comparatively little concern to Perkin since by the early 1860s he was already making eight synthetic dyes derived from coal tar, seven of which he had discovered himself. His first discovery, had, in fact, started a race to produce new synthetics in which chemists throughout Europe took part. In 1859 the purple-red fuchsine was made in France, and named Magenta after the battle in which the French and the Sardinians had fought the Austrians. In 1862 Cherpin produced Aldehyde Green by chemical reaction on the unstable Aldehyde Blue, and a successful way of making Aniline Black, discovered in 1860, was patented in 1863. Coal-tar products also replaced cochineal, the bright red dye made from small insects, 200,000 of which were needed to make one kilogram of the dye.

There also came from the German chemists Karl Graebe and Carl

Lieberman in 1868 the synthesizing of alizarin, the orange-red crystalline solid made from the fleshy part of the madder root. Natural alizarin had been used by the ancient Egyptians for colouring some of the robes of King Tut and a complex method of using madder to produce the fiery Turkey-red had reached England in the eighteenth century via the Levant and France, a country in which some 400,000 acres (160,000 hectares) were given over to madder. The German method of synthesizing alizarin was at first too expensive to be put into commercial production. But the work was published and in Britain Perkin began to seek a viable commercial process. He finally succeeded and on 26 June 1869 tried to patent it – only to find that the Germans had discovered the same method and had beaten him by a single day to the patent office.

During these early days of the synthetic dye industry, comparatively little was known about the organic chemicals which were its raw materials. Initially, 'organic chemicals' described the substances produced by living organisms as distinct from those of mineral origin. The meaning slowly changed to indicate the complex materials such as coal tar and its derivatives which were based on carbon; but the original meaning still held, if more tenuously, since coal had been formed in the geologically distant past by the breakdown of animal and vegetable matter.

Credit for starting to elucidate the structure of coal tar chemicals and their multiple derivatives which were to provide Perkin and others with their synthetic dyes belongs to the German chemist Friedrich August Kekulé. In 1865 he postulated how the six carbon atoms and six hydrogen atoms making up the molecule of benzene were linked together in a hexagonal benzene ring. In this the six carbon atoms formed a six-sided structure from each point of which a carbon atom was attached to a single hydrogen atom. Kekulé's work made it easier to understand the structure of natural dyes and, as a result, the best way of synthesizing them. An early example was alizarin which was shown to have a molecular structure based on anthracene, a coal tar intermediate made up of three joined rings of carbon atoms.

While the Kekulé theory of the benzene ring helped the artificial dye industry, the theory itself would not have been verified without the help of the industry. 'No better illustration of the interdependence of science and industry', Perkin was to say 'has ever been given to the world than this particular example of the action and reaction between theoretical and applied chemistry.'

Despite the German patent success with alizarin, Perkin was the first to devise a viable production machinery. 'Before the end of the year [1869], he later said, 'we had produced one ton of this colouring matter in the form of paste; in 1870, 40 tons; and in 1871, 220 tons, and so on in increasing quantities year by year.'

The madder root industry was killed almost overnight. The price of what had been the almost useless anthracene rose from a few shillings to a hundred pounds a ton, and Perkin's plant outside Harrow was almost swamped with orders.

Chemists throughout the world, but particularly in Germany, were now searching for methods of making a wide range of synthetic dyes. One of those to evade them longest was a substitute for indigo, the product of the indigo plant, of which about 19,000 tons a year were being exported from India during the latter part of the nineteenth century. When Johann Friedrich Wilhelm Adolf von Baeyer began to investigate the chemical composition of the indigo dye in 1865, little was known about its structure. After fifteen years' work Baeyer had not only discovered what this was, but had published a method of synthesizing the dye. But another fifteen years, and the expenditure of a million pounds by the Badische Anilin und Soda Fabrik in Germany, were both necessary before an economically viable way was found for making synthetic indigo.

The organic chemicals which Perkin and his successors were to transform into synthetic dyes were soon to become the foundation of other important industries. Even as colorants, however, they were quickly playing decisive roles completely outside the textile industries where they were being used in such quantities.

Synthetic dyes appeared just as biologists were beginning to understand the cellular basis of life and just as bacteriologists were beginning to appreciate the bacteriological cause of disease. In both fields coal-tar colours proved ideal for staining specimens to be studied under the microscope. Both cells and bacteria were virtually transparent, and thus difficult to detect. The ability to stain them with dyes of a quality and intensity which could be rigidly controlled therefore came at a relevant stage of research. Its usefulness was increased when it was found that some parts of a living cell could absorb certain dyes but that other parts could not.

Using these new staining methods Eduard Adolph Strasburger observed, and was able to explain, the changes in the cells of living organisms during cell division. At the same time another German, Walther Flemming, found that the nucleus of the cell contained material that strongly absorbed dyes. This he christened chromatin, based on the Greek word for colour, while the threadlike objects into which the chromatin coalesced he called chromosomes – later found to contain the physical carriers of heredity.

In the purely medical and bacteriological fields there were other uses of the new dyes which were to be equally important. In 1869 Julius Cohenheim tagged white blood corpuscles with aniline blue and by following their course was able to demonstrate the actual process of inflammation in living tissue. In 1875 Carl Weigart used methyl violet to demonstrate the existence of globular bacteria in tissues and two years later stained the anthrax bacillus with the same dye.

Robert Koch employed methylene blue to help detect the bacilli of tuberculosis and cholera, and between 1875 and 1895 staining agents were used to discover the causes of leprosy, typhoid fever, malaria and brucellosis, thereby making possible the battle against preventable infectious diseases.

While the dyers thus aided the bacteriologists, the latter paid the debt after Paul Ehrlich found that methylene blue and some of its congeners were the only colours that stained living nerve tissue. Ehrlich did not know whether this was due to the dye itself or the presence of sulphur. The resulting experiments led to the discovery of the rhodamine colours, new coal-tar dyes which were to be invaluable to the textile industry. One of Erhlich's most important discoveries with the synthetic dyes was that while most bacteria stained with magenta lost their colour when a mineral acid was added to the solution, tubercle bacilli retained their stain, a useful fact in the study of tuberculosis.

Another, and quasi-medical by-product of the industry which Perkin had originated with 'Mauve' was saccharin, a derivative of coal tar 550 times sweeter than sugar. A Dr Fahlberg, working on coal tar in Johns Hopkins University found on returning home one night that his hands had a sweet taste. Going back to the laboratory he found that one of the glasses with which he had been working had this same taste. When tests showed that the product, 4 pounds (1.8 kilograms) of which had the sweetening power of a ton of cane or beet sugar, was harmless unless taken in very large quantities, the Continental sugar beet industry was threatened, and in some countries its use was limited to medicinal purposes.

The discovery of the sweetening properties of saccharin by no means exhausted the potentials of the organic chemical industry, and in 1868 Perkin announced the synthesis of coumarin, the white crystalline substance previously obtained from the tonka bean and other vegetable sources. Smelling of vanilla, coumarin was used both as a perfume and as a flavouring and became the first of a number of scents which Perkin produced from coal tar products.

The development of synthetic dyes and their by-products had repercussions on many aspects of the chemical industry and of chemical engineering. The great demand for sulphuric acid for use with nitric acids in nitrating benzol, itself led to the development of the contact process for making the acid. Growing understanding of the complex coal-tar derivatives led to the design and building of specialized equipment so that the derivatives could be utilized most easily in industries which grew up as a result of their fresh availability. Insecticides, weedkillers, antiseptics and anaesthetics were only some of the materials which were to be made from coal-tar products, while toluene, one of the scores of by-products, was the basis of trinitrotoluene, the explosive TNT.

While organic chemicals were providing the world's inhabitants with a huge range of new products during the second half of the nineteenth century, another new industry was being developed, the output of which was eventually to permeate almost all aspects of everyday life. This was the plastics industry whose growth well illustrates one fact that regularly runs through the story of engineering advance and technological invention: that advance is invariably prodded on by fresh needs, and that without the demand there would usually have been no progress at all or, at most, a mere footnote to history.

There are various definitions, scientific and industrial, of the word 'plastics'. But it can be said that they are usually man-made materials produced by a variety of chemical and engineering processes; and that while they are stable in normal use they can at some stage in their manufacture be moulded or shaped by heat, or pressure or both. Some, known as thermoplastic materials, became plastic on being heated, lose that characteristic when allowed to cool, but can be repeatedly reheated to restore their plastic quality. Others known as thermosetting plastics, lose their plasticity for good once they have been heated or submitted to pressure. Plastics of both kinds are polymers; that is, materials consisting of molecules which are very large as compared with 'ordinary' organic or inorganic molecules.

The possibility of making an artificial material which could be put to common use had been demonstrated by Professor Henri Bracconet of Nancy in 1832 when by pouring concentrated nitric acid on to cotton or wood fibres he produced a hard water-resistant film which he named xylodine. Neither he nor Professor Théophile Pelouze of Paris, who created a similar substance the following year, followed up their discoveries and more than a decade was to pass before there were any further developments. Then the German-Swiss Christian Schönbein realized that the cellulose nitrate produced when the cellulose in the cell walls of plants was treated with a mixture of nitric and sulphuric acids had a variety of characteristics which depended on the proportions of materials used.

'I have of late', Schönbein wrote to Michael Faraday on 27 February 1846, 'also made a little chemical discovery which enables me to change *very suddenly*, *very easily* and *very cheaply* common paper in such a way, as to render that substance exceedingly strong and entirely waterproof.' He enclosed a specimen and went on. 'My prepared paper can be easily written and printed upon. Paper enjoying the properties mentioned is, to my opinion, a valuable substance and in many respects very superior to common paper, it ought therefore to be manufactured on a large scale.' Thus although plastics were a century later to acquire a misleading reputation for being second-class substitutes for 'natural' material, the first suggestion for their use rested on quality.

However, Schönbein's new material had other characteristics including those of a powerful explosive. Indeed, there is a legend that these were discovered after he had wiped up spilt acid in his wife's kitchen with a cotton apron, started to dry it over a stove, and watched it disappear with a smokeless bang. Certainly in March 1846 Schönbein described the properties of what was to be known as guncotton. At the same time he described the first plastic articles, soon to be produced in quantity. 'To give you an idea of what may be made out of vegetable fibre', he told Faraday, 'I send you a specimen of a transparent substance which I have prepared out of common paper. The matter is capable of being shaped out into all sorts of things and forms and I have made from it a number of beautiful vessels. The first perfect one I obtained is destined to be sent to the Mistress of the Royal Institution [Faraday's wife], as soon as a convenient opportunity will offer itself for doing so and I shall ask the Lady mentioned to preserve it as a sort of scientific keepsake.'

Although plastics were soon found to have significant insulating properties, Faraday appears to have been relatively uninterested in the new material. It was, instead, to be developed by Alexander Parkes, an inveterate inventor whose Parkesine was to be the world's first commercial plastic.

Apprenticed as an art metal worker to a Birmingham firm of brass founders, Parkes discovered a method of electroplating delicate objects and carried out the process on flowers when Queen Victoria visited the works. When Prince Albert arrived Parkes surpassed himself by producing an entire spider's web plated with silver. His patent for this also described a way of waterproofing fabrics with rubber dissolved in carbon bisulphide, a process later sold to Charles Macintosh.

It is not certain what led Parkes to Parkesine. But one of his early patents had involved photography and he had experimented with the idea of producing a layer of collodion – a solution of cellulose nitrate in alcohol and ether – so thick that it would support a photographic emulsion without the aid of glass. Certainly, during the 1850s, Parkes lodged a number of patents covering 'a new material [which] might be introduced into the arts and manufactures and in fact was much required.' The requirement increased throughout the 1850s as the engineering wonders of the Great Exhibition encouraged the belief that there was no limit to which the standards of living might not be raised.

These hopes appeared to be on the edge of fulfilment in 1862 when a case of exhibits made from Parkesine was shown at the International Exhibition held in London. Here were displayed, as a notice described them, 'Medallions, Salvers, Hollow Ware, Tubes, Buttons, Combs, Knife Handles, Pierced and Fret Work, Inlaid Work, Bookbinding, Card Cases, Boxes, Pens, Penholders etc.' The material, it went on, 'can be made Hard as Ivory, Transparent or Opaque, of any degree of Flexibility, and is also Waterproof; may be of the most Brilliant Colours, can be used in the Solid, Plastic or Fluid, State, may be worked in Dies and Pressure, as Metals, may be Cast or used as a Coating to a great variety of substances; can be spread or worked in a similar manner to India Rubber, and has stood exposure to the atmosphere for years without change or composition. And by the system of ornamentation Patented by Henry Parkes [Alexander's brother] in 1861, the most perfect imitation of Tortoise-shell, Woods, and an endless variety of effects can be produced.'

Different proportions of raw material were required according to the role of the finished Parkesine, and by 1865 Parkes was taking out patents to cover specific purposes. In that year he patented an improvement in electric telegraph conductors in which his plastic was to be an insulator. 'The copper-covered steel wire of this invention is preferably insulated with parkesine', ran the application for the patent. 'For this use, the ingredient of the parkesine may be mixed in a suitable solvent, 100 to 150 parts of oil (preferring castor oil) and 50 parts of camphor.' Next year another patent was taken out for moulding Parkesine brush backs, with perforations for bristles, and horizontal holes to receive the wire by which they were secured.

There were intrinsic virtues in Parkesine, and a company set up by Parkes was soon producing an enormous range of goods which included bracelets, earrings, necklaces, knife-handles and boxes. The company nevertheless failed, due partly to Parkes' lack of business sense, to the lack of quality control in his works, and to his insistence on keeping prices down which affected the quality of the raw materials he bought.

Parkes was to be followed by his works manager, Daniel Spill, who set up his own Xylonite Company on the wreckage of the Parkesine Company. Spill took out a number of patents for processes which he hoped would improve the quality of what had been called Parkesine and was eventually to be known as celluloid. One covered the use of alcohol and camphor as solvents which would, it was discovered, limit the shrinkage of the material while it was changing from the liquid to the solid state. This was to lead him into one of the major patent infringement cases which feature so prominently in the early history of plastics – a not unnatural state of affairs since genuine misunderstanding was only too easy.

The case, which was to end disastrously for Spill, arose from the activities of John Wesley Hyatt, a New Yorker who eventually solved many of the problems of celluloid production. Hyatt was drawn into the field by a fact as esoteric as any which have pushed technology or engineering along the road of progress. By the late 1860s elephant-hunters were threatening the supply of ivory and in the process putting up the price of such ivory as was available. Phelan and Collander, the billiard ball manufacturers of Albany, New York, were thereby led to offer $10,000 to anyone who could produce a suitable substitute for the ivory on which their business depended.

The challenge was taken up by Hyatt, a printer who may first have seen collodion in use for covering minor wounds. However, he lodged a number of patents utilizing shellac, paper pulp and ivory dust before patenting in 1869 'an improved method of coating billiard balls, consisting of dipping the billiard balls, made of some suitable composition, into a solution of collodion, which might be given any desired colour beforehand.'

The result was a billiard ball of cellulose-nitrate. Although partly successful, there still remained production problems. There were also other problems as Hyatt once explained after saying that the minimum amount of pigment was added to produce coloured balls which meant that they were in effect covered with a thin film of guncotton. 'Consequently', he said, 'a lighted cigar applied would at once result in a serious flame and occasionally the violent contact of the balls would produce a mild explosion like a percussion guncap. We had a letter from a billiard saloon proprietor in Colorado, mentioning this fact and saying that he did not care so much about it but that instantly every man in the room pulled a gun.'

Hyatt continued to experiment and in 1870 filed a patent covering the solvent action of camphor on cellulose nitrate and the manufacture of celluloid under the influence of pressure and heat. The product, he reported, was 'a solid, about the consistency of sole leather, but which substantially becomes as hard as horn or bone by the evaporation of the camphor. Before the camphor is evaporated the material is easily softened by heat, and may be molded into any desirable form, which neither changes nor appreciably shrinks in hardening.'

His success was too much for Daniel Spill who began what was to be a three-year battle against Hyatt's Celluloid Manufacturing Company, claiming that its processes infringed a number of the patents which Spill had lodged in both Britain and the United States. The defence tried to show that Parkes had discovered the processes concerned before they were patented by Spill, but was unsuccessful. However, when Spill filed his suit for damages, Hyatt succeeded in having the case reheard. This time the verdict went in his favour and the manufacture of celluloid went ahead in the United States without restrictions.

Confident that he could produce a material with almost infinite possibilities, Hyatt now set up celluloid plants in Germany, France and Britain. In all of these countries, but particularly in Britain, the new plastic celluloid became extremely popular in the manufacture of detachable collars and shirt-cuffs which could be washed clean every night. The trend increased when it was found possible to use cellulose acetate instead of nitrocellulose, the great advantage being that the end-product was non-flammable. But the word 'celluloid' was used for material based on cellulose nitrate.

For years celluloid was the most used plastic, but it did not remain the only one for long. Another was the group of casein plastics based on the reaction between casein, the main protein in milk, and formaldehyde. Legend – most unlikely to have any basis – maintains that the story of casein plastics began when the cat belonging to Adolf Spitteler of Prien in Bavaria knocked a bottle of formaldehyde solution into its saucer of milk and the reaction which followed produced a hard water-resistant plastic. Whatever the truth of this, Spitteler and W. Krische, a lithographer from Hanover, announced in 1897 the discovery of the hard hornlike material that could thus be produced. The first casein plastics went into production in Germany in 1900 and throughout the century their output has continued to grow, particularly for such articles as buttons where dimensional stability is not necessary.

More important was the discovery, early in the twentieth century, of the first thermo-setting plastic. This was the eponymous Bakelite, announced by its inventor, Dr Leo Hendrik Baekeland, in 1909. More than thirty years later the German Professor Baeyer noted that when phenol, or carbolic acid, was mixed with formaldehyde, the result was a solid resinous material. The first man to establish the conditions for controlling the phenol-formaldehyde reaction so that it could be exploited was Baekeland who was seeking a synthetic substitute for shellac. When the material was heated, he found, its properties changed remarkably. At first it softened and could be moulded into any required shape. On being heated, to a still higher temperature, however, it hardened again. But once the higher temperature had been reached, it was found impossible to soften the material again by any means. Bakelite was water-resistant, solvent-resistant, was an electrical insulator and could be both cut with a knife and easily machined.

Baekeland discovered that the new plastic could be made in one of three varieties. If the reaction was stopped while the material was hot and liquid, it produced, on cooling, Bakelite A, which was still soluble in certain solvents. If there was no cooling, Bakelite A turned into Bakelite B which remained soft while hot and could easily be moulded. If this was then heated under pressure, Bakelite C was produced.

In 1909 Baekeland read a paper to the American Chemical Society in New York in which he claimed that liquid Bakelite would give wood a coating superior to that of the best Japanese lacquer. 'But I can do better', he continued. 'I may prepare an A, much more liquid than this one, and which has great penetrating power, and I may soak cheap porous soft wood in it, until the fibres have absorbed as much liquid as possible, then transfer the impregnated wood to the Bakeliser, and let the synthesis take place in and around the fibres of the wood. The result is a very hard wood, as hard as mahogany, or ebony of which the tensile, and more especially the crushing strength, has been considerably increased and which can stand dilute acids or steam; henceforth it is proof against dry rot. In the same way I have succeeded in impregnating cheap ordinary cardboard or pulp board and changing it into a hard resisting polished material that can be carved, turned and brought into many shapes.'

The extraordinary variety of new products made possible by the development of plastics was well illustrated by Baekeland in his paper. 'I cannot better illustrate [Bakelite's versatility],' he said, 'than by telling you that here you have before you a grindstone made of Bakelite and on the other hand a self-lubricating bearing which has been run dry for nine hours at 1,800 revolutions per minute without objectionable heating and without injuring the quickly revolving shaft.'

Thermo-setting Bakelite, the early casein-based plastics, and celluloid with its successors would, together with the continuing growth of electrical engineering and its demand for insulating material, have assured the spread of plastics even had World War I not broken out in 1914. As it was, cellulose acetate was soon in great demand for impregnating and then tautening the fabric of aircraft wings.

The growth of plastics after World War I, and its even greater growth after World War II, was due partly to the development of the internal combustion engine, partly to the chemical researches of a handful of outstanding chemists in Europe and the United States. During the first half of the twentieth century the raw materials of plastics were largely agricultural, and phenol and formaldehyde, both obtained from coal. With the growth of the motor car there came a great intensification of research into the crude oil from which petrol was made; and, simultaneously, research into the large number of different chemicals which it was possible to produce from crude oil by cracking, distillation or fractionation.

As research into petroleum products widened out, chemists began to consider more closely the structure of such materials as cellulose, the constituent of the first viable plastic. Cellulose consisted of molecules each containing six

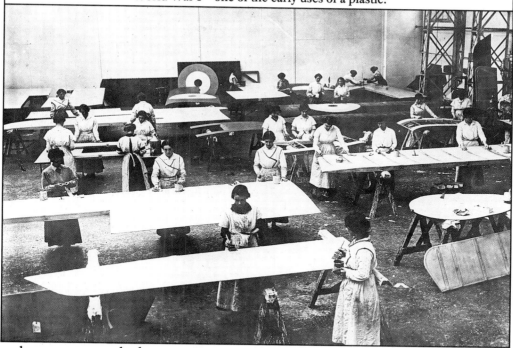

'Doping' aircraft wing during
World War I – one of the early uses of a plastic.

carbon atoms, ten hydrogen atoms, and five oxygen atoms all linked together by complex chemical processes which took place within the living plant which produced the cellulose. Why, it was soon being asked, should it not be possible to produce in the laboratory materials as large as those made naturally in living organisms? Moreover, since the physical characteristics of materials were known to be dependent on their molecular structure, why should it not be possible to tailormake plastics with characteristics particularly suitable for specific jobs?

These possibilities began to be realized by the German chemist Hermann Staudinger and the American Wallace H. Carothers. Their work concentrated on the theoretical basis of polymerization, the process in which two or more molecules of the same compound are chemically united to form a larger molecule. Among those with which Staudinger experimented was styrene, which he polymerized into the thermoplastic material polystyrene. While styrene had a molecular weight of 104 the polystyrene chains which he produced had a molecular weight of about 20,000, each macro-molecule thus consisting of some 200 links.

While Staudinger was investigating in Germany the chemical basis of what was to become the post-World War II plastics industry, a young man, Wallace Carothers, was being hired by the Du Pont Company in the United States to start a programme of basic chemical research. He soon found when polymerizing acetylene that the addition of a chlorine atom produced a synthetic rubber with high tensile strength and better heat – and ozone-resistance than natural rubber. The product was registered under the trade name Neoprene, and was of immense use to the Allies when the Malayan rubber plantations were occupied by the Japanese during World War II.

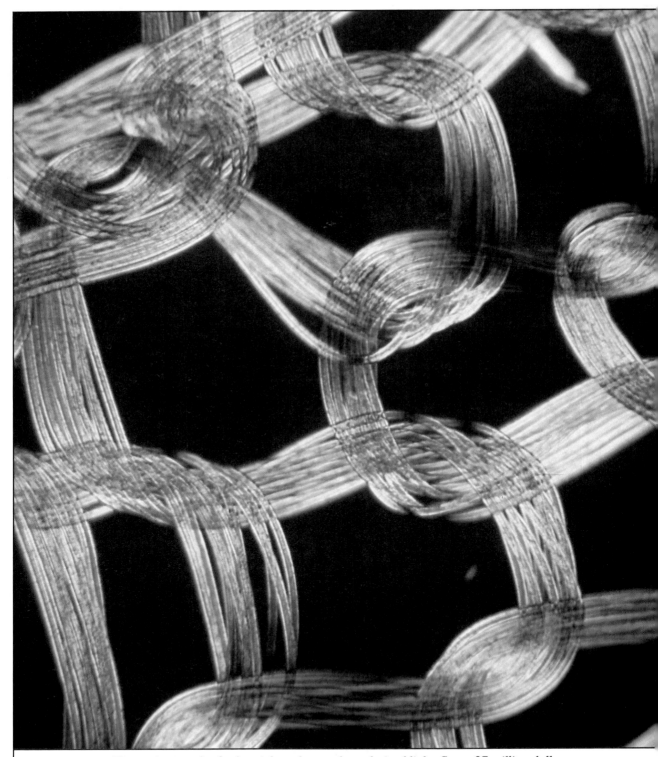

Photomicrograph of nylon tights taken under polarized light. Some 27 million dollars
were spent and 230 chemists worked hard before an economical way of making nylon was discovered.

Plastic spoons. During the first half
of the twentieth century plastic superseded metal in a wide range of domestic equipment.

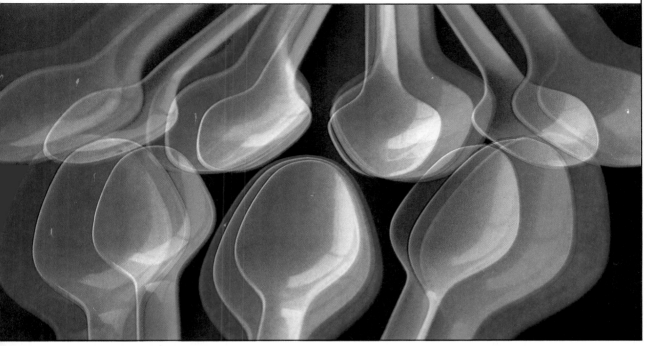

Plastic protection laid over rows of pineapple plants in Hawaii.

However, Carothers' main aim, apart from his theoretical studies, was to discover a substitute for silk, American imports of which came mainly from Japan, a country with whom America's political relations grew increasingly strained during the years between the two world wars. Politics were thus to become a major factor in the emergence and growth of the artificial fibre industry. Half a century earlier Joseph Swan, trying to perfect a filament for electric bulbs, had patented a process for forcing nitrocellulose through minute holes so that it emerged in the form of fibres. They were intended to be carbonized for use in the light bulbs, but the process came to nothing and it was left to Louis Marie Hilaire Bernigaud, Comte de Chardonnet, to produce in 1884 his own cellulose fibres in Paris. These were developed seven years later into 'Chardonnet silk', later christened rayon since its texture appeared to throw off rays of light. Rayon was not a fully artificial fibre, since its raw material was naturally existing cellulose. It was also expensive to make and dangerously inflammable – so much so that it was called 'mother-in-law-silk', since a rayon dress and a lighted match were claimed to be ideal presents for that relative.

Carothers' most important discovery was with diaminohexane, a soluble organic substance, and adipic acid, a white crystalline solid, which he used to form long chain molecules of a substance soon christened nylon. The two raw materials contained only the four elements carbon, nitrogen, hydrogen and oxygen and were arranged in the nylon molecule with the carbon and nitrogen forming the 'backbone' of the chain and the hydrogen and oxygen atoms arranged on either side of it. In the manufacture of nylon, the carbon and hydrogen are obtained from hydrocarbons in coal or petrol; nitrogen and hydrogen come, via ammonia, from the air while the atmosphere provides the oxygen. Thus, as is often stated: 'Nylon is made from coal, air and water.'

Some 27 million dollars were spent and 230 chemists worked hard before an economical way of making nylon was discovered. But the end-product was quickly found more suitable for ladies' stockings than natural silk. Other uses quickly followed and after more than forty years nylon remains one of the foundations of the man-made fibre section of the plastics industry. Sewing thread, bootlaces and climbing rope are all made of it, as are fishing lines, sheeting and surgical thread. Its strength makes it ideal for the manufacture of parachutes, while permanent pleats and creases can be heat-impressed into fabrics produced from it. Shrink-proof, resistant to insect attack and to mildew, nylon has so many good characteristics that other companies in both the United States and Britain soon began to follow the lead that Du Pont had taken, notably Imperial Chemical Industries which in 1941 produced another type of synthetic fibre, 'Terylene'.

Nylon, now almost a generic word for a number of long-chain synthetic polymeric amides, was an end-product of the kind that was hoped for. The emergence of polyethylene – first called polythene – in Britain at roughly the same time was, by constant, purely fortuitous. In fact Sir Michael Perrin one of those deeply involved, has called it, in the magazine *Research*, 'an unusually clear-cut instance of the unexpected results that may come from research, and of the importance of the role of chance in such work.' In 1932 Imperial Chemical Industries began a programme of basic research into the effects of very high pressure on chemical reactions. J.B. Conant, later President of Harvard, had already discovered that the process of polymerization was affected by high pressures, and early in March 1933 an attempt was made to produce a reaction between ethylene and benzaldehyde at a pressure of 1,400 atmospheres and at a temperature of 170°C. 'There was no indication from change in pressure that reaction had occurred', Perrin later wrote, 'and when the pressure vessel was dismantled the benzaldehyde was recovered unchanged. The walls of the vessel were, however, found to be coated with a thin layer of a 'white, waxy solid' to quote from R.O. Gibson's notebook record of 27 March. 'This material was analyzed and found to contain no oxygen, which confirmed the observation that the benzaldehyde, present in the vessel, had taken no part in the reaction. The solid was recognized as a hydrocarbon and, apparently, a polymer of ethylene. A similar result was obtained when ethylene alone was subjected to this pressure but the amount of product formed was always extremely small.'

When the pressure was further raised in a subsequent experiment, the joints and gauges on the apparatus were blown open. Only in December 1935 were the experiments repeated under better conditions and about 8 grams of a new white material in powder form collected. Its properties were soon found to be of great interest. 'It was crystalline and, by x-ray methods, its molecular structure could be accurately elucidated', Perrin has written. 'A high molecular weight polymer of ethylene would be the simplest compound of this type that could exist, consisting only of long chains of methylene ($-CH_2-$) groups. It showed the phenomenon of molecular orientation, or "cold-drawing", that had recently been observed by W.H. Carothers of du Pont's in his original work on poly-esters and nylon.'

Even more important, the new material had characteristics of great use to industry. It did not melt in boiling water, it could be formed into threads and films when heated under slight pressure, it was chemically resistant and it had remarkable insulating properties. First known in Britain as polythene and later officially known as polyethylene, it was soon being made in ounces, then in pounds. The potentials appeared so good that by September 1939 it was being made by the ton.

Within a few years annual production in Britain alone had reached 100,000 tons, due almost entirely to the development of radar whose high-frequency equipment swallowed up vast quantities of polythene as it was still called. World War II was, in fact, the main reason for the transformation of the plastics industry which took place in the 1940s. Perspex hoods for the Spitfires and Hurricanes which won the Battle of Britain and for many parts of the bombers of 1942 to 1945 were examples of what could be made better in plastics than in any other material. When the Japanese overran the Malayan rubber plantations, artificial rubber was produced, following a crash programme of research to make good the deficits of defeat. In fields as diverse as clothing manufacture and the waterproofing of amphibious vehicles, new plastics tailormade for specific tasks were turned out in production-line quantities where only laboratory-size quantities would have been considered possible in a country at peace.

At the time, the impact of the newly grown industry was hardly noticed by a population concentrating on grim tasks. Yet it was the demands of the war which were to make the contemporary plastics world so different from that of the late 1930s. When Carothers was making nylon and Perrin was preparing to make polythene, the general public had little reason to be more than barely aware of man-made materials. Celluloid and celluloid wrappings were familiar and the nylon stocking was a wonder that was to last for years. But other plastics tended to be regarded as poor substitutes for 'real' materials.

Today, after more than three post-war decades of plastics development, it is difficult to consider any home, study or office in which plastics are not evident in abundance. Translucent office partitions, typewriter frames and typewriter keys, light fittings and filing cabinets are only the most obvious examples. In the home, vacuum cleaner housings, picture frames, towel bars and door knobs, telephone handsets and crockery are among the plastic items in universal use. Fibreglass car bodies and plastic car seat covers are a part of everyday life; and so, for that matter, are plastic teeth. The plastic fishing line and the nylon climbing rope have almost become traditional.

The ubiquity of plastics is partly the result of characteristics which are shared by most of them. They are not only light and with good insulating properties. They are also biologically inert – in other words, they do not provide food for insects and they can be used inside the human body. Clothes made from plastic materials are not attacked by insects; neither are the plastic jackets of books or the synthetic resins which in many fields have replaced vegetable and animal glues.

Since the end of World War II the petro-chemical industry has been the main source of new materials rather than the coal tar of earlier years and petro-chemicals are relatively cheap and very convenient. In the 1950s petro-chemical complexes sprang up not only in Britain but elsewhere in Europe with facilities for producing methane, ethylene, propylene and butylene, all chemicals to be used in the plastics industry. To this almost extravagant source of materials there was added an extension of manufacturing processes, and within a few years end-products were being produced by extrusion, laminating, compression moulding, shell moulding and transfer moulding.

Yet new possibilities have opened up during the last decade as the treatment of plastics by nuclear radiations during manufacture has been found to affect their final characteristics. Once again, the potentialities of tailormaking plastics for specific jobs have been enlarged, even though applications are now known to be more limited than was at first expected.

Petro-chemical complexes became
widespread in the 1950s. This one is in Grangemouth, Scotland.

A NEW SOURCE
OF POWER

While the great achievements of the mid-nineteenth century rested mainly on steam power and, later, on the large-scale production of steel, another force was by the 1850s beginning to influence the potentials of engineering. This was electricity, soon to be demanding the power stations and eventually the huge hydro-electric schemes which by the first years of the twentieth century were becoming the most spectacular works of man that engineering had so far made possible.

The early pioneers of electricity differed in one important way from the men who harnessed steam. With few exceptions they were scientists who had little if any connection with industry and who, even after the potentials of electricity had become obvious, still continued to concentrate their energies on solving the riddles which obfuscated the new source of energy. It was hardly surprising, since the forces which enabled amber to attract lightweight objects when rubbed had been an unsolved mystery for centuries. The mystery was if anything deepened when in the sixteenth century William Gilbert, the English physicist, began to investigate the properties of what he named electricity, derived from the Greek word for amber.

The ability to create electricity by friction was enlarged when Otto von Guericke, whose investigations on the powers of a vacuum had surprised the seventeenth century, mounted a globe of sulphur on a shaft, had it revolved, and found that large quantities of electricity could be built up when it was touched by the hand. The accumulations gave forth sparks and shocks when dispersed and were for long used throughout Europe for demonstrations and entertainments. However, it was only with the invention of the Leyden jar, produced about the mid-1740s by a number of experimenters but generally attributed to Pieter van Musschenbroek, that it became possible to store quantities of static electricity – electricity at rest in contrast to the as yet undiscovered current of electricity.

During the second half of the eighteenth century the idea of 'two electricities', plus and minus, was developed. The new phenomenon was found strong enough to kill small animals, while Benjamin Franklin, the American who played such an important role in enabling the Americans to win the War of Independence at the end of the eighteenth century, showed that lightning was nothing less than the sudden discharge of electricity which had accumulated in the clouds. All this was necessary preparation for what was to blossom into the profession of electrical engineering; but it was only preparation. Not until 1800 did there start a succession of discoveries which within three decades laid the foundation of the profession as it is still practised a century and a half later.

In the last years of the eighteenth century Alessandro Volta, the Italian physicist, was investigating the reports of the Italian anatomist, Luigi Galvani, that frogs' muscles twitched whenever they made contact with two different metals.

Galvani, the anatomist, believed that the electricity thought to be involved came from the muscle; Volta, the physicist, believed that the source was the metal, and demonstrated his belief with an apparatus described in a letter to Sir Joseph Banks, the President of the Royal Society.

'It consists', he said, 'of a long series of an alternate succession of three conducting substances, either copper, tin and water; or what is much preferable, silver, zinc and a solution of any neutral or alkaline salt. The mode of combining these substances consists in placing horizontally, first, a plate or disc of silver (half-a-crown, for instance); next a plate of zinc of the same dimensions; and, lastly, a similar piece of spongy matter, such as paste board or leather, fully impregnated with the saline solution. This set of three-fold layers is to be repeated thirty or forty times, forming thus what the author calls his "columnar machine".'

When a wire was attached to the top and the bottom of the 'Voltaic pile' as it was soon called, an electric current passed through it. Volta had made the first electric battery, an apparatus in which a continuous flow of electricity was produced not by friction but by a chemical reaction. A few months later the Englishman William Nicholson built his own battery and by dipping the wires at either end into water had created bubbles of hydrogen and oxygen; he had produced the converse of Volta's demonstration by showing that an electric current could cause a chemical reaction.

Volta's work triggered off a series of experiments throughout Europe, and many kinds of battery were developed. However, it was only in 1819 that Volta's demonstration of the electric current was used to reveal the long suspected link between electricity and magnetism. The experimenter was the Danish physicist, Hans Christian Oersted, who while passing a current through a length of wire brought a compass needle near to it. The needle twitched and then turned to rest almost at right angles with the wire and the current flowing through it. When the current was turned off the needle returned to its former position. And when the current was set going again, but in the opposite direction, the needle again turned almost at right angles to the wire, but pointed in the opposite direction.

This first demonstration that there definitely existed a link between electricity and magnetism led in Europe to a renewed series of experiments and to the work of Dominique Arago and André Ampère which created much of the theoretical basis for electro-magnetism. In America it encouraged Joseph Henry among others into the investigations which were to lead to the electric telegraph.

During these early days of electricity, even more so than in other departments of science, each man advanced with the help of his colleagues' work, reading their papers as published, repeating their experiments and then revising theories which would soon be revised by yet others. Even so, it is difficult not to give to Michael Faraday a major undiluted portion of the credit for first demonstrating the principles of the electric motor and the electric generator on which electrical engineering was to be based.

Following Oersted's experiment, and its repetition throughout the world, there was much speculation on the reasons for electro-magnetism. Investigating this in 1821, Faraday devised two ingenious pieces of equipment which were to have astonishing implications. The first was a bowl of mercury in which a magnet stood up vertically. Above it there hung a length of wire, free to move as it wished and with its bottom dipping into the mercury. When a battery was connected to

A print depicting Faraday's laboratory in the basement of the
Royal Institution, London, which disappeared during alterations made in 1872.

Faraday's Disk Dynamo experiment
carried out in the Royal Institution, London.

mercury and the freely hanging wire, the wire revolved about the magnet. The second piece of equipment was somewhat similar, but in this case the hanging wire was fixed while the magnet was free to revolve, which it did when the current from the battery flowed into and out of the mercury. Both of these instruments demonstrated the same startling fact: it was possible to turn an electric current into continuous mechanical motion. Faraday had invented an electric motor.

Although he now turned to other things, he retained one special ambition: having shown that electricity could be turned into mechanical energy, he asked himself whether it was possible to produce electricity with the aid of the magnetic attraction already shown to be intimately linked with electricity.

He began the experiments which were to answer his question in the autumn of 1831, and did so in two ways. To start with he produced short bursts of current by moving a coil in a strong magnetic field. Faraday had thus produced the converse of his earlier electricity-into-mechanical-motion experiment. He had now shown how, with the help of magnetism, mechanical energy could be used to create electricity. In other words, he had invented the electrical generator.

Faraday's demonstration that electricity could be generated by mechanical means was very soon followed up both in Europe and in the United States where Joseph Henry had discovered self-induction and had made a first primitive electric motor quite independently of Faraday. As early as 1832 Hippolyte Pixii of Paris had produced the first practical generator built on Faraday's principle, a hand-driven machine in which the field magnet revolved with respect to the stationary coils.

During the next decade and more, numerous improvements were made. The current, in the Pixii machine, like that of Faraday's earlier experiments, was what came to be known as an alternating current; that is, after reaching a maximum in one direction it decreased, then reversed and finally reached a maximum in the opposite direction, after which the process was repeated, the number of cycles per second being known as the frequency. Having made a generator Pixii then made a second in which he incorporated a device later known as a commutator. This reversed the connections of the circuit at every half-cycle, thereby converting the alternating current into direct current. The hand-operated generator was soon supplanted by the generator turned by a small steam engine and at the time of the Great Exhibition in 1851 electricity was available in ever-growing quantity for a wide variety of purposes. The first generators were often used to provide current for the treatment of patients by electro-therapy, the practice of giving them small shocks which had been followed from the eighteenth century with static electricity accumulated in the Leyden jar.

From now onwards, however, electricity began to be developed for three different and distinct purposes. One was for communication, another was for lighting and the third was for turning machinery. All three rested on increasing scientific research into the nature of electricity and all presented many similar problems to the electrical engineer.

It had been known from the mid-eighteenth century that the passage of static electricity along a wire could be recorded at the end of the wire. Many of the early attempts at operating an electric telegraph, therefore, involved a number of wires each representing a different letter of the alphabet. In 1832 Baron Schilling introduced a system in which the on/off current could indicate letters by a code in which on/off indicated 'A', on/on/on indicated 'B', on/off/off indicated 'C' and so on throughout the alphabet. There were other similar attempts but it was not until 1837 that the first practical electric telegraph system was experimentally installed by Charles Wheatstone and William Fothergill Cooke on a two-mile stretch of the London and North Western Railway between Euston and Camden Town. The system, utilizing needles which pointed to a letter when the signal was received, was installed on Brunel's Great Western Railway between Paddington and West Drayton 13 miles (21 kilometres) away.

While Wheatstone and Cooke were improving their telegraph system, a revolutionary step was being made on the other side of the Atlantic by an American artist who turned to technology at the age of forty-one. He was Samuel F.B. Morse who had returned to the United States in 1832 in a ship also carrying Charles Thomas Jackson, the American chemist. Jackson had attended lectures on electricity while in Paris and carried with him an electro-magnet. The magnet and its properties were discussed by the two men and Morse argued that if a magnet could reveal the presence or absence of an electric current in a wire, then the current could be used to send information over long distances.

Back in the United States, Morse and his assistant, Alfred Vail, devised an instrument in which current sent through a wire caused a magnet at its far end to attract a piece of soft iron to which was attached a pen or pencil; while the current was on, this would make marks on a moving strip of paper. Morse decided that two units of information should be sent along the wire by the current, a long unit or dash and a short unit or dot, these being separated as required by the cutting off of the current when the pen or pencil would make no marks.

By combining the two units of information in different ways it was possible to transmit any letter of the alphabet – a dot and a dash for the letter 'A', a dash and three dots for the letter 'B', a dot, dash, dot and dash for the letter 'C', and so on. To reduce transmission time as much as possible it was essential that the letters used most frequently should be represented by the shortest codes and Vail began to compute the relative frequency of letters used in English. Eventually, however, he found a short cut by visiting the local newspaper office and discovering the quantities of the different letters required by the compositors. The result was that in the resulting Morse code, the three commonest letters, 'E', 'T', and 'A' are transmitted as a dot; a dash; and a dot and dash respectively; and three of the least used, 'Z', 'Q', and 'V' are transmitted by two dashes and two dots; two dots, a dash and a dot; and three dots and a dash.

Morse obtained an American patent for his sytem on 3 October 1837, but it was 1841 before Congress gave its approval and 1843 before Morse transmitted his first and famous message from the Capitol in Washington to Baltimore – 'What hath God wrought?'.

Various improvements made during the next few years enabled messages to be produced with greater speed, and recorded at the end of their journey in different ways. But the speed of the transmission was still governed by the telegraphist who could tap only about fifty words a minute for any length of time. The handicap was removed in 1846 when Alexander Bain was granted a patent for an automatic transmitting apparatus. A strip of paper was hand-punched with long and short perforations comparable to the dots and dashes of Morse code. The punched tape was then automatically drawn between a metal roller and a metal stylus, the circuit being completed when the perforations allowed. Tapes could in this way be punched up at leisure by more than one operator and then it was possible to send them at speeds of up to four hundred words a minute.

Bain, however, was superseded by Wheatstone whose automatic paper tape transmitter was used, with only minor changes, until the end of the nineteenth century. As the distance over which messages could be sent was increased it was found that one signal would sometimes blur into its successor since the discharge at the receiving end caused an extension of the signal. The problem was overcome by sending through the circuit a current of opposite polarity which swept the circuit clear for reception of the next signal.

While the telegraph was being developed into a system enabling messages to be sent instantaneously between any two places linked by wire carrying an electric current, progress was being made in the use of electricity for lighting. As far back as 1808 Humphry Davy had built a battery, huge by the standards of the time, made up of two hundred metal plates, and had used it to pass an electric current through two charcoal rods. As the rods were drawn apart, a brilliant arc of light came from the white-hot carbon rods, the arc light which was the only form of electric illumination until the independent development of the incandescent bulb by Edison, Swan and others in the 1870s.

Because of the smoke and heat generated by the electric arc, the light was almost entirely restricted to use in the open air. A more important trouble rose from the fact that as the ends of the carbon rods burned away the gap between them had to be continually shortened to keep the arc light burning. Various mechanical devices were contrived to do this automatically, and arc lights were used experimentally in the streets of Paris in the 1850s. One was also erected on the Clock Tower of London's Palace of Westminster – the tower of Big Ben – and used whenever Parliament was sitting at night. Furthermore, in 1858 the South Foreland lighthouse on the Kentish coast, equipped with arc lights, became the world's first lighthouse to use electricity.

However, it was not until 1876 that a major step forward was taken by the invention of the 'electric candle' by Paul Jablochkoff, a Russian engineer living in Paris. The Jablochkoff candle consisted of two parallel rods of carbon, mounted vertically and with a separator between them made of a fine white clay known as kaolin. Since both rods burned down at the same rate and the distance between them was constant, no mechanism was necessary. The year after the arrival of the Jablochkoff candle the Grands Magasins du Louvre in Paris was using eighty, and experiments with them were being carried out both for street lighting and at the West India Docks in England.

During the later 1870s Jablochkoff candles became popular in Europe. But the largest practical size burned for only two hours, smell and smoke remained a problem, and it was obvious that any more convenient method of turning an electric current into light would be quickly exploited. On this occasion the demands of the time produced not one but two men of genius. Thomas Edison in America and Joseph Swan in England tackled the problem with the same determination.

The incandescent lamp had been foreshadowed as far back as 1838 when Jobart sealed a carbon rod inside a vacuum and watched it glow as a current was passed through it. Seven years later J.W. Starr, an American from Cincinnati, patented an incandescent lamp in England. He had his successors and in 1858 Moses G. Farmer lit a room in his house in Salem, Massachusetts with incandescent bulbs. In England Joseph Swan was already passing an electric current through filaments of various materials sealed inside glass bulbs in which a vacuum had been created.

These men, and others such as Maxim and Lane Fox were trying to make a bulb which would burn long enough and constantly enough. Edison believed that the solution to the problem lay in selecting the right material for the filament, and at his headquarters at Menlo Park, 24 miles (39 kilometres) from New York, used burnt paper, wood, corn and a variety of fibres, as well as platinum. For a while he believed the last would be the answer, and patented a bulb with a platinum filament. It consisted of a double helix of wire incorporating a rod which expanded as the filament grew hotter and created a short circuit before the melting point of platinum was reached. When the current was cut off the rod contracted allowing the current to flow – an electrical analogue of Watt's steam engine governor.

The platinum filament was not a success and Edison turned back to carbon. This had burned too quickly in the low vacuum that was all he had previously been able to obtain but late in 1879 he acquired one of the new Sprengel air pumps which enabled him to produce a vacuum of about one hundred-thousandth of an atmosphere. He soon adapted the pump to produce a millionth of an atmosphere and found that his sealed bulbs would retain such a vacuum for long periods and possibly indefinitely.

In October 1879 Edison and his assistant Charles Batchelor watched the first bulb evacuated by the new pump start to glow. Its filament was of cotton thread fitted into a hairpin-shaped groove cut in a nickel plate, and then cooked and carbonized. It burned for many hours and a second one burned for forty. 'I think I've got it', Edison exclaimed. 'If it can burn for forty hours I can make it last a hundred.' But Edison knew that if the problem of creating a viable filament had been solved, there were others apart from opposition from Joseph Swan.

The Englishman had been some twenty years in advance of Edison in making the first light with a carbon filament. His problem, however, had always been his inability to make a bulb which would burn for a sufficiently long time. He finally solved the problem at the same time as Edison, and in the early 1880s the London House of Commons and then the British Museum were lit by Swan bulbs. There was ample room for disagreement and litigation between the two men but they sensibly settled their differences out of court – although only after Edison had failed to get a High Court injunction against Swan – and eventually combined their interests to form the Edison and Swan United Electric Company.

It had been the aim of the early electrical engineers to make electric light available from a central generating station.

In New York it was impracticable to string electric wires overhead, on tall poles, as telephone wires had been strung, and lengthy experiments were made to discover the best way of insulating the wires when they were buried in the ground. A fraud-proof system of measuring the amount of electricity used in any one house had to be devised, as did some way of ensuring that an accidentally large flow of current did not burn out dozens of lights.

While meters, fuses, switches and connecting boxes were being designed and built, Edison was also at work on what was to be the main key to his production of power, the dynamo. Ever since 1831, when Faraday had first shown that it was possible to turn mechanical energy into electric current, engineers had been designing progressively more efficient equipment to do this. E.M. Clarke of London was producing hand-powered generators before the end of the 1830s. In

1856 Ernst Werner von Siemens patented in Britain a satisfactory powered generator. Siemens was a brother of Karl Wilhelm von Siemens (Sir William Siemens), members of the family which set up electrical enterprises in Berlin, Britain and Russia, Sir William Siemens becoming a naturalized Englishman. A decade later Siemens and two British colleagues patented what they called methods of producing 'dynamic electricity'. But although the efficiency of these dynamos continued to improve, the best of them still provided as electric current only about 40 per cent of the energy fed into them.

Edison experimented with armature cores of cast iron, forged iron and sheet iron, and then tested them to find which was the most efficient. The same was done with the armature windings and with the armatures themselves which were built in a variety of different shapes. The result of the experiments was 'Long-Waisted Mary Ann', a 1,100 pound (500 kilogram) dynamo which was the largest that had ever been built. After early tests with the dynamo Edison proudly announced that it showed an efficiency of 90 per cent, a figure met with incredulity by many experts. Eventually, the figure was claimed to be an almost equally remarkable 82 per cent. At one bound Edison had greatly increased the efficiency of the machine that was to become the main method of producing electricity. Once these problems had been solved he could hope to achieve his ambition of operating a central generating station in a big city which would supply hundreds of homes. In New York he chose 255-7 Pearl Street as the site.

But it was in London that Edison's first station was to be operated. It had been built on Holborn Viaduct in 1880 and was soon lighting some 3,000 lamps in the area, including the street lights on Holborn Viaduct itself. The nearby City Temple became the first church in the world to use electric light while the telegraph room of the City's main post office was lit by four hundred lamps. The post office contract had been a personal triumph for Edison since the organization's leading engineer was Sir William Preece who had said that it would be impossible to 'divide electricity' for use in small bulbs. Now he recanted, writing in the *Journal of the Society of Arts*, 'Many unkind things have been said of Edison and his promises; perhaps no one has been severer in this direction than myself. It is some gratification for me to be able to announce my belief that he has at last solved the problem that he set himself to solve, and to be able to describe to the Society the way in which he has solved it.'

The Pearl Street power station in New York opened on 4 September 1882, the culmination of what Edison was to call 'the greatest adventure of my life' and the first firm indication of a technological revolution that was decisively to change human affairs. The *New York Times* which had been wired for the new light, reported the following day: 'It was not until about seven o'clock, when it began to grow dark, that the electric light really made itself known and showed how bright and steady it is. Then the twenty-seven electric lamps in the editorial rooms and the twenty-five lamps in the counting rooms made those departments as bright as day, but without any unpleasant glare. It was a light that a man could sit down under and write for hours without the consciousness of having any artificial light about him . . . The light was soft, mellow, and grateful to the eye, and it seemed almost like writing by daylight to have a light without a particle of flicker and with scarcely any heat to make the head ache.'

The early electrical engineers pushed on in the hope of founding their own profitable empires, but it would be unfair to claim they were not also moved by other motives. Edison was later to say after a tour of Switzerland: 'Where waterpower and electric light had been developed, everyone seemed normally intelligent. Where these applications did not exist and the natives went to bed with the chickens, staying there till daylight, they were far less intelligent.' Men in other fields also saw the advances of the mid-nineteenth century in moral terms, Ferdinand de Lesseps who had built the Suez canal writing: 'After centuries of war and destruction, steam and electricity seem likely to open an era of unlimited progress, by multiplying the means of pacific communications between the people of the earth.'

It was certainly clear within a few months of power being switched on in Pearl Street that the coming of the incandescent bulb was beginning to revolutionize life. By the end of 1882 the station was lighting 3,400 lamps of 231 customers; by mid-August 1883, the figures were 10,000 and 431. Moreover, by this time more than three hundred other individual plants were in operation, lighting cotton mills, flour mills, a music hall, and the *Pittsburgh Times* building. Many steamers had been equipped, including the *Atlanta* of the railway tycoon, Jay Gould, while the Edison system of lighting had soon spread to theatres in London, Berlin and Prague, to factories in France and Germany, and to Government buildings in Melbourne and Brisbane.

It was very probably Edison's absorption in the programme for spreading electric light which prevented him from exploiting one of his few scientific, as opposed to engineering, discoveries. He was engineer and inventor rather than scientist, a fact accounting for his failure to recognize the significance of a phenomenon which became the foundation of the huge electronics industry. This was the 'Edison effect' which he recorded in his notebook in 1883 after sealing a metal wire into one of his light bulbs already fitted with the light-creating filament. When the wire was attached to the positive terminal of a generator an electric current was found to pass between the hot filament and the metal wire. He not only reported the effect in his notebook but wrote it up in the technical literature and patented it the following year. After that he appears to have forgotten it.

Edison was busy at the time with a multitude of inventions. He could see no practical use for the inexplicable passage of electricity; and if he had he would hardly have known how to utilize it since the existence of electrons had not yet been revealed. Not until the first years of the twentieth century did John Ambrose Fleming, who had worked with Edison in England, realize that the electricity was produced by the recently discovered negatively-charged electrons boiling off the hot filament on to the cold plate. The effect only took place when the wire was attached to a positive terminal, which meant that when alternating current was fed into the device only direct current would leave it.

In 1904, working on wireless, Fleming was trying to transform an electrical oscillation into a direct current. 'Thinking over the subject intensely', he wrote in his *Memories of a Scientific Life*, 'I had in October 1904, a sudden very happy thought. I recalled to mind my experiments on the "Edison effect", and in particular my observation that the space between an incandescent carbon filament and a cold metal plate in a bulb exhausted of its air had a one-watt conductivity for electricity. Then I said to myself, "if that is the case we have here the exact implement required to rectify high-frequency oscillations". I asked my assistant, G.B. Dyke, to put up the arrangements for creating feeble high-frequency

Edison's Light Bulb. Thomas Edison in the United States and Joseph Swan in England were largely responsible for giving electric light to the world.

currents in a circuit and I took out of a cupboard one of my old experimental bulbs. . . .'

The experiments were a success, and since the device satisfactorily allowed the current to pass in one direction and prevented its passage in the other, Fleming called it a valve. Two years later Lee De Forest found that if a third element called the grid, was introduced into the device, then a varying but weak electric potential on the grid could be converted into a varying but much stronger flow of electrons to the combination of filament and plate. Thus the valve worked not only as a rectifier but as an amplifier, and the 'Edison effect' could be utilized as the basis for the wireless and electronics industry.

Although Edison had been too immersed in the problems of electric light to note the possibilities of what was to become the wireless valve, he embarked on another enterprise which was eventually developed by others into a series of engineering projects. This was the utilization of electricity for rail transport. Edison had first thought of electricity-powered locomotives during a visit to the Middle West; steam power would be uneconomic to deal with the seasonal wheat trade, he argued, but electric power might do the trick.

He now began to turn his ideas into hardware, no doubt encouraged by the efforts of Ernst Werner von Siemens in Berlin. Siemens, was an expert on submarine cables and one of the foremost electrical engineers of his day. In 1878

The brightly lit downtown
financial area of New York photographed at dusk.

Siemens had been asked by a German coal mine manager to design an electrically powered railway engine that could be used in conditions where smoke and fumes would rule out a steam engine. The result was built and exhibited, with considerable success, at the Berlin Trade Fair in 1879.

Siemens' track consisted of a 900 yard (823 metre) circle of double rails laid on wooden sleepers which insulated them from the ground, while between the rails there ran an iron bar. At the station on the circular track a steam engine operated a dynamo which fed electricity on to the central iron bar. In contact with the iron bar were metal brushes on the train engine; they picked up the current, took it to a dynamo on the engine which was connected to a pair of driving wheels, and then returned the electricity on to the two rails. The engine produced about 5 horsepower and drew at speeds of up to 20 mph (32 kph) four or five carriages carrying up to thirty passengers.

Wet weather reduced the insulating properties of the sleepers and the line operated only on dry days, but with this qualification it was a considerable success. There were certain built-in advantages of the system as Siemens explained to the Society of Telegraph Engineers in June 1880. 'When the motion of the train is slow,' he said, 'the force acting on it is at its maximum, and owing to this it starts with a remarkable energy. When the motion increases, the accelerating power diminishes; so that the driving force regulates itself according to the velocity of the train. On an ascending gradient the speed diminishes and the propelling power increases; on a falling gradient the reverse effect takes place; and in the latter case, if the train runs down by gravity and over-runs, so to speak, the power of the dynamo, the current will serve as a brake to check the speed. In all these particulars the electic motor, therefore, fulfils perfectly, by automatic action, all the functions of the driving power on an ordinary locomotive line.'

So successful was this first electric railway that Siemens applied for permission to build a longer, overhead, line. It was not allowed, but in 1881 a mile and a half long track was built from a station on the Berlin-Anhalt Railway to the Berlin Military Academy and was successfully operated for a number of years. Other similar lines were laid in mines in Saxony and at the Paris Electrical Exhibition of 1881 when in seven weeks 95,000 passengers were carried to the Exhibition Building from the Place de la Concorde.

Edison, who by now regarded his name as synonymous with electricity, felt himself challenged. He believed that he could do better than Siemens, possibly being misled by the rather toy-town impression at first created by the demonstration in Berlin. By May 1880 he had laid out at his Menlo Park laboratories, New Jersey, a track made from streetcar rails that ran in a rough U for a quarter of a mile.

Edison lost no time in patenting a number of features he had incorporated. He had a genuine enthusiasm for the electric railway and, as was often the case with his enthusiasms, he mixed a genuine social motive with the financial ones. In this case he knew the disadvantages of the steam engine. As the *New York Herald* put it, an electric train would be 'most pleasing to the average New Yorker, whose head has ached with noise, whose eyes have been filled with dust, or whose clothes have been ruined with oil.'

Judging by a report in the *Scientific American*, Edison himself enjoyed the tests and trials as he sought to improve his first electric train. 'By invitation of Mr. Edison, representatives of this journal were present at a recent trial of this novel motor,' it said '& had the pleasure of riding, with some twelve or fourteen other passengers, at a breakneck rate up and down the grades, around sharp curves, over humps and bumps, at the rate of twenty-five to thirty miles an hour. Our experiences were sufficient to enable us to see the desirableness of a little smoother road and to convince us that there was no lack of power in the machine.'

Edison continued to improve and to patent. But the lure of electric light remained dominant and it was perhaps because of this that his railway venture came to nothing. He was also, for once, unlucky. Henry Villard, president of the Northern Pacific Railroad, agreed to build 50 miles (80 kilometres) of electric railway track once Edison had brought operating costs to below those of steam. But before Edison could do so Villard's fortunes had crashed while Edison found that some of his own railway patents were being challenged by an engineer with whom he had collaborated. Eventually, at the Chicago Exposition, he ran an electric railway 1/3 of a mile (500 metres) long which carried more than 26,000 passengers a total of 466 miles (750 kilometres). But there was little follow-up to the success, and it was the turn of the century before electric railways, their power usually delivered by overhead wires, began to grow in popularity throughout both Europe and America.

Before power stations began to proliferate throughout Europe there had arrived on the London scene Sebastian Ziani de Ferranti, a young man who in

1886 became engineer to the Grosvenor Gallery Electric Supply Corporation which was then supplying the district with electricity. Ferranti quickly increased the area covered, then decided that the whole of London north of the Thames could be served by a single power station if one were planned for a site on the outskirts where coal for the generators could easily be delivered. The result was the Deptford power station, the prototype for many of the world's major power stations built in the twentieth century.

Ferranti decided that transmission should be at 10,000 volts. But in the 1880s neither generators for producing more than 2,500 volts, nor cables for distributing any higher voltages, had yet been constructed. He finally succeeded in designing his own equipment and his own cables, and current at 10,000 volts began to be distributed in 1889. Regular operation started the same year and some of the Ferranti cables remained in use for more than forty years.

The Deptford station, which demonstrated that it was possible to transmit electricity at high voltages and then reduce the voltage before the current was used, played its part in 'the battle of the currents' as it was known in America, 'the battle of the systems' as it was called in Britain. It was fought round the rival advantages and disadvantages of direct and alternating current, and had been raging almost since the time when Edison and Swan had first revealed the value of electric light. Edison's system, like those of his rivals, had been developed to exploit direct current which flows constantly from the generator to the user and then back to the generator. The current was limited for practical purposes to a maximum of 250 volts, a disadvantage since for long-distance transmission power losses are much less for high voltages than for low. This fact had not mattered so much when distribution was limited to buildings within a short distance of such generating centres as Pearl Street or High Holborn. The situation began to change as smaller groups of potential users, farther from the generators, began to clamour for electricity. Few of the schemes for transmitting high voltage alternating current over longish distances, then reducing the voltage before it was led into houses or factories, had been successful.

They had not been successful, that is, until they were taken up by Edison's rival, George Westinghouse, aided by Nikola Tesla, a Croat engineeer who had for a while worked with Edison. Tesla's main achievement was the design of a practical AC motor, but he also succeeded in building transformers which would raise the voltage of generated alternating current for transmission and would bring it down to a low voltage at its destination. In Britain, the equivalent of the war which Edison began to wage with Westinghouse was fought out between Ferranti, supporting alternating current to distribute electricity from his Deptford power station, and more conservative electrical engineers such as Crompton and Hopkinson. On both sides of the Atlantic alternating current finally triumphed but only after a long war of attrition during which Edison made every effort to stress the alleged dangers of using his rival's system. His company published 'A Warning', bound in red and detailing every accident that could allegedly be attributed to alternating current. And in 1889 Edison bought three Westinghouse alternating machines without revealing that he would be selling them to the prison authorities. When on 6 August 1890 William Kemmer was electrocuted in Auburn State Prison for murder, alternating current was used and, for many people, became synonymous with death.

A few years later Westinghouse won a more important victory. The decision was at last taken to exploit the huge energies running to waste over the Niagara Falls. Westinghouse obtained the contract to supply the motors – alternating – for what was the most important of America's hydro-electric schemes. By 1896 ten 5,000 horsepower pressure turbines were at work in the Niagara Falls scheme helping to provide alternating current for Buffalo 26 miles (42 kilometres) away.

Niagara was, of course, internationally famous, and according to local groups, visitors had been asking for years why Americans 'should allow the unlimited power of Niagara to waste itself away without attempting to divert a fraction of the force flowing by their doors, to increase the material prosperity of their country.' Its hydro-scheme was, therefore, psychologically important, although it was not the first time that water had been used to produce electricity. In 1882, a decade previously, Edison had installed one of the world's first hydro-electricity plants at Appleton, Wisconsin, where, powered by the Fox River, it lit between two hundred and three hundred lamps in the neighbourhood.

The previous year Siemens in Britain had opened the world's first public power station at Godalming, Surrey. 'Siemen's contract', says the firm's historian, 'was a street-lighting one with the Town Council, but power (derived from a waterfall on the River Wey) was also sold to private consumers. Authority to dig up streets for an electrical undertaking did not yet exist [it was introduced in 1882], the cables were laid in the gutters. Godalming was brave, but premature; the citizens did not apply for electricity in sufficient numbers, and in 1884 the streets went back to gas.'

It was from such beginnings that there was to grow, in the last years of the nineteenth century and throughout the first half of the twentieth, the hydro-electric industry and the awesome hydro-dams. They produce a comparatively small proportion of the world's electricity but are impressive examples of the engineer's profession. Niagara, like Lake Victoria and the Rhine Falls in later years, offered the engineer not only a natural source of energy which merely had to be channelled into properly designed machinery, but a source which was available throughout the year. There were exceptions, however. Many of the world's great sources of water energy had to be dammed if they were to be used most efficiently, and the history of hydro-electric power is a history of continuous co-operative research between the structural engineers who build dams, hydraulics engineers who design water turbines and electrical engineers who design generators.

Edison's hydro-plant at Appleton had been a simple affair in which water was carried to a turbine connected to a generator immediately above it. The next step was taken when a rock-filled timber dam on the Willamette River at Oregon City was in 1889 built to hold back water for a small hydro-plant. Four years later a 65 foot (20 metre) high, 1,150 foot (350 metre) long masonry dam was built across the Colorado at Austin, Texas, to supply power for generators producing nearly 15,000 horsepower. The scheme, operated by the city of Austin, supplied electric power for water pumps, electric lights and streetcars, but in 1900 unexpected flooding of the reservoir built up behind the dam caused part of it to give way. The scheme was abandoned although thirty-seven years later it was not only restored and reopened but enlarged by the Public Works Administration.

Other hydro-electric dams were built near the turn of the century both in the United States and in Europe – particularly in France and Italy where run-off from the Alps offered potentially great hydro-power, and in Norway where Scandinavia's high mountains did the same. Dams built for generations of hydro-power are more complex than those that preceded them, since water is normally led to the turbine rotors through circular tunnels embodied in the dam structure and on leaving the turbines is led through a draft tube to an exit below the power-house. 'As the design of the hydro-electric dam progressed', Carl J. Condit has said, 'the passage leading to the turbine was given a spiral form with a diminishing radius in order to accelerate the velocity of flow. The cross-sectional area of the draft tube, on the other hand, expands continuously throughout its length to reduce the velocity of the discharge water and thus minimise the erosion of the stream bed at the tail race. The power-house and the control room came to be housed in a single enclosure that grew in size until it reached a volume comparable to that of the largest urban buildings.'

The dams of the pre-hydro-electric age had been built to impound water either for irrigation or to supply local homes, and many of the hydro schemes also used water for multiple purposes. Once it had passed through the power-house it could, with little extra complication, be used for irrigation or diverted into a city's water supply system. What such multi-purpose schemes did require, however, was ever larger amounts of money and in 1902, the US Congress set up the Bureau of Reclamation as part of the Department of the Interior with the specific task of encouraging irrigation schemes, particularly in the arid areas between the prairies and the western mountain ranges. The Bureau not only took over the organization and building of dams primarily for hydro-electric use but financed the research which led, during the first half of the twentieth century, to a number of new and improved methods of construction tailormade for use on specific sites.

Among the first of the spectacular dams which it sponsored was the Roosevelt Dam, built on the Salt River between 1906 and 1911 near Phoenix, Arizona. A traditional arch-gravity construction in which the pressure of the water behind the dam is borne by its arched shape and also by its sheer weight, it was 284 feet (87 metres) high and 1,125 feet (343 metres) long, and was for some years the tallest dam in the world.

The Roosevelt Dam was followed by a number of others, many of them creating new records for height, length, volume of concrete used, volume of water impounded, or electric power created. Most of them incorporated the latest hydro-electric engineering techniques, and soon showed that under the leadership of the Bureau the Americans had during the first half of the twentieth century became the world's leading dam builders.

The Hoover Dam – formerly Boulder Dam – on the Colorado River where it forms the boundary between Nevada and Arizona, rises for 577 feet (176 metres) and for twenty-two years after its completion in 1936 was the highest dam in the world. Long before 1956, however, the record 4.40 m.yd^3 (3.36 m.m^3) of concrete which went into the Hoover Dam was surpassed by the 10.58 m.yd^3 (8.09m.m^3) of the Grand Coulee, still a major showpiece of American hydro-engineering. Completed in 1943, the Grand Coulee was the key element in an ambitious irrigation and power-generating scheme which is still being expanded. The 550 foot (168 metre) high, 4,175 foot (1,272 metre) long concrete barrier across the Columbia river has created a reservoir 150 miles (240 kilometres) long and has made the Columbia navigable for 350 miles (560 kilometres) upstream, as far as Revelstoke in Canada. The dam itself – the first structure to contain more

Water pouring through the
Dalleo Dam, Columbia River, Oregon, Washington.

The Hoover Dam, on the Colorado River, Nevada.

material than the Great Pyramid of Cheops – is part of a hydro complex that contains two power-houses already generating 2 million kilowatts, a wing dam above the upstream face to protect pumping installations, two pumping plants, two earth and rock-fill dams in the equalizing reservoir, and thousands of miles of tunnels and pipes. Up to 1 million cubic feet (28,000 cubic metres) of water can be discharged by the Grand Coulee, part of this going into the eighteen turbines of the power-houses part into the pumping plant. The Grand Coulee can claim many records. Its pumps, for instance, are the largest that have ever been built while the 40 foot (12 metre) wide pipes forming its penstocks are the largest of their kind.

Since 1911 the size of large dams, and the volume of the water impounded behind them, has continued to increase, with records regularly being leapfrogged by new ones. There are various ways of defining a 'large' dam but the International Commission on Large Dams lists them as those over 15 metres (50 feet) high or a dam between 10 metres (30 feet) and 15 metres (50 feet) high as long as its crest is more than 500 metres (1,640 feet) long, its reservoir holds more than 100,000 m^3 (220 m. Imperial gallons), it can discharge more than 2,000 m^3 (70,000 feet^3) per second and has especially difficult foundations.

Despite this necessarily complicated definition, dams are often graded by their height alone. In this rating, Russia is expected to hold the record soon with the 1,040 foot (317 metre) Nourek Dam on the Vakhsh River in Tadjikstan, USSR, near the frontier with Afghanistan. With a length of 2,395 feet (730 metres) the dam will impound about 2.5 cubic miles (10.5 cubic kilometres) of water which will be used for supply, irrigation and hydro-electric power. Until the Nourek Dam is fully operational, the world's highest dam is the Grande Dixence in Switzerland, whose huge 932 feet (284 metres) face towers over the traveller coming up the Vispthal from the Rhône in the Canton of Valais. The 858 foot (261 metre) high Vajont Dam, rendered useless by a huge earthslip in 1963; the 800 foot (244 metre) Mica Dam on the Columbia River in British Columbia; and other dams more than 700 feet (213 metres) high in Mexico, India and Yugoslavia illustrate how the demands for hydro-power have by now spread across the whole world.

There has also been the development of pumped storage schemes in which water is held at an upper level, led down at times of peak demand through tunnels to turbo-generators which produce electricity, collected at a lower level, and then pumped back to the higher level when off-peak electricity can be used for the task. The biggest of these schemes in Europe is Britain's Dinorwic station in North Wales which in December 1982 began to achieve what will be its full commercial load of 1,800 megawatts. Lying on the northern edge of the Snowdonia National Park, the Dinorwic uses as its upper reservoir Lyn Marchlyn Mawr, 692 yards (633 metres) above sea level. From this natural lake up to 92,400 gallons of water per second can flow down to Lyn Peris at 115 yards (105 metres), passing through one of six turbo-generators on the way. Up to 1,680 megawatts can be generated continuously for five years, and in an emergency the turbines can start supplying substantial amounts of electricity within ten seconds. The station itself, as well as the 10 miles (16 kilometres) of tunnels has been built entirely within the mountain from which 1.94 million cubic yards (1.48 million cubic metres) of rock have been excavated. The machine hall, which is 195 yards (178 metres) long, 26 yards (24 metres) wide and 65 yards (59 metres) deep, is believed to be the most sizeable civil engineering excavation to have been made by man in Europe.

Long before pumped storage schemes gave a new significance to hydro-electricity and the dams that go with them, a hydro-scheme which marked the start of modern irrigation on a new scale was being inaugurated in Africa with the construction of the Aswan Dam on the Nile. It had been decided in 1890 that the Nile should be dammed so that areas larger than those of the Delta could be served by irrigation. The man put in charge was William Willcocks who started by devoting four years to a detailed survey of the river's lower 800 miles (1287 kilometres). At the end of his preparatory work a dam across the Nile was planned at Aswan about 555 miles (893 kilometres) south of Cairo where the river is 6,400 feet (1951 metres) broad. Here, by 1902, there had risen a 65½ foot (20 metre) high dam of conventional masonry construction. However, one feature made the Aswan dam different from all others. The silt brought down by the river was essential to the nourishment of Egypt and the 180 deep sluices were built so that not only water but also silt would pass through them. When they were shut the river above Aswan was turned into a new man-made lake 200 miles (320 kilometres) long.

Water from the Aswan Dam was fed to the Assuit barrage, 350 miles (563 kilometres) downstream, and part of what during the following decades was to develop into a more and more complicated Nile irrigation complex. So successful was the initial operation that in 1905 it was decided to increase the height of the Aswan Dam to 88½ feet (27 metres), thereby doubling the capacity of the reservoir which could be built up behind it. Even so, it was found to be insufficient for Egypt's growing needs and in 1929 the dam was once again heightened. This led on to the construction, between 1960 and 1970 of the Aswan High Dam which will eventually flood the Nile valley for 400 miles (644 kilometres) upstream. Already it has meant the removal to new homes of 90,000 Egyptian and Sudanese peasants and the rebuilding, on higher ground, of the Abu Simbel temples which would otherwise have been destroyed.

The birth and growth of the hydro-electric industry, with the repercussions which it was to have on civil engineering and the use of concrete for massive structures, was not the only by-product of the demand for electricity which continued to increase during the last two decades of the nineteenth century. Another was the development of the steam turbine.

Once the ability of the dynamo to produce electricity had been demonstrated by Faraday in the 1830s, the steam engine had been conscripted to drive the dynamo; the method was the simple one of using the engine to produce rotary motion which was then transmitted by belting to the dynamo shaft. This was the method used in Edison's Pearl Street station in New York in 1881 and it was subsequently used elsewhere. Even lighthouses sometimes used steam-driven dynamos to create their light.

However, there was one limitation to the steam engine working a piston to provide rotary motion which could be taken to the dynamo: in practice, there was an upper limit to the revolutions per minute which could be given to the dynamo shaft. Eventually it was realized that increased revolutions could be utilized and a way of creating them was found by Charles (later Sir Charles) Algernon Parsons. He decided that he would utilize and exploit the principle used by Hero of Alexandria almost 2,000 years earlier, and then by Giovanni Branca in 1629. Richard Trevithick, who had tried out so many engineering possiblities, had in

Armatures designed for small
electric motors show the windings that carry the current.

1815 also built an engine in which high pressure steam was led into a hollow tube from which it emerged to spin the sails of a windmill. But none of these systems was mechanically sound and none was ever developed.

Earlier in the nineteenth century many men had considered directing steam not into a cylinder to move a piston but directly on to blades attached to the perimeter of a wheel, thus producing direct rotary movement with all the benefits of power conservation that that would involve. One thing prevented such ideas being carried from theory into practice: the wheel would have to stand the stresses of extremely rapid rotation and its metal would also have to withstand extremely high temperatures; and until the production of specialist steels no metal had the necessary characteristics.

By the 1880s, with high-duty steel available, only imagination and engineering expertise were required. They were provided by Parsons, in whose turbine high-pressure steam was directed on to specially shaped blades mounted on the rim of a disc attached to an axle which was caused to rotate. The design of the turbine was such that although the steam expanded it did so in a number of successive stages, thus securing the most efficient relationship between steam speed and blade speed.

Parsons' first turbine, made in 1884, turned a dynamo at 18,000 revolutions a minute, more than ten times the previous speed. It was a complete success and by 1890 he had designed turbo-generators for the Forth Banks power station in Newcastle, which became the first in the world to use such equipment. Other stations quickly followed while before the end of the century the German city of Elberfeld bought two Parsons turbines and established his fame on the Continent.

Once the advantages of the turbine had been proven, engineers in America and Europe began to develop and refine its features. Some put their faith in axial flow turbines in which the steam moves parallel to the axis of the rotor; others supported radial flow engines. There were also differences between those who believed in the impulse turbine in which the steam emerged from sets of nozzles on to the blades of the rotating wheel, and the supporters of the reaction turbine in which a ring of stationary blades replaces the nozzles, and the rotor is forced to rotate by the reaction between the stationary blades and the blades of the rotor. Various combinations of steam turbines were installed in generating stations during the first decade of the twentieth century, but it was universally acknowledged that whatever combination was used the steam turbine was far more efficient than the reciprocating steam engine.

The steam turbine had, meanwhile, been used for a different but equally important purpose. In his original patent of 1884 Parsons had proposed that the turbine could be applied to marine propulsion, a commonsense suggestion since the turbine was smaller and lighter than comparable power-producers and thus ideal for installation in the relatively confined space of a ship. When the future of the turbine had been assured, Parsons fitted a model in a 100 foot (30 metre) 44-ton vessel which he renamed the *Turbinia*.

He believed that the turbine could revolutionize propulsion of the Royal Navy's fleet, but he knew also the conservatism of the Admiralty. The one chance of getting such a new power-producer adopted lay in giving a practical, and if possible dramatic, demonstration. The opportunity came in 1897 when Queen Victoria's Diamond Jubilee was celebrated by an impressive Royal Navy review at Spithead, between Portsmouth and the Isle of Wight. The *Turbinia* put in an uninvited appearance and raced down the line of warships at 34.5 knots (64 kph) . She could not be followed since the fastest naval destroyers could barely reach 27 knots (50 kph).

Two years later the Admiralty commissioned from Parsons the first turbines to be fitted into a destroyer, and in 1905 an Admiralty committee on naval design advised that in future all warships should be turbine-powered. The merchant navy was quick to follow and the Cunard Line powered with turbines first the 30,000-ton *Carmania*, and then the *Lusitania* and the *Mauretania*.

One problem still remained. The turbine had been designed to work at the high revolutions per minute needed for generating electricity most efficiently. But when propellers were to be driven as efficiently as possible there was an upper practical speed, which had limited the use of turbines in smaller ships. Parsons dealt with that by designing gearing that brought propeller-revolutions down to a tenth that of the turbines. From now onwards the turbine could be fitted into the world's low-speed tramp steamers and merchant vessels.

The twentieth century thus opened with the prospect of ever-growing uses for electricity, the smoke-free, fume-free power-producer that was eventually to take over from steam. The possibilities had already been seen, and not only by scientists and engineers. 'The applications of electricity', Pedro d'Alcantara, the Emperor of Brazil, had already remarked to Sir William Siemens, 'are becoming each day more wonderful, if one may use that word in speaking of science, and I would mention that I read everything which is published and I am able to obtain regarding the distant transmission of power and utilization of electricity.

'Perhaps it may solve the problem of aerial navigation.'

Electricity pylons, St Julie, Quebec.

MEN WITH WINGS

As they float high in the sky, minute silver specks against the deep cerulean blue of outer space, it is difficult to regard them as works of man. Yet the stratojets carrying humans from one continent to another are the evolutionary end-products of the gaudy balloons which less than two hundred years ago astonished Europe with their promises and threats for the future.

From earliest times man has been fascinated by flight. Legend maintains that in the eighteenth century BC Ki-King Shi owned a flying chariot. The legendary Greek Icarus flew too near to the sun whose heat melted the wax of his wings and made him fall to his death, and in Britain King Bladud, the mythical founder of Bath, came a cropper while trying to fly. In the East, caliphs and sultans fly on magic carpets while the Nihon Shoki, one of the oldest books of Japanese History, published in 720 AD, maintains that the Son of the Heavenly God descended to Japan by the Sky Ship. In the West, witches ride through the air on broomsticks. Magic, throughout the ages, has been invoked to help men emulate the birds.

Only slowly did these fancies give way to drawings and plans which, although never executed, do suggest that men were beginning to see flight as a genuine possibility. Leonardo da Vinci with his design for a helicopter, drawn about 1500, and Francesco de Lana, in 1670, with his plan for an aerial ship that would be raised by four copper spheres which had been emptied of air, were two of those who thought of achieving what most men believed to be impossible.

Yet the conquest of the air started almost by chance when, in Annonay near Lyons, two French papermakers, Joseph-Michel and Jacques-Etienne de Montgolfier, noticed how burning paper rose above a fire. Would it not, they reasoned, be possible to catch in a bag the gas that they believed was being generated and let it raise the bag into the sky? And so, at Annonay on 5 June 1783, a crowd collected to watch a fire of wool and straw set alight under a spherical balloon of linen, 38 feet (11.6 metres) across. When the guy ropes holding the contraption to the ground were released it rose more than a mile in the sky before slowly coming to earth as the air cooled. 'Mr Montgolfier attributed the effect of the machine, not to the rarefaction of the air, which is the true cause, but to a certain *gas*, specifically lighter than common air, which was supposed to be developed from burning substances, and which was commonly called Mr. Montgolfier's gas.' Thus wrote Tiberias Cavallo, who discovered the lifting power of hydrogen and wrote *The History and Practice of Aerostation*.

The dramatic demonstration was followed, five months later, by the ascent of

> Dr Pilatre de Rozier and the Marquis d'Arlandes
> ascending in a Montgolfier from the Bois de Boulogne, Paris, in 1783.

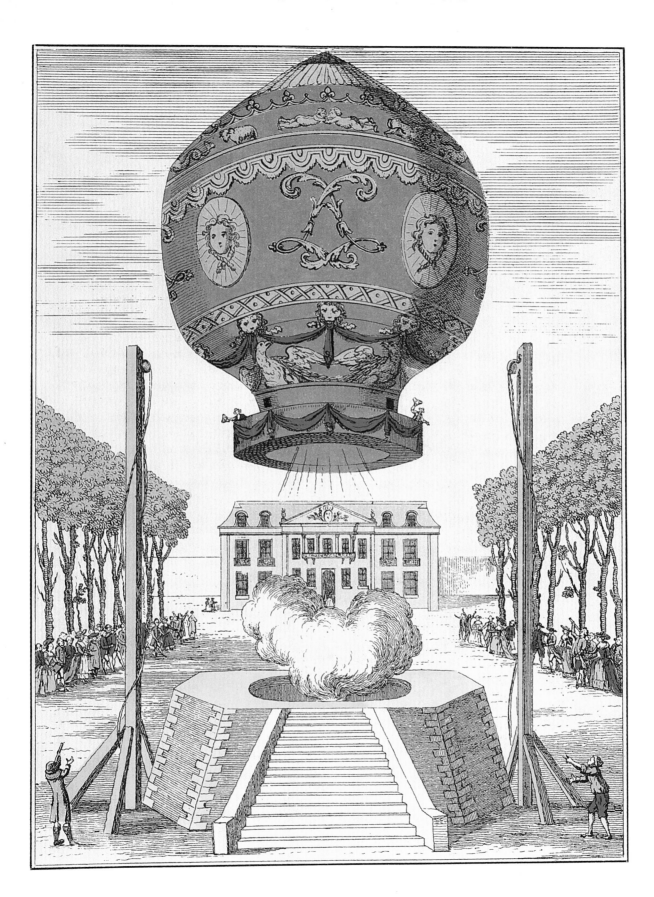

a young doctor, Pilatre de Rozier, and the Marquis d'Arlandes in a 'Montgolfier' sent up from the Bois de Boulogne in Paris. It was on that occasion that Benjamin Franklin, the American commissioner in Paris, heard a bystander question the usefulness of balloons. 'What use', he answered prophetically, 'is a new-born babe?'

This first manned flight in history triggered off the ballooning age, and for some years the skies of Europe were regularly visited by the conveyances of the pioneer aeronauts, frequently brightly coloured and encouraging the earth-bound to speculate on the next wonders to come. In 1784 Vincent Lunardi sailed up in a balloon basket with a dog and a cat from Chelsea in London, landing some twenty miles to the north at Ware in Hertfordshire. The following year the English Channel was crossed by balloon for the first time. The lift given by hot air from a fire could not easily be controlled, the fire itself was a constant source of danger, and before the end of the eighteenth century heated air as a source of lift had been replaced first by hydrogen and then by coal gas.

As ballooning grew in popularity and sophistication – and by 1804 a manned balloon had reached a height of 23,000 feet (7,010 metres) – it became clear that conquest of the air would be of two different kinds. One would be by lighter-than-air craft; the second by machines heavier than air. Each of these two kinds of flying machine presented its own serious problems, although many were common to both categories, and their stories overlap, especially during the second half of the nineteenth century when attempts to give independent motive power to balloons led to the development of the dirigible or directable airship. There were also gliders, which were important for one thing in the complex tale of early aviation since it was largely by their use that the pioneers gained their knowledge of aerodynamics, that interaction between airflow and the movement of solid bodies through the air.

The pioneer in this new field was Sir George Cayley, a Yorkshire baronet who has deservedly been called 'the father of aerial navigation'. Most of his experiments were carried out at Brompton Hall, his home near Scarborough, Yorkshire, and their nature is shown by his Note-Book entry for 1 December 1804 in which he described how he used a Whirling-Arm to determine air resistance. 'I made the following experiments upon the resistance of air to a surface of a foot sq. carried round with an horizontal motion upon an arm suspended upon a delicate hinge. The length of the arm from the pivot to the centre of resistance was 4 feet 9.2 inches. This arm played up & down upon an hinge fixed into the part that was carried round by the upright pivot, & counter-poised by a weight on the other side of the hinge, so that the $\frac{1}{16}$ of an oz. turned it easily from an horizontal position. The whole was braced by a cord over an upright in a part of the frame to give firmness to the structure.

'The motion was communicated by a cord going over a pulley & then down a stair case, with a bag to hold weights at pleasure.'

Such simple equipment led Cayley to see that the basic problem of mechanical flight was 'to make a surface support a given weight by the application of power to the resistance of air.' Flapping wings, he concluded, were useless since, although many men had tried to fly like the birds, 'the pectoral muscles of a bird exceed 7 or 8 times in proportional strength the whole power of a man's arm, when the weight of the bird is compared with the weight of a man.' Having decided that a fixed wing was necessary, Cayley carried on with his experiments, finding that a cambered wing – one having a slight arch or upward curve to it – gave greater lift than a flat one; that the basic technique of bird propulsion was the propeller action of the outer portions of the wing, and that streamlining reduced resistance to the air. These ideas, and many more, he incorporated in a three part-paper 'On Aerial Navigation' published in 1809 in Nicholson's *A Journal of Natural Philosophy, Chemistry and the Arts*. Here Cayley laid down for the first time the principles of heavier-than-air flight and discussed the power required. It was, the historian Charles H. Gibbs-Smith has commented in *Sir George Cayley's Aeronautics, 1796-1855*, 'a literally epoch-making document and is the first treatise in history – and also the first to be published – on both theoretical and practical aerodynamics'.

Cayley consolidated his towering position among the early theoreticians of flight by building a number of gliders. The first, flown in 1804, had the configuration of a modern aircraft. The next, a full-size machine, was flown, unmanned, in 1809. Legend maintains that at least two manned glider flights were made in Cayley's machines: the first by a young boy, the second by Cayley's coachman who is said to have glided across a valley near Brompton Hall, and, getting out of the glider afterwards, protested: 'I give notice. I was hired to drive, not to fly.' In the 1850s, following the suggestion of a young designer named Robert Taylor, Cayley designed the convertiplane. This had four circular discs, mounted in pairs, which were to get the machine off the ground, after which they were to close and form circular wings. Two pusher propellers then took over to drive the machine through the air. Like many of Cayley's ambitious projects, it was never built.

It was not only in the fields of gliders and aircraft that he was awash with ideas throughout a long life. In 1816 he designed a streamlined airship and the following year suggested that such airships should be lifted off the ground not by a single gas-filled envelope but by a number of separate gas compartments contained in a single outer covering. Years later it was Cayley who proposed that they should be equipped with swivelling propellers.

His ideas for directable balloons were as much in advance of contemporary technology as were his ideas for aircraft. Soon after the first balloons had risen into the air, men had put forward various methods of directing their flight. As early as 1799 Cayley himself had suggested 'flappers' that would be operated by crew members pulling on rowing levers. Paddle-wheels turned by hand were suggested, and in 1835 the 160 foot (49 metre) long The Eagle, 50 feet (15 metres) high 40 feet (12 metres) wide, and to be manned by a crew of seventeen, was built in London. The Eagle, said an announcement from the European Aeronautical Society, had been 'constructed for establishing a direct communication between the several capitals of Europe', and the first journey was to be a return trip from London to Paris and back again. But The Eagle never left the ground.

By contrast, Henri Giffard's airship was an engineering success. Giffard had made a fortune from his invention of steam injection and had devoted the money to aerial research. His dirigible, built in 1852, was 144 feet (44 metres) long and with its pointed ends took the form of what was to be the conventional airship. Slung some 20 feet (6 metres) below its cigar-shaped body was a 66 foot (20 metre) spar at the end of which a vertical triangular sail acted as rudder. From this spar there was suspended a container for the pilot and for the single-cylinder steam engine which drove an 11 foot (3 metre) propeller. Delivering between 3 and 5 horsepower, the engine weighed only 100 pounds (45 kilograms) while the all-up weight of the dirigible was only about 350 pounds (159 kilograms). Sitting in the 'pilot's seat', Giffard rose from the Hippodrome in Paris on 24 September 1852 and was successfully carried to Trappes, 17 miles (27 kilometres) away, a journey made at between 4 and 5 mph (6.4 and 8 kph).

While Giffard had achieved the first successful application of mechanical power to flight, and while his dirigible could be manoeuvred to a limited extent, he was unable to navigate in a circle. To overcome the problem he designed a number of larger and potentially more manageable craft, one of them a giant which was to be 1,970 feet (600 metres) long, containing nearly 8 million cubic feet (226,500 cubic metres) of gas, and capable of travelling at 40 mph (64 kph). But Giffard died before these plans could be carried out and it was left to other Frenchmen finally to solve the problem of controlled dirigible flight.

First came the Tissandier brothers, Albert and Gaston, who powered their 92 feet (28 metre) long dirigible with a 1½ horsepower Siemens electric motor weighing 600 pounds (270 kilograms). When the airship was flown from Auteuil in 1883 a measure of control was exercised in spite of bad weather conditions. But full control was obtained only the following year when the 165 foot (50 metre) La France was flown by Captain Krebs, who had designed the airship, and Renard, the engineer who had designed the chromium-chloride batteries powering the 9 horsepower Gramme electric motor. They flew at 13 mph (21 kph) and made a fully controlled circular flight of nearly five miles: but, as seemed likely after the flight of the Tissandier craft the previous year, La France demonstrated that electric engines were too heavy for successful flight.

The last half of the nineteenth century witnessed a great surge of interest in attempts to conquer the air. The Aeronautical Society – later Royal Aeronautical Society – was founded in Britain in 1866. Five years later its members commissioned Francis Henry Wenham and John Browning to build a wind tunnel in which scientific tests could be made. In Austria Ernst Mach, who noted that the nature of airflow over a moving object changes as it reaches the speed of sound – and whose name lives on in the form of Mach numbers for the speed of sound in air – also built one.

John Stringfellow constructed a triplane which in 1868 was tested, unsuccessfully, in the grounds of the Crystal Palace, outside London. Alphonse Pénaud, a Frenchman, built a stable aircraft model powered by rubber bands in 1871 and planned to build a full-scale machine but was unable to get financial backing. In 1876 N.A. Otto produced the first four-stroke petrol engine, and immediately aroused the ambitions of engineers, who began to consider whether it would one day power aircraft. Shortly afterwards, Horatio F. Phillips patented wings of two-surface section, having demonstrated what Cayley had surmised: that most of the lift is produced from a cambered wing by suction above the surface. In the 1890s Clement Ader, another Frenchman, began a series of experiments which used steam to power a succession of aircraft whose wings were modelled on those of the bat, and in the United States Samuel Langley designed and built a steam-driven plane which, unmanned, successfully flew across the Potomac.

Yet if it is possible to trace a direct line of progress from the pioneers to successful powered flight it runs through the efforts of Otto Lilienthal, who with his brother Gustav was brought up in the little town of Anklam on the Baltic coast of North Germany. To Anklam there would come, every February or March, great flocks of storks, slow but confident birds whose deliberately beating wings had brought them from their winter homes in Africa. Every autumn, on what seemed to be a pre-decided day, they would rise from their nests on the Stettiner Haff, or from the boxes the Anklamers had built for them, and after circling and recircling the rooftops would fly south to Africa. Both the Lilienthal brothers were fascinated by the storks. Like good men before them, they at first put their faith in man-made wings which could be attached to a flying machine and moved up and down by the feet, bicycle fashion. All the home-made wings failed, despite the brothers' construction of an ingenious valve which worked between the separate canes making up each wing and allowed the air to pass through the wings on the upstroke but closed to prevent it happening on the downstroke.

One machine, the size of a stork and powered with a ¼ horsepower internal combustion engine, had wings designed as copies of real wings. 'We [also]', Gustav later wrote, 'possesed an apparatus which was fitted with beating wings, moved by spiral springs and which was launched from an inclined plane out of the window of our lodgings on the fourth floor at four o'clock in the morning, so as to avoid being seen.'

But neither brother had succeeded when Otto married and Gustav emigrated to Australia. When Gustav returned in 1886, both men were nearing forty. But each now had money and they embarked, once again, on attempts to fly. Otto, who was the leading spirit, realized by now that man was not strong enough to flap himself into the air. Sufficient power, in light enough form, might be available one day but meanwhile it was necessary to discover what actually happened to wings when they were airborne. To find out, Otto Lilienthal decided, it was necessary to launch oneself downhill into the wind.

The machines which the brothers built to achieve this were elementary hang-gliders consisting of two wings below which there ran a stout iron cross-bar. The pilot gripped the bar and, once airborne, was able to counter the varying wind force by swinging his legs backwards or forwards, to left or to right, and thus changing the centre of gravity. After experiments from the edge of a gravel pit, the Lilienthals had a 50 foot (15 metre) artificial hill built at the Reinersdorfer brickworks near Gross Lichterfelde outside Berlin. Otto, running down the slope, would become airborne and, before long, was covering as much as 100 yards (91 metres) in the air. The brothers then moved to the Rhinow Hills 45 miles (72 kilometres) west of Berlin. And here, in 1895, Otto was soon gliding a distance of 1,000 feet (305 metres) at times soaring above the height from which he had taken off. He now added a vertical tailplane to the glider and constructed another which had two wings, one above the other, instead of one. He also made another vital improvement. 'I am now engaged in building an apparatus in which the position of the wings can be changed during flight in such a way that the balance is not affected by changing the position of the centre of gravity of the body' he wrote. The apparatus was a head harness so adjusted that if the flyer lowered his head it would move a hinged elevator and make the machine rise. Raising the head would cause the glider to dip.

It may have been this early control that brought about the disaster on the morning of 9 August 1896, when Otto became airborne, flew for a while at a height of 50 feet (15 metres), then crashed to the ground. He died next day. His last words were reported to have been: 'Sacrifices must be made.'

Otto Lilienthal in
one of his gliders, 1896.

The first powered free flight at Kitty Hawk, North Carolina,
17 December, 1903; Orville Wright in the plane, Wilbur Wright on foot.

Orville Wright on 'Baby Wright', one of the Wright brothers' early aircraft.

The Lilienthals were important in the story of flight not only for their achievements but for the detail in which they recorded them. Otto Lilienthal's book *Der Vogelflug als Grundlage der Fliegekunst*, 1889 (*Birdflight as the Basis of Aviation*) enabled those who followed him to test and elaborate the theories he had formed. In addition, many of the Lilienthals' flights were photographed so that aeronauts and engineers all over the world were able to study what actually happened in flight.

Among them was Hiram Maxim, an American engineer who had for a while vied with Thomas Edison to light up New York with electricity. While contributing little to the conquest of the air, Maxim illustrates the inventive engineer's willingness to tackle any problem, and he did, in fact, produce a massive steam-powered machine which became airborne even though it failed to fly in the conventional sense of the word. Maxim had been chief engineer to the United States Electric Lighting Company, had emigrated to England and had invented his famous gun, then the most efficient of all automatic weapons, before turning to aircraft in 1889. His aim was not so much flight itself as 'to build a flying machine that would lift itself from the ground,' and he achieved this in 1894 with one of the oddest works of man that the aeronautical pioneers were to produce.

Maxim's machine had two main wings, two sidewings and two elevators, one fore and one aft, giving a total wing surface of 4,000 square feet (370 square metres). Power, transmitted to two propellers each nearly 18 feet (5 metres) long, was supplied by two 180 horsepower steam engines. The engines and their boilers weighed about 1,800 pounds (816 kilograms) and the total weight of the machine – a test-rig rather than an aircraft – was about 8,000 pounds (3,629 kilograms) with its three-man crew. This enormous structure was built in the grounds of Baldwyns Park, Kent, and in July 1894 was set in motion along the 1,800 feet (549 metres) of railway track laid down for it. From each side of the machine two outriggers carried additional wheels which, if the machine rose more than about two feet, came into contact with restraining guards.

There were two test runs. Then on 31 July Maxim raised pressure in the boilers to 320 pounds per square inch (22.06 bars). The propellers began turning and the machine sped along the tracks. It had covered about 600 feet (183 metres) when it became airborne, being prevented from rising further only by the guards. Then a part of one outrigger broke, Maxim shut off the power, and the story of one remarkable vehicle came to an end with a destructive crash. It had cost its designer and builder between £20,000 and £30,000.

The old saying that Everest explorers advance from the shoulders of their predecessors is never more true than in the story of aeronautical engineering. Cayley's early theories were used by the Lilienthals and the Lilienthals' efforts hurried along the Wright brothers to their appointment with destiny at Kitty Hawk. Wilbur and Orville Wright, bicycle makers and local newspaper publishers of Dayton, Ohio, had been fascinated by the possibilities of human flight since their father had given them a bamboo, cork and paper model of Pénaud's aircraft powered by an elastic band. But it was the news of Otto Lilienthal's death, coming into their office via the news agency service, which inspired them to carry on where Lilienthal had left off and seriously to study the problems of human flight.

'Those who tried to study the science of aerodynamics knew not what to believe', Wilbur Wright later said. 'Things which seemed reasonable were very often found to be untrue, and things which seemed unreasonable were sometimes true. Under this condition of affairs students were accustomed to pay little attention to things that they had not personally tested.' And on 30 May 1899, he wrote to the Smithsonian Institution in Washington saying 'I am about to begin a systematic study of the subject in preparation for practical work to which I expect to devote what time I can spare from my regular business'.

Wilbur Wright's letter came at just the best moment, at a time when a practical internal combustion engine was to provide the satisfactory power-to-weight ratio that made powered flight possible. This also led to the engines which made the motorcar a form of transport that carried on from where the earlier steam-driven vehicles had left off. Gottlieb Daimler's invention of the carburettor in 1887 enabled gasoline – petrol – to be used as a fuel, and was among the most important steps which led to the first horseless carriages. There were to follow the multiple improvements of the ensuing decades which capitalized on improvements in rubber technology, electricity and metallurgy, and which together gave via the motorcar a new mobility to huge numbers of the world's population.

It was in the air, however, that the petrol engine most strikingly made possible what had before been impossible by giving the first aviators the power to take flying machines, and themselves, up into the air and keep both airborne. However, that was only one of the problems that had been waiting. Once the Wrights began reading, with the help of the Smithsonian, they realized that a basic problem to be solved was that of keeping a machine stably balanced in the air despite the fluctuating wind eddies which passed across and around it. The Lilienthals had tried to do it by swinging their bodies backward or forward and thus changing the machine's centre of gravity. The Wrights, watching a group of pigeons, found an easier method. They noticed that the birds would at times oscillate their bodies, first tilting up the left wing and depressing the right, and then reversing the process. 'These lateral tiltings,' Wright stated in a patent action years later, 'first one way and then the other, were repeated four or five times very rapidly; so rapidly, in fact, as to indicate that some other force than gravity was at work. The method of drawing in one wing or the other, was, of course, dependent in principle on the action of gravity, but it seemed certain that these alternate tiltings of the pigeon were more rapid than gravity could cause, especially in view of the fact that we could not detect any drawing-in first of one wing and then of the other.'

The Wrights at first considered producing the effect by what were later known as ailerons – flaps on the rear, or trailing, edge of a wing which can be raised or lowered. In the Wrights' day it was considered too complicated a solution and they adopted, instead, a system which came to be called wing-warping, in which wires were attached to the extremities of the wings and to two sticks held by the aviator. By pulling or pushing on the sticks, the left and right extremities of the wings could be made to meet the air at differing angles.

The Wrights tested their wing-warping idea first with model gliders and then with a full-scale machine which they flew on the sandhills adjoining Kitty Hawk, a small isolated fishing community on the coast of North Carolina. 'These experiments', Wilbur Wright later testified, 'constituted the first instance in the history of the world that wings adjustable to different angles of incidence on the right and left sides had been used in attempting to control the balance of an aeroplane.'

Further experiments in the summer of 1901 were discouraging and the Wrights might have given up had it not been for Octave Chanute, who had witnessed their early flights and now invited Wilbur Wright to lecture in Chicago to the Western Society of Engineers. More important than the lecture itself was the work that preceded it: in order to check certain facts and figures, the Wrights built their own wind tunnel in Dayton. Six feet (1.8 metres) long and 16 inches (41 centimetres) square, it was used to check the aerodynamic characteristics of more than two hundred different models of wings.

During September and October 1902, the Wrights made more than seven hundred successful glides at Kitty Hawk in a full-scale machine. And before they returned home to Dayton they had made another discovery significant for the future of flight. This was the value of changing the immovable vertical tailplane which they had fitted to the rear of their machine into a tailplane that could be moved left or right as required, and linking the movement to the wing-warping mechanism. The results were remarkable. Wing-warping produced difficulties but these were automatically overcome if the tailplane was moved at the same time. The glider could, therefore, be directed left or right in a smooth turn. Thus it was not only the problem of balance that had been solved but also that of controlled directional flight.

So far, however, it was flight only, not powered flight. But the trials of 1902 took the Wright brothers across a watershed. They patented their wing-warping and tailplane devices. They began to see controlled powered flight as a possibility of the not-too-distant future. And in the Dayton workshop they began construction of the Flyer, an aircraft with a wing span of 40 feet (12 metres) – 3 feet (0.91 metres) more than that of the World War II Spitfire. It was powered by a four-cylinder petrol engine that could produce 16 horsepower and which, by the use of two chains, one of them crossed, turned in opposite directions two propellers mounted at the rear of the wings. The all-up weight of the engine, including water and fuel, was 200 pounds (91 kilograms) and of the Flyer, 700 pounds (317 kilograms). The engine, which had initially been used to pump air through the Wrights' wind tunnel, was primitive. The petrol was vaporized as it passed over the heated water jacket, and the only way in which the engine speed could be controlled was by adjusting the ignition timing.

During the last days of September 1903 the Wrights left Dayton for Kitty Hawk with the 1902 glider and the Flyer. The two machines had to be uncrated and assembled on arrival. There were delays, due to bad weather. When tests were made in an improvised hangar, some of the propeller shafting broke and had to be replaced. Then a number of successful glides were made with the 1902 machine.

Not until 15 December did Wilbur Wright lie in the Flyer and rev. up the engine. As helpers let go of the wing-tips the Flyer sped along its wooden launching rail and rose steeply into the air. It rose too steeply, developed what would now be called a stall, and plunged to the ground.

Luckily the damage to the machine could be made good in a few days and on the morning of 17 December Orville Wright was in position. At 10.30 am the aircraft sped down the launching rail. About 40 feet (12 metres) from point of release the Flyer rose up to a height of 10 feet (3 metres) and straight into a 25 mph (40 kph) wind. Twelve seconds later Orville eased the plane down. It had covered about 120 feet (37 metres) at a ground speed of about 10 mph (16 kph).

The Flyer was dragged back and used for the world's second powered flight by man, this time with Wilbur Wright at the controls. Both brothers made second flights. Duration and distance continued to increase, the fourth flight lasting for almost half a minute and covering nearly half a mile (800 metres). Then, while they were discussing the morning's events, a gust of wind turned the plane over and badly damaged it. They decided to call it a day.

What they had achieved was the first powered free flight. The phrase 'free flight' is important. 'A power-driven aerodyne' it was later to be laid down, 'is in free flight when, having left the ground, it is maintained in the air by its own power on a level or upward path, for a distance beyond that over which air forces could alone sustain it.'

The Wright brothers' success at Kitty Hawk in 1903 marked the end of one age and the start of another. Three years earlier, Wilbur Wright had written to Octave Chanute: 'For some years I have been afflicted with the belief that flight is possible to man. My disease has increased in severity and I feel that it will soon cost me an increased amount of money if not my life ... It is possible to fly without motors, but not without knowledge and skill. This I conceive to be fortunate, for man, by reason of his greater intellect, can more reasonably hope to equal birds in knowledge, than to equal nature in the perfection of her machinery.'

The Wrights had now proved that flight was 'possible to man'. But that was only the beginning. The following year they built an improved machine and in 1905 produced their famous Flyer III. They succeeded in flying a circular route and remained in the air for as much as half-an-hour. In Europe, Alberto Santos-Dumont, the Brazilian millionaire living in France whose airship had circled the Eiffel Tower with spectacular effect in 1901, made the first powered flights on the Continent in 1906. Before the end of the following year Gabriel Voisin had flown the first of his biplanes, each driven by a 50 horsepower Antoinette motor which turned a pusher propeller.

Despite these, and other European flights, it was only in 1908, when Orville Wright publicly flew one of the brothers' planes in Army trials at Fort Myer and Wilbur toured France, that the implications of flight began to be appreciated. The awakening was to be expected since by the time that Wilbur Wright had finished his tour at Auvours he had made more than a hundred flights, had stayed in the air for two hours at a time, and had carried more than sixty passengers.

All was now ready for 1909, the year when the future of powered flight became assured. At 4.41 on the morning of Sunday 25 July, Louis Blériot took off from Les Baraques, near Calais, in a small monoplane. Of 25.6 foot (7.8 metre) wingspan, carrying a 25 horsepower Anzani engine, the No. XI as it was called was laterally controlled by Wright-type wing-warping equipment while a fixed tailplane carried a rudder and two elevators. Thirty-seven minutes after take-off Blériot landed 23½ miles (37.6 kilometres) away on the fields below Dover Castle. Britain, it was ominously pointed out, had ceased to be an island.

Blériot's demonstration that the English Channel was no longer a barrier against invasion – a demonstration which won him the *Daily Mail* £1,000 prize for the first cross-Channel flight – was followed a month later by the world's first aviation meeting. Held a few miles north of Rheims it was attended by nearly forty different planes, of ten different makes. Lloyd George, the British Chancellor of the Exchequer and one of many politicians present, commented: 'Flying machines are no longer toys and dreams; they are an established fact.' Then he added, self-deprecatingly: 'I feel, as a Britisher, rather ashamed that we are so completely out of it.' Yet it was in 1909 that a significant move was made in England: the War Office's Balloon Equipment Store, in existence since 1878, was split into two wings, one becoming in succession the Farnborough Air Battalion, the Royal Flying Corps and the Royal Air Force, the second becoming H.M. Balloon Factory, the Army Aircraft Factory, and finally the Royal Aircraft Establishment (R.A.E), Farnborough, from which there were to come so many aeronautical innovations during the next half-century.

While Britain was beginning to note the military uses of aircraft, the United States was doing the same. Two years earlier President Theodore Roosevelt had seen a newspaper clipping about the Wright brothers and had passed it to an assistant with the one word: 'Investigate'. Now, in 1909, the United States bought its first warplane, a 36 foot (11 metre) wing-span Wright biplane with a speed of 44 mph (70 kph), a range of 75 miles (120 kilometres) and a ceiling of 3,000 feet (915 metres).

The crucial upsurge of interest in flight came only five years before the outbreak of World War I whose demands were drastically to increase the speed of aeronautical progress. First used mainly for observation – as balloons had been used during the siege of Paris nearly half a century previously – planes soon came under attack, from enemy aircraft and from the ground. This led to the development of fighter planes and to the need for quicker rates of climb, greater manoeuvrability, and higher speeds and operational ceilings. Simultaneously, a demand for bombing aircraft led to a fundamentally different kind of aircraft.

These requirements helped forge a new industry. At the start of the war only a handful of men had been employed in building British planes. By 1918 the number had increased to 350,000. In Germany, and after 1917 in the United States, there were comparable increases. In 1914 the rare hand-built products of the aircraft engineers were still something of a curiosity; five years later they were becoming comparatively commonplace. The practical ceiling had been raised from 7,000 feet (2,135 metres) to nearly 30,000 feet (9,150 metres). The engine weight per generated horsepower had been cut from 4 pounds to 2 (1.8 to 0.9 kilograms) while wing-loadings had been doubled from 4 pounds per square inch to 8.

But war did not only drive aeronautical engineers into producing aircraft that flew higher and faster, and were more manoeuvrable. It also thrust some of them into experimenting with new materials. Until 1914 most aircraft consisted very largely of wooden spars covered with canvas, reinforced to withstand the weight of the engine and the fuel tanks, but not very unlike the Flyer which had flown into history at Kitty Hawk in 1903.

One of the rare exceptions was Ponche and Primard's French Tubavion monoplane which in 1912 became the first all-metal aircraft to fly. But the man who really changed the situation was Hugo Junkers, a German engineer whose business success had earned him enough money for aeronautical research. In 1910 Junkers had patented a thick-section cantilever wing which, alone, would give enough lift for an aircraft. Five years later he incorporated it in his revolutionary J.1. The plane was revolutionary not only because it lacked the external bracing and wires of most sucessful aircraft. It was, sensationally, built entirely of iron and steel. Junkers later explained why he had built the J.1. despite the criticism of friends that he would ruin his business in the process. 'What I

A Bristol fighter F2B and a Sopwith Pup.

wanted, however, as in my other work,' he said, 'was to push into the unknown, to pioneer. I am fully aware that there the likelihood of success is less than along the conventional paths. But why should we not tackle problems that may hold tremendous possibilities? *Must* that be commercially and economically unsound? The criterion, surely, must be the way in which such a pioneering venture is carried out, no matter how extensive it be, or how nebulous the possibilities of success.'

The J.1., nicknamed the 'Tin Donkey', was the first of a series of all-metal planes which were to have an enormous effect on the ideas that aeronautical engineers believed to be practicable. At first the very weight of the planes made them suspect. Then, as steel was replaced with the lighter ribbed duralumin, and as the planes were rigorously tested in battle, it was realized that metal had come to stay.

During the first decade of peace, war-time progress in aircraft design and aeronautical engineering was redirected, with the growth of passenger services, into increased reliability and increased comfort. The Atlantic was flown, first in May 1919 with a stop at the Azores in the NC-4 flying boat, then the following month – the first non-stop flight – by Sir John Alcock and Sir Arthur Whitten-Brown in a Vickers Vimy. Air-to-ground radio, developed during the war mainly to aid the direction of artillery fire, became commonplace, while radio direction-finding made it easier to operate services between one country and the next. Mail was sent by air and by the mid-1920s it had been shown that while flying was still an adventure it was now also a means of civilian travel and communication. In the ten years that followed 1919, miles flown on scheduled routes (excluding those of Russia and China) increased from 1 million to 57 million, (1,600,000 to 91 million kilometres) and passengers carried from 5,000 to 434,000.

Popularization of flight during the two decades which followed World War I went hand-in-hand with numerous technical improvements. Prominent among them was the use of flaps or of slots to increase the efficiency of wings. In Germany Junkers invented his aerofoil wing-flaps. In Britain Sir Frederick Handley-Page a pioneer airman and builder of aircraft, introduced the slotted wing, in which the wing is pierced, near the leading edge, by one or more slots which slope back from the under surface to the upper. A single slot was found to increase lift by 60 per cent without increasing landing speed, while several slots could double lift.

Another important development during the period was the introduction of the variable-pitch propeller. Since the days of the Wright brothers it had been realized that if an engine was to be used to the maximum efficiency, the propeller which it drove should meet the air at one angle during take-off and at a different angle while it was cruising. It had been difficult enough to produce an efficient propeller for one set of conditions, and the problem of designing one whose pitch – the angle at which it met the air – could be varied while actually being driven, for long appeared insurmountable. British and German designers made attempts during World War I but it was not until the 1920s that Dr Hele-Shaw and T.E. Beacham patented a variable-pitch propeller. This was only the first step. A few years later, a constant-speed unit was built into the device so that it could be used to govern engine speeds. Similar propellers were developed to 'feather' the blades in the event of engine failure, thus minimizing drag and providing a useful safety factor. Later still, it was found possible to change propeller-pitch while landing: in other words, to use the propeller as a brake.

The corrugated metal Junker wings brought high drag and were thus aerodynamically inefficient. The disadvantage began to be overcome when another German designer, Dr Adolph Rohrback, started to build smooth metal wings supported by internal metal box constructions, thus allowing the main loads to be borne by the wing surfaces – the start of what came to be known as stressed-skin construction. In its early form the framework inside the wing took bending and shear while the skin transmitted torsion. Later the skin itself was built thick enough to support bending loads in the form of tension and compression. The method was subsequently used not only for metal but for fibreglass wings and, in the supersonic aircraft of the 1950s, for planes having wings of titanium alloys.

Another form of construction first developed between the wars was the geodetic structure used by the designer Barnes Wallis in the fuselages of Wellesley and Wellington bombers. It utilizes large numbers of comparatively short members pinned together where they cross, and pinned in such a way that the compression loads in any member are braced by tension loads in the crossing member. The 'structural redundancy' of such a construction has the advantage that many of the pinned joints can fail without endangering the aircraft – a fact which allowed many of the World War II bombers to survive when any other form of construction would have meant their destruction.

Another, and also revolutionary, method of aircraft building was used a few years later when the 'sandwich' process appeared in the Mosquito, a high-performance plane built almost entirely of wood. Its fuselage was made up of two plywood layers between which there was a layer of extremely light balsa wood, the three being bonded together with strong adhesive. The wings also had inner and outer skins bonded to spanwise spruce stringers.

As planes grew larger and heavier between the wars, and as they operated at higher altitudes, the aircraft industry had to cope with numerous ancillary problems. One was the pilot's difficulty in operating controls unaided. An answer was at first found in the development of servo mechanisms in which a small deflecting force operated by the pilot created a larger deflection in an opposite direction. Gearing was also employed and, towards the end of the 1930s, power-operated controls.

Civilian flight at high altitudes led to the need for pressurized cabins, while as performances were pushed up it became increasingly necessary to measure accurately the strains set up in various parts of an aircraft. Discovering what these were was made easier by the invention of the electrical strain gauge, a device which became one of the most useful auxiliaries of the inter-war years. In this a grid of fine resistance wire supported on a paper base is attached to the surface under test so that any strains in the surface are transferred to the wire. The electrical resistance of the gauge is proportional to the strain and a comparatively simple measurement of the resistance indicates the strain.

The 1920s witnessed not only a large number of small improvements in aircraft design, but the first successful experiments with the autogiro, later to be replaced by the superficially similar but, in fact, very different helicopter. The autogiro, the brainchild of Don Juan de la Cierva, was drawn through the air by a normal engine, but the lift was partly dependent on a rotor, mounted above the aircraft, whose wings were 'autorotated' by the aircraft's passage through the air. Rolling motion brought about by the difference in lift from the advancing and the retreating rotor blades was counteracted by an ingenious method of hingeing the blades. In the helicopter a powered rotor, or rotors, provides both lift and forward movement.

The 1920s also saw the arrival of the monstrous Dornier Do X, the culmination of a line that had begun with Voisin's float-glider flown off the Seine in 1905 and Glenn Curtiss's hydro-aeroplane of 1911. When the Dornier Do X made its maiden flight in 1929 it was by far the largest aircraft in the world, a giant which lifted 150 passengers, a crew of 10 – and nine stowaways. Powered first by twelve Siemens-built Bristol Jupiter air-cooled radial engines and later by Conqueror engines, the Do X had a chequered career but in 1930 and 1931 made a step by step flight from Friedrichshafen on Lake Constance to South America and then to New York. Two more Dornier Do Xs were built for Italy but the giant craft was never a success.

The development of mid-air refuelling and of retractable undercarriages were two other features of the 1920s and 1930s which became permanent items in aviation history. However, the period also witnessed one development which turned out to be abortive: the seemingly triumphant return of the airship. Only towards the end of the nineteenth century had the airship become practicable, largely due to the efforts of Alberto Santos-Dumont, who between 1898 and 1906

Sir Barnes Wallis, pioneer aeronautical engineer
and designer of the 'bouncing bomb' which destroyed the Ruhr dams in World War II.

had made no fewer than fourteen airships, which increased in capacity from 6,350 cubic feet to 71,000 cubic feet (180 to 2,010 cubic metres) and were powered by internal combustion engines of from 3 to 60 horsepower. Santos-Dumont has explained how his obsession had begun with ballooning in Brazil where, at the age of fifteen he had watched an early ascent. 'I, too, desired to go ballooning,' he wrote in *My Airships: the Story of my Life*. 'In the long, sun-bathed Brazilian afternoons, when the hum of insects, punctuated by the far-off cry of some bird, lulled me, I would lie in the shade of the verandah and gaze into the fair sky of Brazil, where the birds fly so high and soar with such ease on their great outstretched wings, where the clouds mount so gaily in the pure light of day, and you have only to raise your eyes to fall in love with space and freedom. So, musing on the exploration of the vast aerial ocean, I, too, devised airships and flying machines in my imagination.'

The most famous of Santos-Dumont's airships was No. VI, 110 feet (33 metres) long, and powered by a 12 horsepower car engine. On 19 October 1901, he took off in No. VI from St Cloud, on the outskirts of Paris, flew to the centre of the city, circled the Eiffel Tower and returned to St Cloud in 29 minutes, 31 seconds. The non-rigid airship, one in which an interior ballonnet could be filled with air to maintain the shape of the envelope despite loss of gas, had arrived.

Meanwhile, on the shores of Lake Constance, Count Ferdinand von Zeppelin, a retired German cavalry officer, had been building the first of his rigid airships. Zeppelin had made his first ascent, by balloon, at St Paul, Minnesota, after serving as an observer with the Northern Army of the Potomac. However, it was more than a decade later that he turned to aeronautical experiments, inspired by 'World Post and Aeronautics', a speech by the German Postmaster-General who urged the development of flight as a method of carrying mail. Zeppelin decided that the answer lay in the directable rigid airship. Passenger ships followed his experimental craft, the passengers being carried in a comfortable central cabin, and between 1910 and the outbreak of war in 1914 Zeppelins made 1,600 flights, completed 3,200 hours in the air and flew 90,000 miles (145,000 kilometres). More than 37,000 passengers had been carried. During World War I the long dark shapes of the Zeppelins hovered over Britain, making raids that did comparatively little damage but forced the British to divert from the Western Front men and planes that were in short supply. Zeppelins were vulnerable to attack by aircraft, and the threat forced a number of improvements on their makers, so that by 1918 their speed had increased to 80 mph (129 kph), their load to 50 tons and their ceiling to 20,000 feet (6,100 metres).

In 1919 the R.34, the latest in a series of airships which Britain had produced during the war, made a double crossing of the Atlantic, and although she broke up over Kingston-upon-Hull in north-east England two years later with the loss of forty lives, it seemed likely that the airship, reconstructed for luxury civilian travel, might be the air vehicle of the future. The view was reinforced in 1929 when the Graf Zeppelin flew round the world, covering 21,500 miles (34,600 kilometres) in twenty-one days, and travelling non-stop from Friedrichshafen to Tokyo.

Then two years later, the huge British R.101, on her maiden flight to Egypt and India, crashed near Beauvais, France. Nearly fifty of the crew and passengers were killed. Plans to complete her sister-ship, the R.100, were abandoned and although inflammable hydrogen was already being replaced as the lifting gas by the much safer helium, the airship appeared to have no future. The belief was strengthened seven years later when the giant Hindenburg burst into flames after hitting the mooring mast as she came into land at Lakehurst, New Jersey.

While the tragedy of the Hindenburg wiped out the hopes of the airship industry, there had already started in Britain a revolution which was to create a new engineering industry, help bring victory in the war already imminent – and make possible the huge expansion of civilian flying in the post-war years. This was the birth of radar, a system in which radio transmissions reflected back from aircraft in flight could pin-point the aircraft's position, its height and even its direction and speed of flight.

The radar industry, which was to draw on the expertise of wireless engineering, was a classic example of an invention sought for and discovered to fill a specific need. Radar might, indeed, have been developed more than a decade before since as early as 1922 Marconi told the American Institute of Radio Engineers: 'In some of my tests I have noticed the effects of reflection and deflection of these [electric] waves by metallic objects miles away. It seems to me that it should be possible to design apparatus by means of which a ship could radiate or project a divergent beam of these rays in any desired direction, which rays, if coming across a metallic object such as another steamer or ship, would be reflected back to a receiver screened from the local transmitter on the sending ship and thereby immediately reveal the presence and bearing of ships, even though these ships be unprovided with any kind of radio.'

Seven years later Edward (later Sir Edward) Appleton, investigating the height of the wave-reflecting ionosphere above the earth, discovered that short bursts of radio energy, echoed back from the ionosphere, could be visually recorded on a cathode ray oscilloscope, an instrument which gives a visual image of one or more varying electrical quantities. Neither Marconi's nor Appleton's observations were put to practical use. The comments by Post Office engineers that radio signals fluttered when a plane passed nearby were similarly ignored. The same thing happened when two Army engineers proposed a primitive method of locating ships with the help of reflected radio signals.

Only in 1935 did Robert (later Sir Robert) Watson-Watt of Britain's Radio Research Board set in motion the events which were within two decades to result in the ubiquitous aerials and bowl-shaped transmitting and receiving equipment of modern radar. In the early 1930s there was no method by which the island of Britain could be effectively warned of air attack from the Continent. Watson-Watt, therefore, asked his assistant A.F. Wilkins to find out what power would be required to obtain a detectable signal from an aircraft at such and such a range. 'I then did two things', Wilkins has said, 'I assumed that the aircraft would have the re-radiating properties of a half-wave aerial, and I took it that the plane would measure about 25 metres horizontally and 3½ metres vertically.'

Wilkins' computations showed that it was theoretically possible to locate an aircraft by means of the radio waves bounced off it. Within months theory was being transformed into practice as experimental transmitters and receivers were set up first at Orfordness, in Suffolk, then at other places around the coast. From the start it had been comparatively simple to find the distance of the aircraft which was reflecting the transmitted bursts of radio energy: the speed of radio waves was known and the time taken for their journey out to a plane and back from it, would give the plane's distance without trouble. It was then found possible to traverse the transmitted beam left and right, a process from which the aircraft's bearing could be determined. Finally, by traversing vertically the aircraft's height could be discovered.

Germany, the United States and Japan had all been experimenting with radio detection to a lesser or greater extent, during these immediate pre-war years. While the British had an early warning system covering south-east England in operation by the time that Germany started the war by invading Poland in September 1939, Germany herself had her own radar defences covering parts of north Germany. The difference between the two countries was that under the proddings of the scientist, Sir Henry Tizard, the British warning system had been so closely tied into Royal Air Force operations that the maximum use could be made of it.

During World War I, flying higher than the enemy had been an aid to survival. But as planes reached ever greater heights the efficiency of their engines decreased. One reason was that the 'thinness' of the air made it more difficult for propellers to do their job. Another was that the rarified air reduced the efficiency of the internal combustion engine. The second disadvantage could be lessened by fitting a supercharger, but this involved additional cost, complications and, above all, weight.

One solution, some engineers believed, might be to make use of a principle that had been known to the ancient Greeks. Air would be drawn into a chamber and compressed; fuel would then be injected into the chamber. The mixture of fuel and air would be fired and ejected through the rear of the chamber. The process would be independent of the density of the surrounding air and the ejected gases would, it was at first proposed, be used to drive a propeller. Dr A.A. Griffith of the Royal Aircraft Establishment at Farnborough suggested the idea in 1920, Charles Guillaume patented a design for a jet engine in France the following year, and shortly afterwards German firms were experimenting along similar lines.

More than a century earlier Sir George Cayley had written of aircraft engines that 'as lightness is of so much value there is the probability of using the expansion of air by the sudden combustion of powders or fluids.' But for two decades after the Wright brothers had first flown, one thing had held back development on these lines – the inability of any metal to operate under the extremes of temperature involved in jet propulsion. It remained a stumbling block even after Griffith had put forward more detailed plans in 1926. A test rig was set up at R.A.E. Farnborough but results were inconclusive.

Only in the 1930s, as new nickel-chrome alloys began to be available, were plans for a significantly different kind of jet engine pushed forward by a young Royal Air Force engineer, Frank (later Sir Frank) Whittle. 'Reciprocating engines are exhausted', he maintained. 'They have hundreds of parts jerking to and fro, and they cannot be made more powerful without becoming too complicated. The engine of the future must produce 2,000 horsepower with one moving part: a spinning turbine and compressor.'

Whittle's 'real jet' dispensed with the propeller of the earlier jet idea and did its job entirely by using the forward thrust produced by the ejected gases – an illustration of Newton's law of equal and opposite reactions. The Germans, working on the same idea, flew the world's first such jet in August 1939, almost as World War II broke out, and for the next few years British and Germans ran a neck-and-neck race to put jet aircraft into squadron service. In German-occupied

France the Germans in 1940 adapted the jet for another purpose: that of assisted take-off for their bombers from runways which were shorter than needed. The British in May 1941 flew Whittle's jet in the experimental Gloster E.28/29 and in 1942 the Germans put into the air the Messerschmitt 262 whose twin wing-mounted jets gave it a ceiling of 40,000 feet (12,200 metres) and a top speed at 20,000 feet (6,100 metres) of 540 mph (869 kph). The following year the Gloster Meteor went into production and eventually reached squadron service a short head in front of the Me.262.

With the jet, as with most other technological jumps forward, national frontiers became virtually non-existent. From the Meteors, shipped to the United States while the war was still being fought, there was developed the Lockheed Shooting Star. The Russians produced their own MiG.15. Before the end of the 1940s aircraft engineers were already entering 'the jet age' in which civilian airliners, led by the Vickers Viscount, powered by four turbo-prop Dart engines, were carrying passengers from continent to continent by the tens of thousand. Significantly, these planes flew at heights not reached before, far above the worst of the weather thus making post-war air travel as different from pre-war as a luxury liner is different from the *Mayflower* in which the Pilgrim Fathers sought refuge in America.

During World War II research had, as in the war of 1914-18, been concentrated on improving the performance of aircraft in battle. Greater speed, manoeuvrability and operating height had been all-important as well as, for bombers, the need to carry the heaviest bomb-load the maximum distance without affecting performance more than was inevitable.

These qualities were not entirely lost sight of after 1945, and while the jet was at first developed primarily for a new level of civilian travel, the military possibilities were not ignored. One of their applications was to be that Holy Grail of aircraft designers, vertical take-off.

Man had so far been airborne with the aid either of lighter-than-air craft or heavier-than-air machines which gained their lift by movement through the air or, in the case of the helicopter, by movement of the horizontally swirling rotors. But if the thrust of a jet engine was directed vertically downwards, and if this thrust were more than the weight of the craft on which it was mounted, then surely the craft would rise vertically in the air? To answer the question Rolls-Royce engineers designed 'the Flying Bedstead', more formally the Rolls-Royce Vertical Test Rig. It incorporated two Nene engines whose jets were directed vertically downwards, one being bifurcated to prevent a turning movement of the rig. The combined thrust was 7,700 pounds (3,500 kilograms) and since the all-up weight of the rig was only 7,200 pounds (3,266 kilograms) the rig rose into the air when full power was developed. Equilibrium in the air was, as expected, a major problem in the early experiments, but this was finally solved by the use of four auxiliary control jets, fore and aft and at either side.

Once the practicability of jet-powered VTOL (Vertical Take Off and Landing) planes had been demonstrated, development went ahead quickly. The ability to take off from or land on a patch the size of a small field, and a patch

A Sea Harrier FRS1 'jump-jet', built by
British Aerospace, landing on HMS *Hermes* during trials in November 1979.

which did not require a concrete runway, gave Service aircraft an immense advantage. The ear-piercing blast of the vertical jets, which limited civilian use, was hardly relevant here.

Three different VTOL systems were devised within the next few years. The first, and simplest, was that in the Hawker Kestrel and later in the Harrier. In these, a Bristol Siddeley Pegasus engine produced a jet stream that was led into four rotatable nozzles, which point downwards as the plane takes off vertically and are then progressively rotated to horizontal positions to give the plane forward flight. The French developed a more complicated system to power the French Dassault-Sud Balzac. This had eight Rolls-Royce lightweight RB.108 engines to provide vertical lift. When height had been gained a Bristol Siddeley Orpheus engine was switched on to give forward movement and the Rolls-Royce engines were shut off. The German EWR VJ 101C used four Rolls-Royce RB.145 engines mounted vertically in pairs at the wing-tips, and a third pair mounted behind the pilot's seat. All six engines were used for vertical take-off and the wing-tip engines were then revolved to a horizontal position as the fuselage engines were shut down.

Vertical take-off, developed in a variety of ways, was to be equalled in importance in the post-war years by the innovation of variable geometry wings whose angle to the aircraft could be changed to make the best of conditions at take-off and landing, and in high-speed flight, and by the solution of the problems involved in ultrasonic flight.

Soon after the end of the war planes began to approach the speed of sound in level flight – about 750 mph (1,200 kph) at 5,000 feet (1,525 metres). During the war supersonic speeds had occasionally been reached by planes in a steep dive. Within eight months of peace, jet planes raised the speed on a 62 mile (100 kilometre) closed circuit from 497 to 605 mph (800 to 974 kph) and in October 1947, the American Bell X-1, a rocket-propelled research aircraft, became the first manned aircraft to exceed the speed of sound in level flight, reaching Mach 1.06 or 700 mph (1,120 kph). In roughly six years, that speed was almost doubled.

Supersonic flight had presented two main problems. One concerned the high temperatures produced on the aircraft's structure, a problem eventually solved by the use of special duty metals. The second was the need to deal with the shock-waves which began to form on the plane's wings and fuselage as the speed of sound was approached. In subsonic flight the shock-waves were, in effect, thrust ahead of the plane; but as the sound barrier was reached the plane caught up with them before they dispersed, thus greatly increasing the drag involved and inhibiting the plane's performance. The answer here was found to lie in the delta-shaped wing, a response to the problem that has already drastically altered the shape of some aircraft and is likely to alter more in the coming years.

The world's first supersonic bomber was the delta-wing Hustler, built by General Dynamics, flying in the late 1950s. But competition centred on supersonic civilian airliners for the next ten years. The Russians' Tupolev 14 was, on 31 December 1968, the first to fly but its performance, marred by a disaster at the Paris Air Exhibition, was overshadowed by that of Concorde.

A view from the traffic controller's
window in the modern international airport, Schipol, Amsterdam.

The first investigations into the potential plane began in 1956 when aircraft engineers from private British companies and from the Royal Aircraft Establishment at Farnborough began to discuss the possibilities of a supersonic passenger plane. One technical problem arose almost at once. If aluminium alloy were used, the plane would have to fly at no more than 1,450 mph (2,330 kilometres), since at higher speeds heat caused by friction with the air would result in unacceptable weakening of the fuselage. Stainless steel and titanium appeared to be the only alternatives. But the cost would be so high that a partnership with another country would have to be arranged. Thus it came about that Britain and France agreed to build the plane together. The contract was signed in 1962, but two years later it was announced that the initial plans for the aircraft had been changed. Its wing-span was increased, as was the engine power, and the number of potential passengers was increased to 118. The first two aircraft flew at Toulouse, France and Filton, England, in March and April 1969 respectively, powered by four Rolls-Royce Olympus turbojet engines, and with a maximum cruising speed at Mach 2.2 or 1,450 mph. The plane has now for years been flying successfully on the Europe-New York run as well as to South America.

The problems of supersonic flight, as exemplified by Concorde. are no longer technical problems but those of economics and of environmental pollution. Cost of design and construction, as well as of fuel, severely limit the use to which such airliners can be put. So does the sonic boom whose great nuisance value – a phrase which many would claim to be a gross understatement – limits its use over densely populated areas.

Overcoming these difficulties still remains a task. Others are provided by the ever-larger 'jumbo jets' carrying their hundreds of passengers. Here the problems are not so much those of aeronautical engineering as of 'processing' the army of passengers and their luggage on arrival at a busy airport. As far as aeronautical engineering is concerned, the successful operation of supersonic jets appears to have brought the profession to the end of one phase of its history. What follows Concorde lies in the speculative future.

BURROWING THROUGH AND BUILDING UP

Since the start of recorded history – and almost certainly before it – tunnellers have belonged to a proud race of men doing skilled and dangerous jobs. Through the years the dangers have hardly changed, and although they are minimized by modern safety precautions tunnellers are still at risk from rock falls and the explosions frequently needed to blast a way forward; from fire, fumes, and the commonest of all tunnelling hazards, the huge volumes of water that can pour into the working areas from the most unexpected of sources. 'Every man seemed to possess the miraculous power of Moses, for whenever a rock was struck, water sprang out of it', said the *Manchester Guardian* of one group of tunnellers working in the 1890s.

The most obvious reason for spending the large sums that tunnels have always claimed is to ease communications. Canals, railways and roads have in turn through the centuries enabled men to take shorter routes by moving through hills and mountains rather than taking a longer route round them. The development of hydro-electric power stations has demanded the building of tunnels both to help collect water behind hydro dams and to carry it from the dam to the electricity-generating works. Water supply has itself used tunnels; so have drainage and sewage disposal schemes. The building of underground caverns for storage or the enlargement of existing ones – for the safe housing of valuables in war time – has exercised the tunnellers' art and craft, while the extraction of minerals by mining has done much the same, although mining and tunnelling have required rather different skills and techniques despite certain similarities.

Tunnels can very roughly be divided into those which are made through hard ground and those which are made through soft. 'Hard' ground can itself range from granite to certain kinds of sandstone while soft ground covers a wide range of possibilities, each of which has to be tackled on its merits. Whatever the material through which the tunnellers have to drive a way, and whatever method of lining the tunnel is used, the end-product is frequently of circular cross-section. In theory, if the vertical pressures are greater than the horizontal pressures, then a tunnel should be wider than it is high; if the horizontal pressures are greater, then it should be higher than it is wide. In practice, and as illustrated by the underground railways with which most people are familiar, tunnels are very frequently circular in section.

Few details are known about the tunnels of ancient times while the oldest to have been recorded was, tantalizingly, built by a monarch, Queen Semiramis, whose very existence has been called into question. This was the pedestrian tunnel built about 2170 BC below the 600 foot (183 metres) wide Euphrates to link the Royal Palace in Babylon with the Temple of Jupiter on the opposite bank. The tunnel was mentioned by Diodorus Siculus, the Sicilian historian who lived during the first decade of the Christian era. The tunnel may have been built after the main stream of the Euphrates had been diverted; however, it was certainly a considerable achievement and more than another 4,000 years were to pass before the next tunnel was built on a river bed – the Thames tunnel which Marc Isambard Brunel constructed between Wapping and Rotherhithe in the first half of the nineteenth century.

Eupalinos' water supply tunnel on Samos, built in the sixth century BC, was at the time considered one of the wonders of the Greek world. More than 3,000 feet (915 metres) long, it had a section of about 19 square feet (1.76 square metres) and a trench in its floor which carried clay water pipes and incorporated inspection pits. Despite these refinements the line of the tunnel was badly sited. Started simultaneously from either side of the hill which it pierced, it was found when nearing completion to be 16 feet (4.8 metres) out of true at its planned junction, a mistake rectified by a right-angle bend.

The Romans, as already noted, built numerous tunnels in the construction of their aqueducts, and in some cases built them also to carry their roads. Among the latter is the 3,000 foot (914 metre) long tunnel taking the road from Naples to Pozzuoli through Posilipo Hill. It is cut through volcanic 'tufa', a calcium deposit, but many of the Roman tunnels penetrated hard rock and were laboriously constructed by one man holding a chisel to the rock face while one or more colleagues sledge-hammered the head of the chisel and eventually cut out a block.

One of the most famous Roman tunnels was the 3½ mile (5.6 kilometre) length excavated during the reign of the Emperor Claudius about 50 AD to drain Lake Fucino in the nearby Liris Valley. Described by Pliny, Tacitus, and other Roman writers, it was begun simultaneously from forty shafts bored down through Monte Salviano to the line which the tunnel was to take. Due to earth movements, it had to be repaired at least twice during succeeding centuries and eventually fell into disuse. However, when it was decided to reopen it during the middle of the nineteenth century some sections, by this time 1,800 years old, were successfully incorporated in the new scheme.

In siege warfare, tunnels were used long before Roman times though certainly exploited during them. A tunnel carefully lined with timber would be dug, under concealed cover if possible, until it stretched beneath the walls of the fortress being besieged. After the excavators had withdrawn the timber support would be set on fire and the tunnel would collapse, bringing down with it the undermined fortifications.

However, this stratagem could sometimes be dealt with, as is described by Vitruvius in Book X during his account of the siege of Apollonia in 214 BC.

'Again, when Apollonia was besieged, and the enemy designed by digging tunnels to penetrate unsuspected within the walls', he wrote, 'this was reported by spies to the citizens of Apollonia. They were panic-stricken at the news and their spirits failed them in their lack of resource. For they did not know the time or the place for certain where the enemy were likely to emerge.

'But at that time Trypho of Alexandria was the architect in charge. Within the walls he planned tunnels and, removing the soil, advanced beyond the wall to a distance of a bowshot. Everywhere [along the tunnel] he hung bronze vessels. Hence in one excavation which was over against the tunnel of the enemy, the hanging vases began to vibrate in response to the blows of iron tools. Hereby it was perceived in what quarter their adversaries purposed to make an entrance with their tunnel. On learning the direction, he filled the bronze buckets with boiling water and pitch overhead where the enemy were, along with human dung and sand roasted to a fiery heat. Then in the night he pierced many openings, and suddenly flooding them, killed all the enemy who were at work there.'

The use of tunnelling for military purposes continued after the disintegration of the Roman Empire, and thoughout much of the Middle Ages it was the sapping of castle walls for which tunnelling engineers were mostly employed. Some of the few exceptions were in central Europe where mining for minerals involved the same laborious work with hammers, wedges and drills.

Then came gunpowder, the first innovation in a series which was to change tunnelling as much as it changed warfare. The difference that explosives were to make to the profession can be gauged from Gösta E. Sandström's verdict in *The History of Tunnelling* 'An underground working advanced with hammer, wedge and bar is a healthy place of work', he says. 'The temperature is uniformly pleasant, there is ample room at the face for men to operate. The dust raised by gad or pick is not enough to matter, and the air is clean. With some exceptions the working postures are natural and do not induce deformities of back or limbs. It is quiet work, during their six-hour shifts. It was also skilled work; the Saxon miner was a recognised member of a mining society at its technical peak.'

The situation was changed first when the burning of wood to heat rock and its subsequent douching with cold water began to replace the hammer and chisel and then, in the mid-seventeenth century, by the use of gunpowder. At first the practice was to drill a hole, charge it with about 2 pounds (0.9 kilograms) of gunpowder, plug the hole with wood in which a gap was left to light the charge, wedge the wood tight with a piece of metal, and then light the match. The hazardous procedure was slightly improved by the use of clay tamping of the charge, introduced towards the end of the seventeenth century, but for more than another century opening up tunnels by explosives was still dangerous.

Their use became practicable just as the European canal age was starting and clay tampons were applied when it was found necessary to blast a 515 foot (157 metres) tunnel on the Languedoc Canal, the success of which encouraged the Duke of Bridgewater to start his own canal-building operations. Opened in 1681, the tunnel, situated at Malpas, a few miles from Béziers, was 22 feet (6.7 metres) wide and 27 feet (8 metres) high. It may well have influenced the Duke who introduced tunnels at both ends of the Bridgewater Canal; and, via the Duke, it may have had its effect on Brindley who, it will be remembered, not only incorporated five tunnels in his Grand Trunk Canal but burrowed 2,888 yards (2,641 metres) through the Harecastle Ridge. Here the traditional enemy of the tunneller – water – was channelled by Brindley to fill the summit level of his canal. But windmills, watermills and eventually a Newcomen steam engine were needed to keep the workings open.

For the rest of the eighteenth century the canal builders were the main employers of the tunnellers, to be followed in the first half of the nineteenth by the railway engineers. Methods of using explosives improved and so did those of lining the tunnels after they had been built. When this was necessary, brick had at first been used. Then with cast iron, it was found quicker, and more economic, to bring in, behind the men working at the face, cast-iron segments which could be put in position and then joined up to produce the circular – or oval-shaped linings similar to those of contemporary underground railway tunnels.

What had not been built by the start of the nineteenth century was any successful under-river tunnel. The first major attempt was started in 1807 by Richard Trevithick who had turned aside from his railway work to begin boring beneath the Thames at London. The plan was first to drive a pilot tunnel from the south bank, only 5 feet (1.5 metres) high and from 3 to 2 feet (0.91 to 0.61 metres) wide; when completed, it would be enlarged to 16 feet by 16 feet (5 by 5 metres). At first all went well. The pilot tunnel was satisfactorily lined with timber, and pumps dealt adequately with the water that seeped in. After twelve months the pilot tunnel was 950 yards (869 metres) long and the north bank of the river was less than 50 yards (46 metres) away.

Then the river broke in. Trevithick and four companions struggled back through the tunnel with the water swirling round them and barely escaped with their lives. Within a few days he had located in the Thames the exact spot where the water was coming through and had sealed it with bags of clay. Gravel on top of the clay consolidated the seal and the flooded tunnel was pumped out. Within a week the work was again being pushed forward.

Once more the Thames burst in. Once more, everyone was lucky to escape. But this time the authorities refused to allow the hole to be plugged on the grounds that this would interfere with navigation. Trevithick's plan to carry on by using a series of coffer dams was turned down and the project was abandoned. It remained so for more than a decade, and then revived and brought to a successful engineering but disastrous financial conclusion by Marc Isambard Brunel.

By now the Napoleonic wars were over. Nearly 4,000 passengers were being rowed every day across the river between Wapping and Rotherhithe, while vehicles travelling between the Essex and the Kentish shores had to make a long detour westwards to cross London Bridge. The need for a tunnel was growing. The man for the job was Brunel, now financially recovered after the end of the war with France when the Government had refused to pay for the 80,000 Army boots he had made.

While working in Chatham Dockyard, Brunel had picked up a length of naval timber being attacked by the destructive 'ship-worm', the *Teredo navalis*, a mollusc which gnawed its way through the toughest wood and was the scourge of shipbuilders, both naval and civilian. Brunel noticed that while the creature was at work it was protected by two strong shell-plates and was thus able to gnaw or bore its way forward while its flanks were guarded. Why, Brunel asked himself, should it not be possible to adapt the principle to human tunnelling operations? By 1818 he had produced and patented plans for 'a casing or cell to be forced forward before the timbering which is generally employed to secure the work.'

Brunel's daring plan for a Thames tunnel was to utilize a new material patented only the previous year and to experiment with a new method of construction, using his skill, which eventually spread throughout the world. The new material was Portland cement, in the use of which Brunel was to be a pioneer. When the London sewerage system began to be built, Portland cement was used in quantity. Slower-setting and more expensive than earlier cements, it was stronger and before the end of the century nearly 3 million barrels a year were being exported to the United States. Then exports declined as the Americans began to make their own.

In the Thames tunnel project Brunel embedded into the cement ties of hemp, reed, laths of fir and birch, as well as hoops of iron to give additional strength. In a yard on the south side of the Thames, he built without centring, a 60 foot (18 metre) long half-arch of cement counterbalanced by 62,700 pound (28,400 kilogram) weights and reinforced first by fir ties and hoop iron. 'To many', wrote Brunel's first biographer, Richard Beamish in *Memoir of the Life of Sir Marc Isambard Brunel*, 'the tunnel itself was not a greater object of interest than was this remarkable structure; and though its application, in its entirety, was never adopted in practice, yet the introduction of hoop iron bonds is now common to a variety of engineering and architectural structures.'

Brunel's experiment of using tensile reinforcement in concrete or masonry reconstruction was slow to be followed up. In 1867 Joseph Monier in France patented a process for reinforcing garden pots and tubs with iron mesh. Another eleven years passed before Thaddeus Hyatt took out the first important patent for the process in the United States and not until 1889 was the first reinforced concrete bridge, the Alford Bridge, built in the United States.

Brunel's main innovation was the Brunel shield, an 80 ton cast iron structure consisting of twelve separate frames, each housing three 3 feet by 6 feet (0.91 by 1.82 metre) cells one above the other. The shield was divided vertically into three parts, each of which could be moved forward separately by large screw jacks pressing on the tunnel lining erected behind the shield. The rubble hewn out by the thirty-six workers was passed back to a working stage at the rear of the shield – much as the *Teredo navalis* passed wood back through its body. The shield was also used by the masons lining the tunnel walls.

Brunel had so impressed the Institution of Civil Engineers with his plans that

Men at work in the Shield while excavating
the tunnel for the Great Northern and City Tube Railway, London, in 1901.

in 1824 a new company was formed to carry out the project Trevithick had abandoned more than a decade earlier. Brunel was retained as engineer at £1,000 a year for the job which was expected to take three years, was awarded £5,000 for the patent covering his shield and promised a bonus of £5,000 when the tunnel was completed.

Work started on 2 March 1825 with the construction on the Rotherhithe shore of a 42 foot (12.8 metre) high brick tower, set on an iron curb which was sunk a few inches each day by excavation from within the tower. By August the excavation was 50 feet (15 metres) deep and before the end of November the shield was assembled at its foot. Geologists had forecast that Brunel would meet only strong blue clay under the river, and with this prediction in mind he had planned a double-road tunnel 37 feet 6 inches by 22 feet 4 inches (11.4 by 6.8 metres) to be driven 14 feet (4 metres) below the river bed.

At first all went well. But when excavating below the Rotherhithe tower Brunel had found pockets of gravel which made him anxious about what he would meet under the river. As he had feared, there was gravel here too and in January 1826, then again in February, water broke through the roof of the tunnel, the situation being saved only when bags of clay were dumped into the river above the tunnel's line of advance.

Brunel became increasingly uneasy. As early as 10 February 1826, he had written in his diary, 'went very early to the Tunnel for the purpose of giving directions to prop up the back of the staves, which, for want of weight at the new shaft, might be overbalanced by the pressure of the ground at the back. I could not rest a moment until it was done, for the consequences might have been fatal, at this moment in particular. What incessant attention and anxiety! To be at the mercy of ignorance and carelessness! No work like this.'

The tunnel had now become the wonder of the country and, against Brunel's advice, the company ordered that sightseers should be admitted at a shilling a head. By April seven hundred visitors a day were being taken down the Rotherhithe entrance and through the lined portion of the tunnel. As if to emphasize the success of the enterprise a concert was held below the Thames.

By mid-May he was growing seriously worried and on the 13th wrote: 'During the preceding night the whole of the ground over our heads must have been in movement, and that, too, at high water. The shield must, therefore, have supported upwards of six hundred tons! It has walked many weeks with that weight, twice a day, over its head! Notwithstanding every prudence on our part a disaster may still occur. May it not be when the arch is full of visitors.'

Three days later the tunnel was visited by Lady Raffles and a large party. 'I attended [her] to the frames, most uneasy all the while as if I had a presentiment. . .' Brunel wrote.

A few hours after the party had left, a portion of the roof gave way and a torrent of water and mud swept into the tunnel, pushing everything before it. William Gravett, one of the assistant engineers, later gave a vivid account of the scene. 'We stayed some time below on the stairs, looking where the water was coming in most magnificently', he said. 'We could see the farthest light in the west arch. The water came upon us so slowly that I walked backwards speaking to Brunel several times. Presently I saw the water pouring in from the east to the west arch through the cross arches. I then ran and got up the stairs with Brunel and Beamish, who were then five or six steps up. It was then we heard a tremendous burst. The cabin had burst and all the lights went out at once. There was a noise at the staircase, and presently the water carried away the lower flight of stairs. Brunel looked towards the men, who were lining the staircase and galleries of the shaft, gazing at the spectacle, and said, "Carry on, carry on as fast as you can!" Upon which they ascended pretty fast. I went to the top and saw the shaft filling. I looked about and saw a man in the water like a rat. He got hold of a bar, but I afterwards saw he was quite spent. I was looking about how to get down when I saw Brunel descending by a rope to his assistance. I got hold of one of the iron ties, and slid down into the water hand over hand with a small rope, and tried to make it fast round his middle, while Brunel was doing the same. Having done it he called out "Haul up". The man was hauled up.' Incredibly enough, when a roll call was held it was found that all the men had succeeded in scrambling to safety. The next day the Rotherhithe vicar described the accident in his sermon as 'a just judgment upon the presumptuous aspirations of mortal men.' Brunel merely noted in his diary: 'The poor man.'

Once more, hundreds of bags of clay were dumped to plug the hole. But it was November before the tunnel was drained and the damage repaired – largely due to the efforts of Marc's son Isambard Kingdom Brunel. To celebrate the resumption of the work, a dinner for fifty special guests, and for 120 of the workers, was held in the tunnel with music provided by the band of the Coldstream Guards.

But Brunel remained unhappy and early in 1827 admitted: 'Every morning I say, Another day of danger over!' The company directors, he regretted, insisted on paying the men piece-rates, which he feared would cause accidents by over-haste. 'None, I feel confident, would occur if all idea of piece-work were abandoned', he wrote. 'It always operates as a stimulant, a very dangerous one. Obliged to drive on, on account of expense, we run imminent risks indeed for it. That a work of this nature, under such circumstances, should be thus carried on, is truly lamentable.'

Only a few months were to pass before the earlier disaster was repeated, with the loss of six men drowned. No one could be found to support the company, whose funds were by now insufficient to carry on, and the project was abandoned.

Seven years later, the company was granted a Government loan of £246,000, the old shield was replaced by a new and stronger one weighing 140 tons, and work was begun once more. This time Brunel was successful. But it was not until 12 August 1841 – more than sixteen years after the project had started – that the shield reached the north shore of the Thames. The tunnel was officially opened on 25 March, 1843 and within twenty-four hours more than 50,000 people had used it. But in completing the work the company had virtually run out of money, and was able to pay Brunel only £1,700 of the promised £5,000 bonus.

Despite the improvements in engineering equipment made in the following decades, the difficulties and dangers of subaqueous tunnelling remained, as was shown in the 1870s when the constructors of the Severn Tunnel faced what at times seemed to be insurmountable problems. Travellers on Brunel's Great Western Railway between London and South Wales were faced at the Severn with making a long detour north through Gloucester or taking the time-wasting ferry across the estuary. A viaduct had been proposed but abandoned; abandoned, too, was a scheme for a tunnel put forward in 1865. It was only in 1872 that a Severn Tunnel Act was passed and only in March 1873 that work was begun.

A banquet held early in the
nineteenth century in the tunnel being excavated beneath the Thames.

The length of the tunnel was just over 4½ miles (7 kilometres), but this included long approaches and only half of the tunnel was actually below the estuary. However, the first near-disaster came not from the estuary but from faults in the ground to the west. By mid-October 1879, the headings being driven from either shore were within 130 yards (119 metres) of meeting each other. Then, on the 16th, fissures above the tunnel being driven through the west bank let in a torrent of water. Further inland, 5 miles (8 kilometres) of a local river dried up, as did many wells.

Luckily, no lives were lost. But it was clear that emergency equipment would be needed to pump out the flooded workings, and special pumps were brought from Cornwall. But first one pump failed, then the second did the same. When they were at last in operation again it was found that to complete the job of clearance a steel door in a section still flooded had to be closed. A diver named Lambert made three attempts, the last in a self-contained diving suit, to close the door and turn one vital valve. Only later was it found that the valve was fitted with a left-hand thread and had, in fact, been closed from the start. Lambert's dangerous task had been followed by his opening rather than closing it.

Walls were built to deal with the flood from the fissures, and the depth of the tunnel was lowered. But in April 1881 water broke in from the estuary and was stopped only when bags of clay were used to plug the hole – as they had been used by Brunel half a century earlier. Then, in October, water began pouring in through the faults which had caused trouble in 1879. Once more, it was Lambert who came to the rescue and shut the metal safety door.

These trials and tribulations were to be compounded by a stroke of ill fortune from a fresh direction. The Severn is noted for its exceptionally high tides, and during the spring they can rise 32 feet (9.7 metres) to produce the famous Severn Bore. But now, when the main dangers appeared to have been dealt with, there came a tide 10 feet (3 metres) higher than the highest on record. 'Tidal wave' was the local description of the huge wall of water which inundated the Welsh shore, swept through pumping stations and put out the boiler fires. Once again, emergency machines were called for. They successfully dealt with the flooding, and earth barriers were built up to guard against a second similar disaster.

Eventually, on 5 September 1885, the first train steamed through the tunnel; but another fifteen months passed before fresh pumping and ventilation equipment was installed and the tunnel was opened to regular traffic.

While British engineers were grappling with the problems of driving tunnels beneath rivers, their contemporaries on the Continent were showing that they could follow up their earlier success of driving a tunnel through the Alps. The Semmering Tunnel was towards the eastern extremity of the range where heights were much less than elsewhere; the tunnel itself was only a mile long and its building had presented none of the difficulties which faced engineers trying to pierce the higher parts of the range. Whichever area of the Central Alps was to be pierced, a tunnel would have to be 7 or 8 miles (11 or 13 kilometres) long whereas the longest in existence by the 1850s was Brunel's 1.8 mile (2.9 kilometre) Box Tunnel on the Great Western Railway. Above any practicable line there towered mountains thousands of feet high, a fact which ruled out the driving of vertical shafts normally used to carry out two essential tasks; making it easier for the planned line to be followed accurately, and ventilating it.

Little was known in mid-century of the rock which would have to be pierced.

The knowledge available was discouraging, and when Robert Stephenson visited Switzerland he ruled out any chance of burrowing through the Alps. Another British engineer, James Swinburne, took the same view.

Yet as early as 1838 G.F. Medail, a young Frenchman – or, technically, young Sardinian, since Savoy then formed part of the Kingdom of Sardinia – had worked out a possible route from Modane in the Arc Valley on the north of the Alps to Bardonecchia in the south and then down the Dora Riparia to Susa. The tunnel between the two villages, or small towns, would pass below Mont Fréjus but, as there was already a road across the nearby Mont Cenis Pass, the tunnel on which work finally started in 1857 was frequently called the Mont Cenis.

Medail was dead by this time and construction was put in charge of another young Sardinian, Germain Sommeiller, already working on a railway being built up the Susa Valley from Turin. Sommeiller was faced with driving a 7½ mile (12 kilometre) tunnel, large enough to contain a railway engine, through rock of which comparatively little was known, and the task was at first expected to take twenty years. If the time sounds exceptional it should be remembered that although explosives were available the holes into which they had to be packed had in the early days of the project to be drilled much as the Romans had done their drilling some 2,000 years earlier. One man would hold the 'jumper', a drill with a simple chisel-shaped head, while a second man would strike the head of the 'jumper' with a hammer. The hole would be driven in for a distance of between 18 and 36 inches (46 and 91 centimetres), depending on the nature of the rock. The holes would then be filled with powder and plugged before the explosive was set off by a fuse. Not only was the work demanding but the rate of progress was slow – an average of 9 inches (22 centimetres) every twenty-four hours.

Although the Mont Cenis Tunnel was being driven from the Modane and the Bardonecchia ends simultaneously its progress was due largely to Sommeiller's insistence that any form of mechanical help should be tested, and, if useful, immediately put into operation. In the first half of the nineteenth century many engineers had experimented with mechanical rock drills. James Nasmyth, inventor of the steam hammer, was one of them. The American Singer brothers were others. T. Bartlett was yet another, and his percussion drill was for a while tried out by Sommeiller on the Mont Cenis project. All these machines had certain disadvantages which were only overcome in 1849 when J. J. Couch of Philadelphia patented a steam-operated rock drill. The drill operated inside a hollow piston; having been thrown against the rock, the drill was caught on the rebound by a gripper and thrown forward again by the stroke of the piston. Couch's drill was improved the following year by Joseph Fowle who had made the original tool and who now had the drill automatically rotated. Perhaps more important, it was operated by compressed air rather than steam.

Sommeiller's problem at the Mont Cenis was to provide the compressed air. He first used hydraulic compressors but found that they did not stand up to continuous operation. He then devised what he called his water-spout machines. These utilized a 164 foot (50 metre) head of water to drive water wheels which in turn compressed air in a vertical cylinder.

Progress speeded up once the compressed air drills were operating. In 1862 the best daily advance reached 3.41 feet (1.03 metres); two years later it had reached 9.75 feet (2.97 metres). Up to nine drills were mounted on a wheeled carriage which could be brought up to the face. Eighty holes were drilled simultaneously and were then fired to produce what was in effect a pilot tunnel. This would be expanded to the full dimensions of the tunnel and the roof would be supported, if necessary, by a masonry arch or by timbering. Meanwhile, further back, the permanent lining of the tunnel would be erected.

Ventilation continued to be a problem, although it was eased when the compressd air used for working the drilling machines was directed to clearing fumes from the tunnel. Heat rose to 90°F (32°C) which although bad enough was less than geologists had predicted. Safety precautions were considerable for the times and between 1857 and 1871 casualties numbered only fifty-five, of which twenty-eight were fatal.

The meeting of the headings from north and south came on Christmas Day 1870 when a drill pierced the thin layer of rock between them. On the following day a charge was fired which brought down the remaining barrier. Only then could the success of the engineers be seen: the difference in height between the floors of the two tunnels was only about one foot and the alignment where they met was less than 18 inches (45 centimetres) out of true.

If the construction of the Mont Cenis Tunnel was almost a model of what tunnelling should be, its successor, the St Gotthard, came near to being a disaster. There had been a road of sorts across the St Gotthard at least since the time of the Romans and it was obvious that the route, lying as it did on the direct path from Zürich to Milan, would be followed by a railway as soon as it was practicable to build one. Surveys in the 1860s showed that a summit tunnel of about 9¼ miles (14.9 kilometres) would be required and that under the main St Gotthard ridge its height above sea level would be nearly 3,800 feet (1,158 metres). Yet the incentive provided by Mont Cenis was so great that when in 1872 the Swiss Central Railway Company invited tenders for the project seven companies submitted them.

The contract was won by Louis Favre, a well known tunnelling engineer of Genoa who was given the target-date of 1 October, 1880 for completion. From the first, therefore, he was working against time. Blasting started in the autumn of 1871 at both ends of the tunnel, at Goeschenen in Switzerland on the north, and at Airolo in Italy on the south. A variety of drilling machines, British, French, Swiss and Italian were tried out. The River Reuss in Switzerland and the Tremolo in Italy were diverted to work stations providing compressed air for the drills, while Favre was able to use as explosive Alfred Nobel's recently invented dynamite instead of gunpowder. The initial pilot heading was soon being pushed forward at 13 feet (4 metres) a day while back behind the working face hundreds of men excavated the tunnel to full size and built a masonry lining.

However, all this was achieved in appalling conditions and with a casualty rate that was finally to claim 310 men killed, 877 seriously injured and hundreds more wrecked by disease. From the start, water poured into the tunnel from fissures in the rock. In the northern heading it reached an inflow of 2,400 gallons (10,910 litres) a minute; in the southern heading, more than 3,000 gallons (13,638 litres). At times the dynamite put into the drilled holes as powder ran out as yellow sludge, and it became necessary to encase the explosive in metal tubes before it was used. If the men found it difficult to work in water up to their knees, conditions were made worse by rock dust and by the heat which at times rose to 122°F (50°C). Men found it impossible to continue working underground after three or four months. About thirty horses and mules used to pull debris from the workings, died every month.

'As we rushed by dripping walls', wrote an American journalist who visited the tunnel, 'and saw here and there ghoul-like figures with dim lamps hiding behind rocks or in deep niches, I involuntarily recalled what our conductor had said of a glimpse of the bowels of hell. It was impossible to speak and be heard. I might as well have addressed myself to the granite wall of the tunnel as to have attempted a word to either of my companions.

'The air was so thick that lights could not be seen twenty yards ahead of us, and we all walked close together for fear of being lost or tumbling into some subterranean hole.

'Far ahead of us we heard the dynamite explosions, sounding like heavy mortars in the midst of battle. In some places where we were walking the water was nearly a foot deep, and again it came through crevasses about our head like April showers . . . [The men] are contented to receive their forty or fifty cents a day for hard work, if they can only escape wounds and death from the bad gases and the thousand accidents to which they are liable every moment of their lives in the tunnel. Alas! they do not escape, for every week records its disaster, either from explosions and flying rocks, falling timbers and masonry, or railway accidents, breaking machinery etc.'

By 1879 further troubles arose. One was caused by the dearth of water serving the Italian end of the tunnel. An aqueduct nearly 2 miles (3 kilometres) long had to be built to tap the Ticino but until this was done the lack of water to drive the air compressors meant that hand-drilling was necessary and the daily advance of the tunnel dropped from 13 feet (4 metres) to less than a foot. The problem was compounded by the fact that the northern workings were hampered by layers of unusually hard serpentine rock.

It was on the northern side, moreover, that the most difficult and dangerous problem was encountered. More than a mile and a half in from the Goeschenen entrance, the tunnel had been driven through a layer of decomposed felspar and gypsum. But contact with moist air turned the material into a plastic the weight of which broke through the tunnel lining. The problem was so acute that for a while it was thought that this portion of the tunnel might have to be redug along a different line. Eventually the situation was saved by building 8 foot (2.5 metres) thick granite supporting walls carrying a 4 foot (1.4 metres) thick granite arch. Despite the difficulties in the main tunnel, there were few in building the minor tunnels which had to be cut for the approach lines. This was surprising since some of them followed a helical route in the rock, entering it at one level and leaving it as much as 311 feet (91 metres) higher.

The ordeal was too much for Favre who collapsed of a fatal heart attack in the tunnel in July 1879 and was buried in the cemetery at Goeschenen, which already held the bodies of so many workers who had died in the St Gotthard. Seven months later the tunnellers from Airolo broke through the wall separating them from those coming from the north. The floors of the two tunnels were only 2 inches (5 centimetres) out, and the error in alignment only 13 inches (33 centimetres). Seven months still remained before the contract date would be reached; but that date was for completion of the lined tunnel and it was only seventeen months later that the work was eventually finished.

Other sub-Alpine railway tunnels followed the St Gotthard: notably the Simplon and the Loetschberg on the line between Berne and Milan, and the Arlberg in Austria between Feldkirch and Innsbruck. In the twentieth century they too were followed by the road tunnels, first being the Great St Bernhard and the San Bernardino and then, in the 1960s, the astonishing tunnel under Mont Blanc.

The longest sub-Alpine road route, the Mont Blanc Tunnel's 7 miles 350 yards (11.6 kilometres), presented special problems to its engineers. Little was known about some of the rock which had to be dealt with, there were ventilation problems created by the three hundred vehicles an hour which were expected to use the route, while the contrast between the heat of the tunnel and the Alpine cold at the two entrances was so great that a room of intermediate temperature was built in the tunnel for warming up or cooling down the workers.

Horace Bénédict de Saussure, the Genevese naturalist who towards the end of the eighteenth century encouraged men to make the first ascent of Mont Blanc, had prophesied that a road would one day be built through the mountain linking the French valley of Chamonix on the north with the Italian valley of Aosta on the south. However, no serious attempts were made for more than 150 years, and it was not until 1946 that Count Dino Lora Totino, a Turin merchant, financed the survey of a possible route and then formed a company to build a tunnel linking Courmayeur in the Aosta Valley with Chamonix.

However, the ambitious Count had started work without official permission. The project was stopped by the Italian authorities and only in 1949 was a group of Italians, French and Swiss set up officially to plan and carry out the project. The Swiss had been brought in since the people of Geneva hoped that their city would become a major clearing house on what would become the main road between Paris and Rome. The French had at first protested against the tunnel on security

The entrances to two tunnels being built above Wassen, Switzerland, for a new two-lane road under the Alps near the St Gotthard Pass.

grounds, just as the British were to protest against plans for a Channel tunnel, but the Swiss acted as honest brokers and work was started early in January 1959.

The task was tackled in different ways by the Italians and the French, the Italians relying on light manual drills for the boring, and on road vehicles both for moving equipment forward in the tunnel and for moving out the debris as work progressed. The French, by contrast, used heavier equipment, typified by their 75-ton three-deck gantry carrying fifteen heavy Ingersoll-Rand boring machines, and used railway trucks to move back the debris.

Soon after the work had started, the Italians began meeting unexpectedly difficult rock and there were two serious rock falls which held up their advance into the mountain. In April 1962 they faced another problem when their camp at the southern entrance to the tunnel was hit by two avalanches. For their part, the French at times had trouble holding up the roof of the tunnel. They dealt with this by extensive bolting, using as much as one million yards of metal which were driven into the rock as supports.

Both Italian and French tunnelling methods worked as predicted and the break-through between north and south took place as planned on 14 August 1962. More work had to be done but after a further eighteen months the 'White Way' was opened to traffic.

Earlier, in the second half of the nineteenth century, the Americans had been progressing with the intractable Hoosac Tunnel in Western Massachusetts, the first major United States project of its kind and one which, although financially disastrous, enabled American tunnellers to lay the foundations of their craft. The first idea for a tunnel through the Hoosac had come in 1825 when after the success of the Erie Canal it had been proposed to build a canal between Boston and the Hudson. Nothing came of the plan, the railways began to take over, and when in 1854 the first money was granted for tunnelling through the Hoosac Mountain it was planned to carry a railway from Boston to Troy. There were disagreements with the contractors and when the State Commission took over nearly a decade later more than 20,000 feet (6,100 metres) of the tunnel's 24,000 feet (7,320 metres) still had to be excavated.

Construction was stopped for a year but a research engineer was sent to Europe. On his return he wrote a series of reports on drilling machines, the use of explosives and other technical matters. The resident engineer, Thomas Doane, designed his own mechanical compressor to work rock drills and introduced both nitroglycerine in the place of gunpowder, and the electrical firing of the charges. Eventually Doane left, and other contractors took over. A decade later Doane once more appeared on the scene but only after the two headings of the tuunnel had met. Despite the changing of contractors the error in alignment was less than an inch and the error in level only a few hundredths of an inch.

In Europe discussion continued about the most controversial tunnelling project of all – a tunnel under the English Channel, only 23 miles (38.8 kilometres) wide between Dover and Calais. The 'Chunnel' as it has been called for years had been considered a theoretical possibility since 1750 when the University of Amiens announced a competition for engineering designs. Napoleon expressed an interest in plans drawn up by a French engineer, Albert Mathieu-Favier, and in 1857 and 1867 respectively a French doctor, Thome de Gramond, and a Scottish engineer, William Lowe, put forward definite proposals. With the development of increasingly effective drilling machines, and the accumulation of experience from all over the world, the 'Chunnel' began to look a practical possibility and in 1872 a number of companies were formed to build it. By driving the heading from just west of Dover to Sangatte, 5 miles (8 kilometres) west of Calais, it was believed that the work could be carried out entirely in the Lower Chalk and that no major technical difficulties would be met. Investigation showed that the material was, in fact, chalk mixed with clay, a mixture so impervious to water that no lining to the tunnel would be needed. Furthermore, the mixture was easy to deal with and a new tunnelling machine invented by Colonel Frederick Beaumont was able to cut a 7 foot (2.13 metre) diameter bore at the astounding rate of 16 feet (4.9 metres) an hour.

Problems were financial rather than engineering and it was only after another company had been formed in 1880 that 7 foot (2.13 metre) diameter pilot tunnels began to be dug from both sides of the Channel. There were to be two parallel tunnels, eventually enlarged to a diameter of 14 feet (4 metres) and through them trains would be drawn by compressed air which, as well as powering the engines, would be used to ventilate the tunnels.

By the summer of 1882 the British heading had been driven for more than a mile beneath the Channel. It was visited by Ferdinand de Lesseps, still basking in his Suez triumph, who with five colleagues was lowered in a skip to the bottom of the shaft which had been dug to the west of Dover's Shakespeare Cliff. 'At the bottom of this shaft', wrote *The Times*, '163 ft. below the surface of the ground, the mouth of the tunnel was reached, and the visitors took their seats on small tramcars that were drawn by workmen. So evenly had the boring machine done its work that we seemed to be looking along a great tube with a slightly downward set, and as the glowing electric lamps, placed alternately on either side of the way, showed fainter and fainter in the far distance, the tunnel, for anything one could tell from appearances, might have had its outlet in France. A journey of some 17 minutes, however, not counting a stoppage for refreshments when 1,000 yards had been traversed, the workmen drawing the cars on the down grade at a fast walk, brought the party to the end of the boring – 1,900 yards from the shaft and about 150 ft. below the sea.'

At last it seemed likely that a Channel tunnel would be successfully built. This in itself was enough to arouse opposition in Britain, mainly on the grounds that it would make a French invasion more easy. The Government had no need to adjudicate on the question when Parliament was petitioned to ban the 'Chunnel' since further work was stopped by the Board of Trade on the grounds that the tunnel passed through Crown property – all that ground between high-water mark and the 3 mile (5 kilometre) limit.

Between 1883 and 1929 no less than thirteen attempts were made to have Parliament raise its objections. All failed, and not until 1957 were the first moves made to lift the ban. Since then, the prospects for the 'Chunnel' have yo-yo'd and in the early 1980s continue to remain uncertain. In the nuclear age, it is difficult to maintain that the defence risk is relevant. What matters more are the economics of operation; the effect of a tunnel on existing cross-Channel shipping services; and, perhaps most important of all, the changes in the life of south-east England, and its transport system, that the influx of 'Chunnel' traffic would make inevitable. As elsewhere in the western world, tunnel development no longer rests on the solution of engineering problems.

Since 1964 the Japanese have been constructing the longest tunnel in the world, the 33.47 mile (53.9 kilometre) Seikan Tunnel which dives beneath the Tsugaru Straits to connect Hokkaido Island to the north with Japan's central Honshu Island. The tunnel, to be used by high-speed trains which will cut travel times between Tokyo and Hokkaido's capital city of Sapporo from twenty hours to less than six, was first discussed almost half a century ago. But it was only after the end of World War II that technological improvements made it possible to deal with the extraordinary difficulties presented by ground beneath the Straits. The Seikan engineers, cutting a pilot tunnel to study soil conditions, a working tunnel for transport of the construction materials, and finally the main tunnel, have been working their way through comparatively young strata, composed largely of fragmentary rock and layers of volcanic ash and lava. To complicate the situation still further, the line of the tunnel unavoidably passes through nine separate zones containing faults created by major movements of the earth's crust. And while it is being bored 109 yards (100 metres) below the sea bed, the Straits are up to 153 yards (140 metres) deep, so that a great weight of water has to be sustained.

Boring has been carried out by a 95-ton 441 kw machine whose cutters work across a diameter of 4.3 yards (4 metres) and which can advance at 2 metres an hour. Two methods are used to reinforce the strength of the tunnel as it is cut. As the cutting machine advances radial holes are driven into the roof and walls and into them there is injected a mixture of cement, liquid glass and special chemicals; this reacts with water to form a solid material strong enough to withstand the pressure on the tunnel. In addition to this grouting, cement is sprayed on the walls and roof just after excavation and in places reinforced by steel supports.

Despite these precautions there have been two occasions when the sea began to break through, much as the river had done with Brunel's Thames tunnel almost a century and a half earlier. In February 1969 water started to pour in at 12 tons a minute and was finally stopped only by an enormous injection of concrete after work had been held up for seven months. In May 1976, another break-through took place with water coming in at 40 tons a minute, but control was regained after only two months.

Both these incidents happened in major fault zones. But they appear to have been the last and the rest of the $1.8 billion project continued without delay.

The Japanese use of fresh methods is typical of the tunnelling industry where new technology is constantly being tried out and, if successful, incorporated into accepted practice. One such is pipe-jacking in which steel, concrete or plastic piping is inserted into the circular hole to form a ready-made tunnel. The use of large diameter tunnel-boring machines has expanded and is continuing to do so, while what is known as the new Austrian tunnelling method – although used elsewhere – involves lining tunnels with reinforced concrete or with similar material which then forms, in effect, an integral constituent of the tunnel wall.

If man has been intrigued since earliest times in burrowing through the earth for severely practical reasons, he has also been tempted to build up towards the stars whenever and wherever possible. The Tower of Babel is the most famous example from the ancient world. The Pharos of Alexandria, and the towers on which men of later centuries put the warning flares of their lighthouses, are others. The campaniles of medieval churches were built high so that the bells they carried could be heard over a great distance. When public clocks were built it was natural that they should be set on towers built as high as was practicable.

All these early examples of high buildings, like the spires of the great cathedrals, were constrained by the limitations of the materials used. Brick, masonry and even the artificial masonry which came into use during the nineteenth century following the invention of Portland cement, had to be of a certain minimum thickness to support what was built on top of it. As height and weight increased so did the thickness of the walls. However ingenious the architect, the height was limited by the practicable strength and width of the foundations and the walls of the lower storeys.

During the second half of the nineteenth century the situation was radically changed by the use first of iron, then of steel, as a building material. At the same time another obvious barrier to high buildings was removed by the development of the safe lift or elevator. No longer was it possible to rule out the potential skyscraper on the ground that no one wanted to walk up a dozen or more storeys to get to their home or office. Elisha Otis is often credited with being the one man, above all others, who made the skyscraper a viable proposition, and it is true that in 1854 he showed at New York City's Crystal Palace that elevators were both workable and safe. But Otis died when his new device had been installed in only two buildings. Progress was taken up by Henry Baldwin Hyde, the insurance magnate, who in 1872 put an elevator in a new business building, an example which encouraged others to do the same.

The combination of iron or steel with the lift transformed the position in the United States. But it was a piecemeal process, and one of the first skyscrapers, the Home Insurance Building of Chicago, had granite base piers and brick party walls. The first five storeys were framed with wrought-iron beams and the next

The elevator shaft in a modern water tower, Chicago, Illinois.

five with steel. Two further storeys were added sometime later on.

Chicago had been built on comparatively water-laden soil and thus presented engineers with more complex problems than did New York whose Manhattan Island offered them solid rock for their foundations. Another fact which made the skyscraper attractive in New York was, of course, the confined area offered for development in Manhattan. With the great expansion that followed the end of the Civil War, New York was virtually compelled to build upwards while Chicago had the space to spread outwards. Nevertheless, both cities went ahead on parallel lines. 'The question whether the skyscraper originated in New York or Chicago is still a matter of controversy', Carl J. Condit observed in 1968, 'for the answer depends on whether we define it in terms of size and economic function or in terms of structural and architectural character. If we think of the skyscraper as a high commercial building whose height greatly exceeds its horizontal dimensions and which grew out of economic exigencies arising from intensive land use, then the form may be regarded as the creation of New York builders. On the other hand, if we hold that the structural system of wind-braced steel or concrete framing expressed in an organic architectural form is an essential characteristic, then the skyscraper was a Chicago achievement. Indeed, the structural techniques used by the New York builders were for many years thoroughly conservative and lagged well behind the level reached in Chicago.'

In both cities the compromise use of masonry and iron continued for some years, and it was to the 230 foot (70 metre) Western Union Building and the 260 foot (79 metre) Tribune building, both built in this manner, that the British Thomas Henry Huxley pointed when he sailed into New York in 1876 with the words 'Ah, that is interesting; that is American. In the Old World the first things you see as you approach a great city are steeples; here you see, first, centres of intelligence.'

However, it was to be the work of an Old World engineer and architect that was to give a great boost to the metal-for-skyscrapers movement in the next decade. He was Gustave Eiffel, who in the early 1880s was given the task of supporting internally the 151 foot (46 metre) high Statue of Liberty that was to be erected at the entrance to New York harbour. Steel was used for the chief bearing members, the first time that it had been specified in New York for any structure other than a bridge.

Gustave Eiffel was already one of Europe's most famous engineers, renowned for engineering successes across the world. At Tan-an in Cochin China he had erected a bridge of 262 foot (80 metre) span without scaffolding by building out the structure piece by piece, a method of which he was the pioneer. He had constructed viaducts using iron piers instead of masonry. At Garabit he had thrown an arch of 541 feet across the Truyère 400 feet (122 metres) below. At Nice he had conceived and built the 75 foot (23 metre) diameter dome of the Observatory, weighing more than 100 tons but floating within a circular trough so that it could be moved with negligible effort.

His work on the Statue of Liberty was one pointer towards the Eiffel Tower. Another was a series of investigations aimed at discovering the extreme limits to

---

The 1,000-foot (305-metre) tower
planned by Gustav Eiffel for the 1889 Paris Exhibition.

which the metallic piers of viaducts could be pushed with safety. These showed that a 981 foot (299 metre) tower was a practical possibility and the Council planning the 1889 Paris Exhibition formally accepted Eiffel's plans for a 1,000 foot (305 metre) tower. Various sites were proposed, including Courbevoie and the Trocadéro, among a number. Eventually it was decided to build at the northern end of the Champ de Mars, the great open space where the Exhibition was to be held, and which runs from the River Seine to the Ecole Militaire.

Protests came quickly, and a petition from French authors, painters, sculptors and architects asked whether Paris would 'allow itself to be deformed by monstrosities, by the mercantile dreams of a maker of machinery?' The Tower was compared to 'a gigantic and black factory chimney' and it was claimed that Parisians would see for twenty years – the estimated life of the tower – 'stretching out like a black blot, the odious shadow of the odious column built up of rivetted iron plates.'

From the four massive foundations for the tower, standing at the corners of a 330 foot (100 metre) square, there arose during 1887 the network of girders supporting the successive storeys forming part of the structure. Once the first storey had been completed a light-weight railway was laid on its flooring; further girders were then drawn up by crane through a hole in the flooring, and then moved by train to the positions from which they could be rivetted into position. Above the second storey the inward curve of the four columns forming the tower restricted the working space to ever narrower limits. But the girders were progressively less massive and the tower was continued without difficulty to a height of 896 feet (273 metres). Here, surrounded by an observation gallery, there was a laboratory, from whose four corners there rose latticed arched girders supporting a giant lantern, reached by a spiral staircase: a lantern, moreover, which was to shed white, blue and red beams across the Exhibition filling the Champ de Mars below.

The maze of intercrossing girders, linked together by 2,500,000 rivets, was a sight which made the Eiffel Tower a wonder of the late nineteenth century. Escalators and two different kinds of lifts brought visitors to the top from which they had a unique view of the country surrounding the capital. And on the first stage, a tribute to the French dedication to food, there was a French restaurant, a Russian restaurant, an Anglo-American bar and a Flemish beer saloon, each of the four capable of holding up to six hundred visitors.

The Eiffel Tower was to be used during the following years for meteorological observations while in the first years of the twentieth century the fall of objects between the various storeys, the heights of which were exactly known, provided information on air resistance so badly needed by the infant aircraft industry. But the Tower was of limited use until the loftiest wireless transmitter in the world was erected on top of it after World War I.

The Eiffel Tower not only became famous throughout the world but convinced Americans that metal was now fully taking over from masonry. In Chicago the steel frame carried out all the structural roles in the second Rand-McNally building. Other similar buildings followed in both Chicago and

---

The C.N. Tower, Toronto, 1,822 feet (555 metres) to
the top of the transmission mast, and the world's tallest free-standing structure.

New York while a different boost for the skyscraper came in 1904 when Eugene Freysinnet pioneered the use of pre-stressed concrete.

Although the development of the skyscraper was due mainly to the incorporation of iron and steel, reinforced concrete was also used. Its real popularity came, however, only after Freysinnet had found that the use of very strong steel rods, pre-tensioned – or stretched – before being incorporated in a building's framework, would greatly increase structural efficiency. This was done by creating a controlled stress during construction which would counteract the stresses created when the working load was put on the building.

As progress continued in both these fields, skyscrapers grew both in numbers and in height. So much so that in more than one American city tall buildings began to turn main roads into canyons. Environmental restraints had to be imposed and in 1916 the New York City Zoning Ordinance made it necessary for buildings to have specific set-backs as they rose in height. The regulations put architects and engineers on their mettle and the outcome was a number of ingenious building complexes, notably the Rockefeller Centre in New York.

Meanwhile, throughout the years, heights have continued to increase. New York's 102 storey Empire State Building, whose 1,250 feet (381 metres) – 1,469 feet (448 metres) to the top of T.V. tower – was finished in 1931, held the record for some years. Today it is claimed by the 1,454 foot (443 metre) Sears Tower of Chicago. Outside the United States, skyscrapers have gone up in increasing numbers since the end of World War II, but nowhere have American heights been neared. Moscow's State University building rises to 787 feet (240 metres), the Pirelli building in Milan to 414 feet (126 metres), and the Vickers Tower in London to 387 feet (118 metres).

New records have also been established in the post-war era for tall towers as distinct from buildings. The Eiffel Tower's 985 feet (299 metres) later raised to 1,052 feet (321 metres) was over-topped by Great Britain's Independent Broadcasting Authority tower at Emley Moor, West Yorkshire which in 1971 rose to 1,080 feet (329 metres). Four years later this was far exceeded by the 1,822 foot (555 metre) C.N. Tower in Toronto's Metro Centre. The 130,000 ton reinforced, post-tensioned concrete structure has a 416-seat revolving restaurant at 1,140 feet (347 metres).

Since 1945 the television towers which have been superimposed on many of the world's tallest buildings have tended to confuse height records, while a completely new category of engineering construction has been created by the radio and television masts of which many countries can now boast. Poland leads the way with its 2,117 foot (645 metre) radio mast at Plock, followed by the 2,063 foot (629 metre) T.V. and radio mast at Fargo, North Dakota. The United States can claim another five masts more that 1,500 feet (457 metres) high. Britain has a 1.269 foot (387 metre) stayed steel tower at Belmont, Lincolnshire, and a 1,080 foot (329 metre) free-standing radio tower outside Huddersfield in Yorkshire.

In the building of masts, as of skyscrapers, there are few indications that engineering or technological limits have been reached. The potentialities of new materials, created with a steadily increasing knowledge of what their molecular structure makes possible, suggest that buildings a mile or more high are steadily moving from the field of fantasy to that of engineering possibility. As in many other departments of engineering, the limitations are financial and environmental rather than technological.

THE RISE OF THE NUMBER-CRUNCHERS

After World War II, as after World War I, every field of engineering began to utilize the new ideas, the new materials and the new processes that had been forced on, hot-house-wise, by the demands of the conflict. In addition, the decades after 1945 witnessed the growth of the two new disciplines of nuclear engineering and of space flight. Both were dependent for success on a family of man-made machines that were among the most remarkable that engineers had ever built: the electronic computers which could handle instructions at the rate of millions per second and thus make possible engineering, scientific and mathematical operations quite beyond the unaided human range.

While the calculations needed for building nuclear weapons and nuclear power stations, as well as for planning the orbits of spacecraft, could not have been made without 'electronic brains', the uses of number-crunching machines have in the last two decades revolutionized commercial applications and statistical work, the study of social questions and of defence. In fact, it is justifiable to claim that few works of man have infiltrated his life as deeply as the modern computer.

Artificial aids to mental computation go back to the abacus of ancient times, a frame along whose parallel rods beads can be moved as an aid to counting and which is still widely used in the Far East. More important calculating aids came in the seventeenth century with Napier's Rods, a device invented in 1614 which allowed multiplication to be carried out by a series of simple additions. They were followed some years later by the slide rule, devised by William Oughtred, the mathematician who introduced the signs for multiplication and proportion. William Schickardt, Blaise Pascal, Gottfried Leibniz and the 3rd Earl Stanhope were others who followed with mechanical calculating machines.

Although Schickardt, with a machine made in the 1620s, was the first of them, Pascal occupies a unique position from having invented, a few years after Schickardt, ten progressively improved models of what he called La Pascaline. Each had eight interlocking geared wheels, the first six dealing with single numbers, tens, hundreds, thousands, tens and then hundreds of thousands, and the last two covering twenties and twelves. Numbers were put into the machine by means of dials resembling a telephone dial and the results could be read through windows in the top of the machine. Apart from addition and subtraction, multiplication could be carried out by a simple system of additions, and division by a comparable series of subtractions.

Men continued to develop mechanical means of calculation but it was only in 1805 that a crucial step forward was taken when Joseph Marie Jacquard used perforated cards to control the operation of the new loom which he had invented in Lyons. The idea of using cards with holes in them to control the movements of a calculating machine was taken up first by the Englishman Charles Babbage who, in 1832, prepared plans for an Analytical Engine, a steam-driven machine which processed information and instructions taken from such cards. The cards passed into the 'store', the equivalent of the electronic computer's memory unit, and were so processed that the results could be either printed or set up in type.

Babbage's Analytical Engine was never built, but his work encouraged others and before the end of the nineteenth century a number of 'difference engines' had been constructed to supply tables for navigation, for insurance and for astronomy. Sir William Thomson, later Lord Kelvin, designed with his brother a comparable machine which simplified the mathematical work of producing tide-tables.

With the exception of Babbage's unbuilt steam-driven machine, all attempts to speed up computation relied on hand power, and this remained true until 1890 when the US Census Office held a competition for the most efficient census-taking system. It was won by Herman Hollerith who used electricity to record information from punched cards fed into his machines. *The Electrical Engineer* in an article headed 'Counting a Nation by Electricity' noted: 'This apparatus works unerringly as the mills of the gods, but beats them hollow as to speed.'

The next important step was taken four decades later by Vannevar Bush, the American electrical engineer. Trying to solve equations associated with power failures he contrived, in 1930, a 'differential analyzer'. It was the first analog computer – one accepting data as a continuously varying quantity and solving problems by physical analogy, usually electrical – which could be used for a wide range of problems, and soon led on to the development by Bush of more complex machines.

This work in the late 1930s marks the watershed between the older, manually or mechanically operated computing machines and the modern marvels capable of carrying out a million or more operations a second – the electronic computer that for most practical purposes is today described merely by the single word 'computer'.

The vital difference between the old computers and the newer ones relying on electric or electronic movements lay in the ease with which fresh calculations could be carried out on answers which the machine had already produced, and the speed with which this could be done. With a mechanical calculator it is necessary to put the first answer back into the machine and then start a new operation. In the new computers that were now being born it was possible by giving the correct instructions at the outset to have such operations carried out automatically. Moreover the instructions would be carried out by electricity whose effect operates at the speed of light. So the message instructing what the machine had to do next would be given along a few feet of wire at the speed of 186,281 miles (299,782 kilometres) per second. This fact alone brings some understanding of how electronic computers can carry out their vast numbers of operations per second.

As with aircraft after 1914, it was the demands of war which in both Britain and the United States led the way to the modern computer. Although the two countries moved forward on parallel lines, Britain was a short head in front, although the fact was not revealed until many years later, due not only to reasons of commercial secrecy but because of the security demanded by war. For the British effort was concentrated on computers to help in the decoding of intercepted messages sent by the Germans on their Enigma coding machines.

The first was built by the British Post Office Research Station for use by the cryptographers working at the Foreign Office station at Bletchley Park. Named 'Heath Robinson' after the English comic artist who drew wildly impractical machines for doing outlandish tasks, it was followed by 'Peter Robinson', 'Robinson and Cleaver' (two London stores), and 'Super Robinson'. The machines, using 1,500 valves and producing a tape which could be read at the rate of 2,000 characters a second, were followed by an improved series of computers, the 'Colossi', which could be read at a rate of 5,000 characters a second. The first 'Colossus' filled a large room in one of the Bletchley Park huts, incorporated many improvements on the 'Heath Robinson' series and had a typewriter output. The next 'Colossus' used 2,400 vacuum tubes, was five times as fast as its predecessor, and was completed by June 1944 in what was considered to be an impossibly short time in order to cope with the glut of messages which it was known would follow the Allied D-day landings in Normandy. Before the end of the war another half-dozen 'Colossi' were built, each one an improvement on the last.

Meanwhile work had started independently in the United States on construction of a computer to replace the electro-mechanical devices used to calculate gun trajectories. The result was ENIAC, the Electronic Numerical Integrator and Computer. The progress of the next three decades and more can be judged by comparing contemporary desk-top computers with figures for ENIAC. It weighed more than 30 tons, occupied a room 30 feet by 50 feet (9 by 15 metres) and contained about 18,000 vacuum tubes, 70,000 resistors, 10,000 capacitors and 6,000 switches.

The pioneer ENIAC had certain disadvantages. If a valve failed it could take eight hours to locate the fault. The heat generated during operation was considerable, the machine was normally run for not more than an hour at a time and it was said that when it was switched on all the lights in the district went dim. Nevertheless, computing a sixty-second trajectory that demanded twenty hours work from a skilled operator and fifteen minutes on a differential analyzer, required only thirty seconds on ENIAC.

Whereas the analogue computer received a continually varying electrical pulse and with this produced results by analogy, ENIAC was a digital computer taking in numbers that were subjected to mathematical manipulation to produce results. Thus while the analogue computer might be compared to a slide rule whose numbers were analogous to length, the digital computer was comparable to an abacus in which numbers were used for calculating. ENIAC, like all digital computers, used the binary code. While numbers are most frequently expressed in the code using ten symbols from 0 to 9, the binary code uses only two symbols, those of 0 and 1. In the decimal system a 1 is put in a second column when 9 has been reached in the first column, and that first column filled; in the binary system, a 1 is put in the second column, to represent the figure 2. To represent the figure 3, a 1 is also put in the first column, while to represent 4, a third column is started with a 1, the other two being filled with 0s. This practice of 'carrying' when 2 rather than 10 is reached would be complex in everyday life, but in a computer it has one great advantage: any number can be represented by a succession of electric pulses representing '1' interrupted by no-pulse intervals representing '0'.

ENIAC, like virtually all computers, stored the data which was put into it and then dealt with this data by means of a programme of instructions; but when a new problem had to be solved a different set of instructions had to be prepared if a considerable rearrangement of the computer's switches was to be avoided. The difficulty was first overcome in 1948 when a small computer holding both programme and data in the same store was built at Manchester University. It was followed the next year by the Cambridge computer EDSAC, the letters standing for Electronic Delay Storage Automatic Computer. Although requiring only 3,000 valves it worked six times faster than ENIAC and both input and output was carried on paper tape.

The Manchester University machine was developed by Ferranti and Professor F.C. Williams, and in 1951 became, as the Ferranti Mark 1 computer, the world's first computer to become commercially available.

From this date onwards computers began to proliferate in both England and America, while in 1951 the Ukrainian Academy of Sciences constructed Russia's first computer of the ENIAC type. Basically, all consisted of an input unit, which could be compared to the ears and eyes of a human being; of a processor which could be compared to the thinking and control functions of the brain: a memory, comparable to memory-storage in a human brain: and of an output unit, comparable to human hands and mouth. Each complete system solved the most complex mathematical problems – the essential job of the computer – by subjecting the data stored in the 'memory' of the computer to the long set of instructions embodied in the programme. Progress made throughout the years lay in the continually increasing amount of information which could be stored in the

The Manchester University Mark I Computer,
photographed in 1949 but showing the computer as working in June 1948.

'memory', the speed with which it could be handled, and the variety of methods which could be used for putting data or programmes into the computer and for taking out the completed results of any operation.

From the early days many methods have been used for both of these latter tasks. Most of them have involved some kind of magnetic recording, among the most common being the magnetic core. Here the memory store consists of hundreds of thousands of ferrite rings, each only a few hundredths of an inch in diameter. Wires pass through the cores, each of which is magnetized with a polarity depending on the direction of the current. Thus every core can represent either 1 or 0. Other forms of storage include magnetic drums, magnetic discs and magnetic tape; in all of them the figures which the computer will use to carry out the operations it is ordered to carry out by the programme are obtained by reading off the binary code figures from the memory. Speed of memory-use has been constantly increasing, and speeds of 10 million read-write cycles per second have been reached with modern semi-conductor store.

Meanwhile, the size of computers has been decreasing at a truly remarkable rate. Much of this has been the result of two revolutionary developments incorporated in the second and third generation of computers built from the mid-1950s and mid-1960s respectively. They are the transistor and the integrated circuit, two devices which have led to the 'desk-top' computer and a series of advances in miniaturization which have transformed the computer into a totally different animal from its predecessor of three decades ago. Progress in both fields has been spurred on by the need to produce electronic equipment small enough and robust enough to be fitted into ballistic missiles; however, the impact on the world of civilian computers cannot be over-stressed.

Each of the 18,000 valves in ENIAC was bulky by modern standards. Each demanded considerable quantities of electricity and they became inconveniently hot in use. It was, therefore, something of a revelation when in 1948 William Bradford Shockley, working for the Bell Telephone Laboratories, found what could be done with semi-conductors such as germanium. Semi-conductors have many curious properties, one of which is that their resistance decreases with rising temperature and the presence of impurities, which is contrary to the action of normal conductors. In addition, they allow an electrical current to pass much more easily in one direction than another. They act, therefore, as rectifiers, so that if alternating current is led through a semi-conductor, it emerges as a direct current. This had long been known, and pre-valve radio sets were, in fact, known as crystal sets since it was a crystal that they used. In 1948, however, Shockley discovered not only that germanium crystals containing certain impurities were far better rectifiers than those of a few decades earlier but that they could be used to amplify current. In other words, operations similar to those which had taken place inside the glass-enclosed vacuum of a valve could be made to take place inside a crystal of germanium. Not only was this minute in size compared with a valve; it also did its job without creating the heat of a valve and was virtually immune to shock and vibration.

Other solid-state devices – 'solid-state' because the movements of the electric current took place in the solid crystal rather than in the vacuum of the valve – soon came into being, and by the end of the 1950s transistors, a fraction of an inch long had largely replaced the cumbersome valve. Some thousands of transistors, each designed to perform a specific electronic function, could easily be incorporated in a computer, where they were joined together by soldering.

The second revolution, the rise of the integrated circuit which came in the 1960s, obviated this final process since the integrated circuit was nothing less than the making on a single crystal of a number of different circuits, already joined up to perform the tasks they would carry out in the finished computer. The idea of creating such circuits was outlined in a paper by a British radar expert, W.A. Dummer, in 1952, and the first integrated circuit came from the Texas Instrument Company six years later. Transistors had been more sturdy than valves, and integrated circuits were even more so. In addition, the circuits reduced still further both the size of the computer and the power needed to operate it. In 1960 a transistor occupied the space of a 5 mm. cube, but four years later, integrated circuits allowed a hundred to be fitted into the same space. By 1975 the figure had risen to 10,000. Two years later Robert N. Noyce underlined the position in the *Scientific American*. 'Today's microcomputer', he said, 'at a cost of perhaps $300, has more computer capacity than the first large electronic computer, ENIAC. It is twenty times faster, has a larger memory, is thousands of times more reliable, consumes the power of a light bulb rather than that of a locomotive, occupies $\frac{1}{30,000}$ the volume and costs $\frac{1}{10,000}$ as much. It is available by mail order or at your local hobby shop.'

It is the integrated circuit, in fact, which is the key not only to the desk-top computer, but also to a multitude of devices – wristwatches being an obvious example – in which micro-miniaturization has been taken to what seems almost miraculous lengths.

Of the world's semi-conductors, which include germanium, silicon and selenium, silicon is the one most frequently used in the manufacture of integrated circuits, a process which relies on photo-engraving at microscopic levels of accuracy. Silica, or silicon dioxide which occurs in sand and rocks, is the second most abundant element in the earth's crust, and the manufacture of most integrated circuits starts with the reduction of silica into silicon which is 99.9999999 per cent pure. This degree of purity, abnormal though it is, typifies the standards which permeate every phase of integrated circuit manufacture. The purified silicon is then raised to its melting temperature of 1420°C, being covered during the process by a layer of purified inert gas, necessary both to prevent oxidation and the taking up of impurities by the purified silicon. The conducting characteristics of the silicon are closely controlled by the nature and quantity of known impurities such as boron or phosphorus, and these are now added to the liquid silicon before the next stage of the operation.

This stage involves the insertion into the liquid of a single crystal of silicon. The crystal grows naturally in the liquid and is allowed to develop until it is three or four inches in diameter and a few feet long. It is pulled from the liquid, its sides ground to a smooth surface and then sliced with a diamond saw into wafers about ½ mm in thickness. The process has to be carried out in conditions of astonishing cleanliness, since even a single particle of dust can stop the operation of the circuits that are now to be built on to the wafers. Workers wear special clothing and the air around them is continually filtered and recirculated. A typical wafer

A Transam Tuscan 100 Computer
Processor Board showing its banks of memory chips.

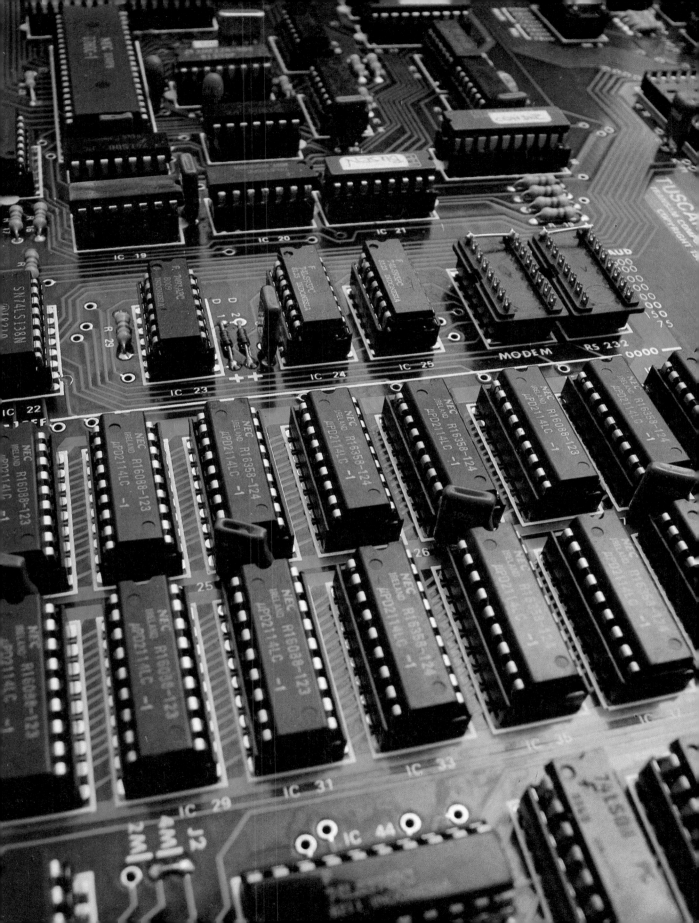

fabrication plant harbours per cubic foot fewer than 100 particles a micrometre [a millionth of a metre] or more in diameter. For the purpose of comparison, the dust level in a modern hospital is 10,000 similar particles per cubic foot.

Each wafer, ½ mm thick and in the region of something like 7 or 10 centimetres (3 or 4 inches) across, will eventually be cut into chips which are in some cases as small as 2 mm square; a single wafer can thus produce hundreds of chips each of which is an integrated circuit. The transformation of silicon wafer into miracle micro-chip is brought about by a series of chemical operations each carried out in conditions as clinically clean as those in which the wafer itself was produced.

The wafer is first exposed to steam which produces on its surface a film of oxide. Next a light-sensitive material is added on top of the oxide, and light is shone through a patterned mask on to it. Where the mask lets the light through, the material becomes hardened; where the mask stops the light, the material remains unhardened and in the next stage of the operation, an etching process, the unhardened areas are removed to reveal the silicon below. The wafer, with its hundreds of potential chips each bearing a pattern of exposed silicon, is now treated with chemicals which transform the patterns on the silica into electricity-bearing circuits whose characteristics are governed by the nature and quantity of the chemicals used. Once the first layer of circuits has been 'grown' on to the wafer, the process is repeated, then repeated again until a succession of circuits, layered above each other, make it possible for each of the potential chips to carry a huge amount of information. As a final stage the wafer is cut up into individual chips or circuits.

The minute size and the great potential of the integrated circuits is the result not only of the initial small size of the chip itself but of the size-control that is possible during the chemical process of building up the circuits. Thus the depth of the oxide film produced during an early stage of the operation can be controlled to within one ten-millionth of a metre.

The miniaturization which integrated circuits have made possible is the most spectacular advance of the last two decades. But there have been others, assimilated into new machines as manufacturers have vied with each other to make computing faster, more flexible and cheaper. One of the first problems was to provide input material, or to design methods of handling output material, at a speed fast enough to cope with the computer's processing. Even in the early days of electronic computing, data was being produced at nearly 100,000 characters a second while a conventional printer worked at some 1,500 characters a second. Systems of presenting data on cathode ray tubes helped to solve the output problem while input was speeded up by the use of magnetic tape.

New methods were evolved for holding ever greater amounts of information in a computer's memory while it was found possible to share the use of a computer between different operators, and to enable one group of computers to 'talk' to another.

These latest developments have come at a time when the computer has infiltrated huge areas of scientific, business and commercial life. The scientific applications are the most easily understood. Once calculations can be carried out at rates made possible by the electronic computer it is obvious that operations in the stratosphere of mathematics, once practically impossible because of the time taken for the human mind to carry them out, then become feasible. The working out of shell trajectories was quickly followed by use of the computer to elucidate the complex interactions which followed nuclear fission and to provide information on the movements of vehicles in space.

At other levels, computers have enlarged engineering possibilities. The increased analytic power has enabled engineers to construct more complex structures – such as the very tall skyscrapers of New York and Chicago – than could easily have been designed without their aid. Another, and perhaps more revealing example has been given in *Science* by Professor Herbert Simon, Professor of Computer Science and Psychology at the Carnegie-Mellon University, Pittsburgh. There are, he says, two different ways in which a computer can assist an engineer in designing electric motors. On the one hand, the engineer can design the motor using conventional procedures, then employ the computer to analyze the prospective operation of the design – the operating temperature, efficiency, and so on. On the other hand, the engineer can supply the computer with the specifications for the motor, leaving to it the task of synthesizing a suitable design. In the second, but not the first case the computer, using various heuristic search procedures, finds, decides on, and evaluates a suitable design.

These esoteric applications of the computer can be matched by the more mundane functions which in the last three decades have progressively changed life in office and factory. Commercial applications came first and it was in the 1950s that Lyons, the British catering firm, installed Leo, one of the first computers tailormade to do the work being carried out by clerical staff. Leo not only handled stock control, ensuring that supplies of the hundreds of products sold by the firm's retail outlets were always available, but worked out the salaries to be paid weekly or monthly to the hundreds of staff. From this comparatively simple start the use of computers to deal with routine office operations has steadily expanded, so that management can today have a virtually up-to-the-minute account of stocks in hand and operations in progress. But here, as in engineering applications, the computer can have more than a passive role. It can take note of past records over any period and on that basis can 'decide' when reordering is necessary and how large orders should be.

Computers can carry out the more complex operation of monitoring the progress of work in factories. In a large chemical works, for example, production may involve more than a thousand measurable variables, on whose quantities may depend the most efficient control of some hundreds of valves. A computer, programmed to operate the works at maximum efficiency, will note the variables and either give directions for operating the valves or, if necessary, operate them without human intervention.

A comparable use can be found in the airline pilot's 'performance advisory system'. Until its introduction, the pilot could only calculate his most fuel-efficient altitude by using a variety of charts. Now the new system notes air temperature, engine pressure ratios and a number of other factors before automatically giving the most efficient altitude – and, it is claimed, saving up to 5 per cent of the fuel.

More familiar is the use of micro-computers in many everyday situations. In hundreds of shops computers enable the cost of a specific amount of food at any of a huge range of prices to be shown automatically on a counter-top weighing machine. Watches that not only tell the time but give the date and trigger a light or an alarm at any set time rely on micro-computers that have already made great inroads into the traditional Swiss watchmaking industry.

Photograph of the earth superimposed to scale
on a Voyager spacecraft computer-enchanced photograph of Saturn's rings.

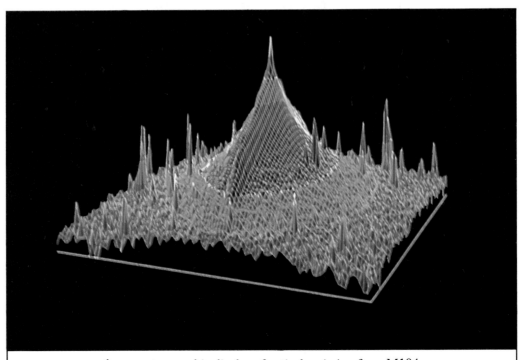

A computer graphic display of optical emission from M104
(Sombrero Galaxy) demonstrating the use of computers in astronomical research.

What is there left for the computer innovators? Long ago they introduced time-sharing, a system which enables a central data processor to be linked by cable to a number of input-output terminals and to process information from any terminal as and when required. And long ago they began to tailormake computers for special tasks or industries. What remains is the reduction in size to dimensions that would have seemed impossible even a decade ago, and operation at speeds which it is still difficult for the human brain to grasp.

Much of this progress is likely to come from increased knowledge of cryogenics, the study of materials and phenomena at temperatures close to absolute zero −273.15°C. Great things, for instance, are expected from use of the Josephson Junction, a device incorporating an insulating layer 15 atoms thick set between two superconducting films at −269°C. This forms the basis of ultra-fast switching circuits that are capable of processing 70 million instructions per second. And it is expected that before long they may enable the largest of today's computers to be contained within a single cubic foot.

However, the most important impact of computers in the future is likely to be in a sphere far removed from the wonder-world of what can be done in what space in what time. Doubts have already been raised that the growing use of computers by pupils in schools is eroding their mathematical ability. However, misgivings can perhaps be removed on the grounds that electronic methods of calculation for routine operations give a child more time to understand the fundamentals of mathematics. More important still is the change that computers are already beginning to bring about in man's concept of himself in the universe. 'The definition of man's uniqueness', Professor Herbert Simon has pointed out in *The New Science of Management Decision*, 'has always formed the kernel of his cosmological and ethical systems. With Copernicus and Galileo, he ceased to be the species located at the centre of the universe, attended by sun and stars. With Darwin, he ceased to be the species created and specially endowed by God with soul and reason. With Freud, he ceased to be the species whose behaviour was – potentially – governable by rational mind. As we begin to produce mechanisms that think and learn, he has ceased to be the species uniquely capable of complex, intelligent manipulation of his environment.'

Thus, in their fashioning of computers, and of the possibilities which they still hold out, engineers may have affected the future almost as decisively as they did when they succeeded in the most potentially apocalyptic of all their endeavours – releasing the power within the atom.

TOWARDS THE
ULTIMATE POWER

Most wars present engineers with new problems, tend to revolutionize their ideas, and often set them on the road to entirely unexpected goals. This was never more true than of World War II which not only brought into being the electronics industry – without which the tightly interlocked routing of international air traffic based on radar control would be impossible – but also helped to create the world of nuclear energy. More than any previous subject, nuclear power, first for use in weapons then as a provider of electricity, demanded the closest collaboration between engineers, technologists and scientists. This was hardly surprising in view of the extraordinary problems to be solved before the prospect of an 'ultimate weapon', could be translated from idea into blueprint and then into working example.

When Otto Hahn and his colleagues first split the uranium atom in Berlin in the winter of 1938 there was immediate speculation that the world might soon have access to a new sort of energy. The reason for optimism was simple. It had quickly become clear that when the nucleus, or core, of an atom of the heavy element uranium absorbed an unwanted neutron, the uncharged particle contained in all elements, then a dramatic process took place. First, a comparatively huge amount of energy was released; secondly, the process released one or more further neutrons from the split uranium nucleus. These in turn could, in the right circumstances, be used to split further nuclei in a block of uranium, thus starting what was described as a chain reaction.

From the early days, two things were realized. The first was that the release of energy, involving the transmutation of an atom into something different would, weight for weight, be on a scale vastly greater than that of chemical combustion in which, for instance, carbon atoms in coal were combined with oxygen atoms in the air. The prognostications of theory were eventually to be confirmed in practice. A ton of uranium would, if completely utilized, provide as much heat as 3 million tons of coal; an inch cube of uranium had as much energy locked within it as ninety wagon loads of coal. 'Completely utilized' is the conditional phrase. But although problems did, and still do, prevent the use of more than a smallish percentage of the nuclear fuel, it still offers options incomparably greater than those of conventional fuels.

Secondly, it was also appreciated that using nuclear energy in a weapon would require virtually instantaneous release of the energy whereas for industrial purposes the release would have to be controlled by some means so that it took place slowly. Thus there arose, in the early days, two lines of research, one leading to 'the bomb', the other aimed at 'the boiler' as it was called. However, before either of these got firmly under way it seemed likely, for a while, that nuclear energy might be a genie that was never to be released from its box.

Early in 1939 comparatively little was understood about the mechanics of nuclear fission, as the splitting process had been christened. It was known, however, that natural uranium consisted of at least three isotopes of the element – atoms containing the same number of protons and electrons but different numbers of neutrons – all apparently inextricably mixed together. The nuclei of each isotope contained 92 positively-charged particles known as protons; each nucleus was surrounded by 92 negatively-charged electrons, forever circling it. But while more than 99 per cent of uranium atoms had 146 neutrons locked to the 92 protons within each nucleus, giving each such atom an 'atomic weight' of 238, a small minority were different. About $7/10$ of 1 per cent of uranium atoms – about 1 in every 140 – had only 143 neutrons and were thus known as uranium 235 atoms, while an even smaller percentage were known to contain only 142 neutrons and were therefore known as uranium 234.

Now all these uranium atoms, obtained from uranium ore only after lengthy and costly processing, are identical in chemical properties and differ in physical properties only in those that depend on the mass number, that is, the number of particles in each nucleus. The different isotopes had, it is true, been separated in a few laboratories, but only in the most minute quantities and only after lengthy processing. It was, therefore, a shock when in February 1939 Niels Bohr, one of the world's most distinguished physicists, maintained that only uranium 235 was easily susceptible to fission and that most uranium 238 atoms hit by a neutron would merely absorb it.

This development appeared to rule out the prospect of nuclear energy for any purpose at all. The feeling was reinforced by the belief that an enormous amount of uranium 235 would be needed to start a chain reaction – the Frenchman Francis Perrin estimated 40 tons – and that, in any case, the complicated physics of the matter might result in a nuclear bomb fizzling instead of exploding. These fears were dispersed when, early in 1940, two scientists in Britain, Rudolf Peierls and Otto Frisch, worked out that the amount of uranium 235 required would be measured not in tons, not in hundredweights, but in pounds. Their calculation showed, furthermore, that a nuclear explosion would indeed be immense. But first the all-important problem of isotope-separation on more than a laboratory scale had to be solved.

The physicists and engineers soon realized that there might be three ways of solving the problem of separating the uranium 235 atoms from the rest. The first, gaseous diffusion, was based on the fact that if a gas containing natural uranium is diffused through the right kind of barrier, the lighter 235 atoms will pass through more quickly. Another method would be to subject uranium to electric and magnetic fields; each isotope would be deflected differently and it would thus be possible to collect them at different points in the equipment. But the magnets would have to be 100 feet (30.5 metres) long, and enormous amounts of electricity would be required. Thirdly, it would be possible to circulate a gas containing uranium between two concentric pipes, one being steam-heated and one water-cooled, the lighter isotope becoming concentrated around the inner pipe.

In Britain, where the decision to build the world's first nuclear weapons was taken in the summer of 1941, it was decided that gaseous diffusion should be the first option, and the problem was handed to Imperial Chemical Industries. The specification was for metal barriers in which there were 400 holes to the linear inch, or 160,000 holes to the square inch. The holes were to be 3 ten-thousandths of an inch in diameter and there was to be a tolerance of only 10 per cent in the mean hole size. The appallingly formidable problems which this demand presented were increased by two things: the barriers, 2 or 3 feet square, had to be produced in millions of square feet, and the gas which was to be filtered was uranium hexafluoride. Little was known about 'hex' except that it was very corrosive.

Eventually, the first diffusion barriers were made by adapting a printing process in which the printing plate consists of stainless steel; on it there is formed in a dotted pattern, by electro-deposition, an extremely thin film of copper, photographically laid down in much the same way as it is in a half-tone block. The process was to be adapted, readapted and then readapted again. It was to be of great use almost a decade later when supplies of uranium 235 were needed for Britain's peacetime nuclear reactors.

Before this, however, Britain's nuclear research was, due to the dangers of attack from German-occupied Europe, removed to the United States. And it was in the States that, under the control of the Manhattan Project, there arose the huge new factories where nuclear fuel was produced, and the research establishment at Los Alamos where the first nuclear weapons were fashioned. The factories were of a size and a complexity quite outside normal engineering practice. Thus the American Gaseous Diffusion plant at Oak Ridge, Tennessee, contained hundreds of acres of diffusion barriers, covering more than 50 acres (20 hectares) of factory floor, connected by many hundreds of miles of piping, and incorporating hundreds of joints and valves.

Before they were built it had to be confirmed that a nuclear chain reaction was possible in practice as well as in theory. The need was to produce a work of man that was in some ways the most significant he had created in all his long history – the world's first nuclear reactor, or 'pile' as it was originally called. Since 1939 it had been known that if the neutrons released during fission could be slowed down, then the chance of their creating further fissions would be increased. The reasons for this were not known but it appeared that if a chain reaction was to be brought about, then some form of moderator was required: some substance in which neutrons would bounce around until their speed was reduced to the level at which they were most likely to produce further fissions. It was important, however, that the neutrons should only be slowed down by the moderator and not absorbed by it; if they were captured by the moderator they would not, of course, be available to create further fissions. In the early days water and carbon were both suggested as moderators, and it was graphite, a form of carbon, which was used by the Italian physicist Enrico Fermi and his team in the world's first nuclear reactor.

The site was a squash court under the west stands of Chicago University's Stagg Athletic Field and here Fermi's team began to assemble in November 1942 nearly 6 tons of uranium metal and uranium oxide, some 40,000 blocks of graphite, of clinical purity and finished to the closest engineering tolerances, and ten long rods of the metal cadmium. The cadmium absorbed neutrons and as long as the rods were imbedded in the core, the nuclear reaction would not start.

If the purity and accuracy of the components gave a foretaste of the problems which had to be solved when nuclear energy entered the industrial world, other features were of a more rough and ready kind. 'The frame supporting [the reactor] was made of wooden timbers', one of the staff has commented. 'Gus Knuth, the millwright, would be called in. We would show him by gestures what

An atomic explosion in Nevada,
1951, one of a series of controlled nuclear bursts.

we wanted, he would take a few measurements, and soon the timbers would be in place. There were no detailed plans or blueprints for the frame or the pile.'

By December 1 more than fifty layers of uranium and graphite had been piled on each other – the origin of the word 'pile' which like 'tank' in World War I was an effective camouflage. The following day Fermi ordered the electrically operated control rods to be withdrawn. From the counters, each with a dial resembling a clock face with a single hand showing the neutron count, there came the clicking which indicated the rate at which neutrons were being produced.

It was early afternoon before the critical moment was neared. 'At first', wrote Herbert Anderson who had taken part in the pre-war nuclear research, 'you could hear the sound of the neutron counters, clickety-clack, clickety-clack. Then the clicks came more and more rapidly, and after a while they began to merge into a roar; the counters couldn't follow any more and their operating range had to be re-set. That was the moment to switch on the chart recorder and switch off the sound. But when the switch was made everyone watched in the sudden silence the mounting deflection of the recorder's pen. It was an awesome silence. Everyone realised the significance of that switch; we were in the high-intensity regime and the counters were again unable to cope with the situation any more. Again and again, the scale of the counters and the recorder had to be changed to accommodate the neutron intensity which was increasing more and more rapidly.'

Suddenly Fermi raised his hand. 'The pile has gone critical.'

For the first time in the history of the world a man-made self-sustaining nuclear chain reaction was running steadily.

Fermi waited for a couple of minutes, then ordered that an emergency safety rod be lowered into the reactor. The watchers celebrated with Chianti, drunk from paper cups. The news was telephoned to J.B. Conant, the US nuclear physicist in Harvard, in guarded words. 'The Italian navigator has just landed in the New World. The earth was not as large as he had supposed [meaning that the necessary pile of uranium and graphite was smaller than anticipated] so he arrived earlier than expected.'

The success in Chicago demonstrated that the physicists' theories had been correct. It demonstrated, also, that it should be possible to make a second nuclear explosive. This was plutonium, an element virtually non-existent in nature, which American physicists had deduced might be created by a chain reaction started with uranium. Calculations on both sides of the Atlantic showed that plutonium would be fissile and Fermi had now demonstrated that the nuclear pile for making it was a practical option. The outcome was two of the greatest industrial engineering plants ever built, the huge Clinton plant in Tennessee, later renamed Oak Ridge, and the 1,000 square mile (2,590 square kilometre) Hanford Engineering Works on the Columbia River north of Pasco.

From these plants there came the plutonium and the uranium 235 which at Los Alamos, New Mexico, were assembled into the weapons which destroyed Hiroshima and Nagasaki.

Despite the overriding need to make a nuclear weapon before the Germans did so, both the United States and Britain had by 1945 an awareness of the great changes which could be brought about by nuclear power. So, for that matter, had France whose Collège de France had in 1939 been a leader in nuclear research, and the USSR whose scientists had managed to continue with her own nuclear research even after the German invasion of 1941.

All the power-producing nuclear reactors which today dot the world have much in common, although it has been estimated that there are no fewer than 135 general classes of reactor and that about a thousand different reactors 'could be listed for theoretical consideration.'

With virtually all of them, the first problem is that of choosing an acceptable site, and the choice is almost always a difficult one. Three conditions have to be met, the first of which is that there must be constantly available a water supply of up to 1,200 million gallons (5,445 million litres) a day. This in itself rules out anywhere that is not on the coast, on a broad estuary or on a sizeable inland lake. Secondly, the sub-soil must be capable of bearing the multi-thousand ton weight of the nuclear reactor and its surrounding concrete shield, a fact which itself tends to make impracticable many riverine situations. The third essential is that the site should be as isolated as possible from large centres of population. In the early days it was decreed in Britain that no power station should be built within 5 miles (8 kilometres) of any town or village with a population of more than 10,000 inhabitants. There is no chance of a nuclear reactor exploding like a bomb, but the extremely remote chance of a release of radioactivity due to a serious accident can never be ignored.

These three requirements are comparatively easy to meet in large countries like the United States and the USSR but in Western Europe, and particularly in the small island of Britain, the nuclear energy industry has been forced on to many hitherto unspoilt areas. 'We have only to mention a possible site', one electricity official has said, 'and it becomes a beauty spot overnight.' Natural as the feelings of conservationists are, it must be admitted that similar protests have been made over the centuries against canals, railways, roads, airports and telegraphs. Sir John Cockcroft, once the head of Britain's civilian nuclear energy programme, has in fact commented: 'These new power stations can appeal by their scale and sense of power derived from the innermost sources of nature. We have a challenge to fit them into the landscape, so that future generations will look on them with admiration mixed with awe, just as earlier generations looked on Bamburgh Castle or the Forts of the Saxon Shore.'

The world's different nuclear reactors all fall into one of two basically different groups. There are the thermal reactors in which neutrons slowed down by a moderator produce energy by splitting atoms of uranium 235. And there are fast reactors in which neutrons which have not been slowed down by a moderator release energy by splitting atoms of plutonium, itself a product of uranium 238.

Thermal reactors, which comprise the great majority of the world's power-producing nuclear reactors, can again be split into two groups based on the moderator which they use. For various reasons, scientific, engineering, and financial, there are only two substances which can be used to slow down neutrons. One is graphite, the other is water. Graphite is used as the moderator in the Magnox reactors which are supplying energy from the first generation of Britain's nuclear power stations, and in the Advanced Gas-Cooled Reactors which are providing energy from the second generation. Ordinary (light) water is used both as the moderator and coolant in Pressurised Water Reactors (PWR) and Boiling Water Reactors (BWR) while in Canada's CANDU (Canada Deuterium Uranium) heavy water is used not only as the moderator but also as the coolant for transferring the heat from the reactor core to the boilers which produce steam to drive the turbines which in turn create electricity. Heavy water, more correctly known as Deuterium Oxide (D_2O), is the best moderator available but also the most expensive.

Certain features form part of all nuclear reactors. They include a controlling system which allows the nuclear reaction to be slowed down or speeded up as required, and a coolant which carries away heat from the reaction so that it can be used to boil water, to produce steam, and operate a turbo-generator which generates electricity.

The engineering problems posed by building reactors can be estimated from a few figures. The concrete foundations, weighing many thousands of tons, have to be laid to an accuracy of $1/100$ of an inch. The reactor itself can weigh 50,000 tons while the shielding necessary to contain the radioactivity produced in the core can add another 150,000 tons. The pressure vessel housing the reactor can be 50 feet (15 metres) across, while the boilers which generate the steam to drive the turbines can be some 80 feet (24.4 metres) high. And at one typical power station each boiler contains more than 26,000 feet (about 5 miles or 8 kilometres) of tubing.

The very size of some components have created engineering problems of their own. At the Hunterston nuclear power station in Scotland, for instance, a 1,200 ton giant crane had to be designed to lift 350 tons and then installed so that there was beneath it a huge working space 200 feet (61 metres) wide and 200 feet high. At other British power stations special cranes capable of lifting 400 tons have been installed.

The concrete shielding, pierced in dozens of places by steel tubes enclosing gas ducts, steam pipes and electric cables, makes its own special problems. The need to ensure that the concrete sets firmly round the tubes without forming air pockets means that it must have a high water-to-cement ratio; but this itself would mean that as the concrete dried out it would shrink unduly and thus cause trouble. So a special compromise usually has to be worked out.

Inside the concrete shield there is a steel liner and a thermal shield which decreases the amount of heat lost through the concrete. At the centre of the reactor there is the steel pressure vessel containing the moderator and the nuclear fuel. In the Berkeley nuclear power station, for example, the pressure vessel was 80 feet (24.4 metres) high and 50 feet (15 metres) across, built up from scores of 3-inch-thick (7.6 centimetres) steel plates. Each piece had to be accurate to $\frac{1}{16}$ of an inch (1.6 millimetres) , and was made in the manufacturer's works from a template itself accurate to $\frac{1}{64}$ of an inch (0.4 millimetres). The separate parts of the pressure vessel were welded together on the site and the men picked for the job attended a month's course before they began work. Even then, each man was able to weld only 6 inches (15 centimetres) of the 3-inch-thick plate in an eight-hour shift. Total time for welding the pressure vessel reached more than a million man-hours.

Only after all the welds had been tested by X-ray radiography could the pressure vessel be stress-relieved. During this operation it was heated to 620°C and kept at that temperature for ten days, accomplished by swathing it in thick insulation and then heating it from the interior by miles of special electrical heating elements.

Today, the problem is slightly less difficult since it has been found possible to combine both the shield and the pressure vessel in a single structure of pre-stressed concrete lined with ductile steel. This is only one of many cases in which the demands of nuclear engineering have pushed ahead the progress of other technologies on which they are dependent. Another example followed the erection of the Calder Hall station in northern England. It was considered impossible, on the Calder Hall site, to weld to the stringent specifications, steel plate more than 2 inches thick (5 centimetres). This in turn limited the pressure inside the pressure vessel to about 114 pounds per square inch (7.9 bars); and the amount of heat which can be carried away from the reactor core in the pressure vessel is in turn limited by the pressure. The demand eventually brought about new methods enabling men to weld plates up to 4 inches thick (10 centimetres). The result was that subsequent stations had gas passing over the reactor at more than two and a half times the previous pressure, and thus producing considerably more electricity for every pound of fuel used.

Inside the pressure vessel lies the heart of the thermal nuclear power station, the reactor itself, consisting of the nuclear fuel, the moderator which slows down the released neutrons, and the control rods by which the speed of the fissioning process can be adjusted. These items, their sizes and configurations, naturally vary according to the kind of reactor, but they have some characteristics in common – notably the great weight involved, the extraordinary high standards of accuracy to which they must be made and the clinical cleanliness – a proper phrase since the degree of cleanliness has to be as high as that of a hospital – with which they must be assembled.

Whatever the fuel and whatever the moderator, the modern reactor will probably weigh more than 80,000 tons, and supporting it adequately presents both mathematical and structural problems. For the diagrid on which it rests must bend

The control rod drive in a high temperature gas-cooled nuclear power station – Peach Bottom, Pennsylvania.

no more than a small fraction of an inch while the pressure vessel itself expands and contracts as a result of the changing temperature inside it. Design of the diagrid for Britain's Calder Hall station was so complex that a computer was used to solve the thirty-five simultaneous equations involved.

On top of the diagrid lie thick steel plates and, in a graphite-moderated reactor, the components of the reactor are assembled on them: graphite blocks and the uranium fuel contained in long thin cans. The graphite is prepared by a carefully controlled chemical process in which petroleum, coke and other raw materials are bonded together by pitch or tar and are then heated in a kiln to about 1,500°C. The resulting substance is then pressed into long columns which are automatically chopped into lengths of a few feet and put into another kiln where they are brought up to about 3,000°C. There can be up to 100,000 graphite blocks in a reactor, and their total weight may be almost 4,000 tons. Each has to be machined to an accuracy of a few thousandths of an inch before it is marked with an identifying number, cleaned with a small vacuum cleaner, sealed in a transparent wrapping, and then brought to the reactor site. Here the thousands of blocks have to be positioned accurately to within 1/500 (0.05 millimetres) of an inch in the three dimensions.

Into what is by now a honeycomb of graphite blocks there have to be placed not only the fuel rods but also up to a thousand thermo-couples which will automatically measure the temperature at different points within the reactor, and the all-important control rods, of which there may be as many as a hundred. These rods are of boron, cadmium, or some other strong absorber of neutrons, encased in stainless steel or aluminium. When let down further into the reactor core they slow down the nuclear chain reaction; when drawn up they allow the nuclear reaction to increase.

The processing of the uranium ore and the separation and packaging of the uranium into metal containers, present a good illustration of the great complexities of nuclear power production and of the engineering problems that have to be overcome. The uranium ore is first dissolved in nitric acid, the impurities are filtered off, and the liquid is submitted to a series of complicated processes which finally yield a green powder, uranium tetrafluoride. This is mixed with shredded magnesium and compressed into 7 pound (3 kilogram) billets which are then heated in a special vessel. From the resulting billets of pure uranium, the fuel rods are cast and machined accurately to size. Their size and number depend on the design of each power station, but at the Berkeley power station on the River Severn, 42,445 rods, each 19 inches (48 centimetres) long and 1.1 inches (2.8 centimetres) in diameter, are stacked, in columns of thirteen, into more than 3,000 channels in the reactor core. At Bradwell, on the east coast, 20,970 36-inch (91 centimetre) rods are stacked, eight high, in 2,837 channels.

Each fuel rod is individually enclosed in a tight-fitting can made from a magnesium-aluminium alloy called Magnox. The can forms a physical barrier between the uranium and the carbon dioxide gas which is circulated throughout the reactor to carry away the heat to the heat-exchanger units, and it prevents the radioactive fission products – some of which are gaseous – created when the reactor is operating, from mixing with the coolant and being transported to the heat-exchangers.

The fuel of the fast reactors, the latter so-called because they use fast neutrons whose speeds have not been slowed down by means of a moderator, is primarily plutonium which is produced as a by-product in thermal reactors. However, the unique feature of the fast reactor is that it can be designed to produce more fuel than it uses. This is done by surrounding the core of the reactor with a blanket of uranium 238 – depleted uranium which has had most of the fissionable uranium 235 used up in the thermal reactors. This non-fissile isotope of uranium is, on capturing a fast neutron, transmuted into the fissile isotope plutonium 239. If the reactor is designed in the correct way it can not only supply power but at the same time breed more plutonium from the uranium blanket than it uses in the core and, in so doing, make use of the uranium 238 that was of no use in thermal reactors.

The outcome in increased energy production is staggering. Some years ago it was estimated that in Britain alone there was available some 20,000 tonnes of depleted uranium 'left over' from thermal reactors. But used in fast reactors this could deliver energy equivalent to some 40,000 million tonnes (metric ton = 2204.6 pounds) of coal, or about three hundred years' supply at the present rate of extraction. This is equivalent to 23,000 million tonnes of oil, or about twenty-five times the amount that is expected to be extracted from the North Sea oilfields.

The nuclear engineers have not found it easy to exploit the advantages of the fast reactor. Since no moderator is required the reactor itself can be smaller than a thermal reactor, and this itself makes for obvious economies. But the rate of heat production is extraordinarily high because of the higher power density, and the need for devising some method by which this heat could be drawn off quickly enough was one of the engineers' first problems. It was solved by the use of liquid metal as a coolant – usually sodium or various alloys of sodium and potassium.

Serious consideration of the fast reactor's engineering problems was begun in Britain in the early 1950s. The first small experimental fast reactor in Britain was built at Harwell in 1954 and a second one built in the following year. These reactors were used to make practical checks on the physics of the fast reactor system. Results were so encouraging that a start was soon afterwards made on what was to become one of the most photographed works of the nuclear age. This was the Dounreay Fast Reactor, built in the far north of Scotland on the bleak shores of the Pentland Firth and enclosed in its huge white spherical container. The problems were immense and the feelings of the reactor engineers when the project was sanctioned are shown by an extract from a review of them. 'At first sight this fast reactor scheme appears unrealistic,' it stated. 'On closer examination it appears fantastic. Scientists solve the problems they can. Engineers solve the problems they have to.' The DFR, as it became known, began operating in 1959, was quickly brought up to its maximum electrical output of 14 megawatts, and since then has pumped more than 500 million units of electricity into the Scottish grid before being closed down in 1977. Before this happened a full-scale Prototype Fast Reactor, (PFR) capable of generating 250 megawatts of electricity had been built and had reached full power in 1976.

The whole concept of the fast reactor has been attacked on the grounds that it uses plutonium, the most dangerous material in the world as it has been called. So much so that the possible precautions necessary for preventing hijacking of the material by terrorists, and the safeguards that might become necessary at fast reactor sites, have nourished fears of what has been called 'the plutonium society'. Nevertheless, both the USSR and France have for many years been operating fast reactors comparable to the larger of the two built at Dounreay. Germany and Italy have collaborated in building a 1,200 megawatt fast reactor. The Americans long ago built the world's first commercial power plant using a fast reactor, the Enrico Fermi Atomic Power Plant outside Chicago.

Of the many hundreds of reactors which are appearing in increasing numbers throughout the world, across the breadth of the United States, on the shores of Japan and throughout France, the majority are designed to generate electricity. But other potential uses of nuclear power became apparent in the aftermath of nuclear fission's discovery. One was in the field of air transport, although here engineers had to grapple with the formidable problem of designing shielding whose weight did not make any design impracticable. This difficulty, as well as the potential dangers from a crashed nuclear-powered aircraft, finally seems to have ruled out this application, although two nuclear-powered planes, a ramjet and a turbojet, were for a while under development in the United States. The same difficulties of shielding and danger have so far ruled out nuclear railway engines and nuclear road vehicles.

However, the development of nuclear-powered ships has been a different story. Shielding is, of course, quite as important on sea as on land; space is confined, while the rolling of a ship presents particularly difficult problems. The control rods in the reactor, for instance, must be so designed as to be virtually unjammable and the reactor itself is so designed to continue working just as efficiently in a heaving vessel as it does on the rock-steady concrete foundation of a power station site. These problems were overcome and as long ago as 1954 the United States launched the US *Nautilus*, a nuclear-powered craft which has been called the first true submarine as distinct from a submersible boat since it could

The UK Atomic Energy Authority's
Prototype Fast Reactor at Dounreay, northern Scotland

stay at sea for more than a year without surfacing to charge its batteries. The *Nautilus*, the first ship to sail under the North Pole, used 2.8 pounds (1.27 kilograms) of uranium 235 for its first 62,000 miles (99,800 kilometres) of travel, instead of the 720,000 gallons (3,273,000 litres) of oil a conventional submarine would have needed. The costs are estimated to have been ten times greater than those of an oil-propelled ship but that was, and is, of little importance when measured against the great strategic advantages of a vessel with such freedom of action.

Five years after the launch of the *Nautilus* two other nuclear-powered vessels were at sea, very different in their aims but each demonstrating some of the advantages of nuclear propulsion. One was the NS (Nuclear Ship) *Savannah*, an American merchant ship launched in 1959 whose 3,650 ton nuclear power-producing equipment had the distinction of being lighter than would have been the comparable oil-fired boilers and bunkers. As important was the *Lenin*, a USSR ice-breaker specifically designed for operations along the USSR's Arctic coastline, launched in 1957 and fully operational two years later. Powered by two reactors, with a third held in reserve, the *Lenin* is rated at 44,000 shp., an extraordinarily high rating for a 16,000-ton ship, and one which allows her to plough through 6 feet (1.83 metres) ice at 2 knots (3.7 kph). With an ice-free speed of nearly 20 knots, and capable of travelling all-out for a year without refuelling, the *Lenin* has transformed maritime operations along the north coast of her country. The pressurized water reactors each independently produce steam for the turbines; these drive electric generators whose current is fed to electric motors which in turn operate the propellers.

Nuclear-powered surface ships have not been developed in the last quarter-century with the speed that might have been expected. Operating costs are likely to be the main reason although they are more than counterbalanced by the specialist advantages of the *Lenin*. As for submarines, nuclear power has continued to take over from oil and some scores are now estimated to be in operation throughout the world.

Meanwhile, nuclear engineers have been developing miniaturized reactors of a size inconceivable at the start of the nuclear age. The Soviets have built a 'pocket power station' which can produce one megawatt for two years from a few hundred pounds of fuel, compared with the 4,000 tons of diesel fuel needed for a conventional engine. In America the SNAP (System for Nuclear Auxiliary Power) scheme has introduced numerous miniature reactors for specific tasks, some of them as small as 14 by 13 by 18 inches (36 by 33 by 46 centimetres). Remote sites such as early warning posts in the Arctic are only among the more obvious of the places where they can be used to good effect, and SNAP devices have been used in American and USSR spacecraft.

In the United States the increase in the number of nuclear power stations has dramatically slowed down as fear of accidents has grown, but elsewhere they continue to be built and France, Sweden and the USSR as well as Britain, have ambitious plans for the rest of century. Apart from engineering and safety problems, as well as the question of waste disposal, which the nuclear power industry is perpetually claiming as on the verge of solution, one riddle remains under controversial discussion. Do nuclear stations produce electrical power more, or less, cheaply than those using coal or oil? 'It all depends' is the almost inevitable answer, and this is true whatever country one is talking about.

On some of the factors involved there can be little argument. The grid pylons and cables, the transformers and switchgear – all the plant needed for the distribution of electricity – is the same whether the electricity comes from a coal-, an oil- or a nuclear-powered station. However, the reactor and the heat-exchangers will be a great deal more expensive than anything required in a conventional station.

Compared with these relatively simple factors there are many which are more complex. It is not easy to estimate what the cost of coal, oil or nuclear fuel is likely to be in the future years when a power station is operating. Cost per unit of electricity will depend on the length of time during a power station's life when it has to be closed for maintenance. The comparatively high capital cost of a nuclear station will make it more vulnerable than a conventional station to a country's financial position; and technological advances in energy production from oil, coal or nuclear fuel, made during the years when a station is being built or operated, can play havoc with even the most carefully worked-out estimates. Thus the position today is basically the same as that described some twenty years ago by a British expert in nuclear power. 'Estimates of the cost of nuclear power', he wrote, 'at present contain many elements of conjecture and should be treated with reserve. Estimates of capital components depend upon, among other things, assumptions about interest rates and life of plant, and estimates of fuel replacement costs involve assumptions about enrichment, irradiation, and value of spent fuel; all of these are subject to wide variations and for many of them there is so far little supporting evidence from experience.'

If nuclear fission has demanded much of engineers, it has made a return which even today, almost half a century after it emerged, is less appreciated outside the engineering professions than it should be. The internal combustion engineer is able to determine just how quickly piston rings wear by incorporating a radioactive isotope in their metal and measuring the minute amounts that appear in lubricating oil filters. Makers of steel sheet – or of paper – are able accurately to control the thickness by measuring the amount of radiation passing through the sheet from a known radioactive isotope. Leaks in underground water or oil pipes can be traced with comparative ease by passing a radioactive tracer in the form of a liquid through the pipes and using a Geiger counter to detect the radiation emitted by the leaking liquid.

These are but a few examples of the ways in which radioactive isotopes – frequently called radioisotopes – are being used in engineering, industry, agriculture and medicine. Their value can be realized when it is appreciated that the radioactive version of an element is chemically identical with its non-radioactive isotope, but that the position of the former can be found, by the use of an instrument such as a Geiger counter. Radioactive versions of most elements can be obtained by irradiating them in a nuclear reactor and about a thousand radioisotopes of the naturally occurring elements have been produced in this way. About six hundred of them have a sufficiently long half-life – the time taken for the activity of a radioisotope to decay to half its original value – to make them potentially useful, and many hundreds of them are obtainable from commercial suppliers.

Measuring wear – of motorcar tyres or of piston rings – is only one of the more obvious engineering uses. Radioactive gases have been pumped along underground telephone cables, as well as along pipelines, to help trace breaks.

Radioactive tracers have also been used to follow the movement of mud in rivers and harbour entrances, thus enabling the dredging engineer to prevent silting-up. The movement of radioactive pebbles has been tracked to help stop coast erosion. British engineers have even solved an important problem in the Middle East by putting radioactive tracers in a number of desert streams, waiting for the materials to seep down and reappear on the far side of a range of hills, and thereby reveal whether a certain site could, or could not, be used for a big new reservoir.

There is also a growing number of uses for radiation in the plastics industry. One of the most used materials here is polyethylene which can be found in a wide variety of household goods. For years polyethylene was made by treating ethylene at a temperature of about 200°C and a pressure of about 1,500 pounds to the square inch (103.42 bars), a process which causes the ethylene molecules to polymerize, in other words to form long-chain molecules. It was then found that if the ethylene was first irradiated, polymerization would take place at room temperature and at about a tenth of the previously required pressure. This not only reduced the complexity and cost of the process but resulted in a final product with more convenient properties. Irradiation was also found to affect cross-linking, the joining up of polymer molecules of which rubber vulcanization is one example. If the process is induced by irradiation, the vulcanized product has a number of new and useful properties, including resistance to brittleness at high temperatures.

In commerce, radiation can be used to check whether packets on a high-speed production line are being properly filled, while for both agriculture and medicine the by-products of the world's nuclear reactors provide tools that had never before existed. The manner in which growing crops utilize fertilizer had always been a dark mystery despite all the efforts of botanists. But after some of the fertilizer had been irradiated in a reactor it became possible with the help of a Geiger counter to follow its course through the living body of the growing crop. In the same way it is now possible to follow the track of the normal 'food and drink' of plants and trees merely by ensuring that a small percentage of such food and drink is radioactive. In the same way a suitable radioactive preparation can be followed during its passage through a human body to reveal exactly how such organs as the liver and kidneys are functioning.

Radioisotopes have also been found of great value in food processing, an area where engineering, commerce and science all meet. Much food 'goes bad' by the multiplication in or on it of minute bacteria, a popular example being fish on which a single bacterium, doubling its number every half hour can turn into more than 64,000 in about nine hours. Irradiation can stop the process, just as it can stop the sprouting of potatoes in storage. The ability of radiation to kill off micro-organisms has been exploited in Australia to stop the growth of anthrax bacilli in carpets, and in Britain the method has been found ideal for sterilizing syringes, scalpels, and other medical instruments.

In all these cases the dangers of radioactivity, against which engineers have to take such complex preventive measures in their reactor designs, have to be avoided – by ensuring that living organisms are shielded from dangerous radiation. However, the potentially damaging effect of radiation on living organisms has been turned to good account in a variety of ways. Genetic change caused by radiation has brought about new varieties of flowers and crops 'tailor-made' to flourish in specific areas and climates. More dramatically, radiations have sterilized large batches of insects which, no longer able to propagate, have brought success to major insect-control programmes.

But it is in medicine that some of the most spectacular benefits from nuclear research may yet emerge. Since the early years of the century it has been known that the radiations emitted by such elements as radium could in some cases kill off cancerous growths in the human body. Knowledge of how this could be brought about was accumulated only slowly and was, until almost the start of World War II, limited by the cost of radium, the most practicable radioactive element to use at that time. The situation was transformed by the post-war ability to create radioactive materials which would throw out their radiations at just the required strength for just the right time. A classic illustration of what is now possible can be cited from Sweden. There, gamma rays from a number of radioactive sources of the same type were directed on to a tumour in the brain of a patient. Where the radiations passed individually through different parts of the skull, they had little effect; where they met, they killed off the tumour.

While the engineering techniques and standards deployed in the construction of nuclear power stations were of an order that would have been unattainable before World War II, some of the most remarkable nuclear work of the post-war world has been carried out for research rather than power production. Much of the early information about atoms had been obtained by the use of accelerators which speed up atomic particles and direct them on to a target-particle. If the speed is great enough the target-particle is shattered and from the speed, trajectory and weight of the fragments it is possible to extract information about the nuclear structure.

The first accelerator was the work of Van de Graaf at Princeton in New Jersey in 1929. John Cockcroft and E.T.S. Walton built another with which they artificially split the atom in Cambridge in 1932. In the same year, Ernest Lawrence of Berkeley, California, built the first cyclotron, an accelerator in which he used a magnetic field to make the particles travel a spiral path between two hollow semicircular D-shaped electrodes called 'dees'. At each half-revolution between the dees, the particles received an additional burst of energy from an oscillating voltage, the last addition being made just before the spiral route ended at the target. James Chadwick built Britain's first cyclotron shortly before the outbreak of war in 1939.

Within a few years nuclear physicists were beginning to realize that there were more kinds of particle than the positively-charged proton, the negatively charged electron and the non-charged neutron. About a hundred have now been found, largely due to of speeding-up particles in accelerators to ever greater velocities.

The accelerators are usually housed in a circular underground tunnel on account of the radioactivity they can produce. Some of the tunnels – including those at the CERN (Conseil Européen pour la Recherche Nucléaire) laboratories outside Geneva, the Brookhaven National Laboratory on Long Island and at Stanford University, California – are more than a mile in circumference. In a different kind of accelerator at the Fermilab laboratory, Batavia, Illinois, the tunnel is 4 miles (6 kilometres) long and particles traverse it 50,000 times per second before hitting targets with an energy of some 500 billion electron-volts.

The engineering problems which have to be solved when building such equipment can be judged from the figures of the synchro-cyclotron at CERN, the organization set up in 1953 and now co-ordinating much of the nuclear research of more than a dozen European countries. The yoke of the 2,500 ton electro-magnet consists of eighteen magnetic steel plates 12 yards (11 metres) long, 1.64 yards (1.5 metres) high and 14 inches (36 centimetres) thick. The magnet is excited by two enormous 55-ton coils made up of nine pancakes of aluminium conductors measuring 19 square centimetres, in cross-section. Moving the coils from Belgium where they were made, on barges up the Rhine and then through Switzerland on a special trailer to Geneva, was a major operation in itself. The 18,500 gauss magnetic field set up by the magnet − a figure that compares startingly with the earth's 1 gauss magnetic field − is applied across a vacuum tank between the magnet's poles. In the tank the vacuum is roughly 76 million times less than that of the earth's atmosphere, a pressure created and maintained by two large vacuum pumps. The measures to stop radiation are formidable. Thus the walls which enclose the synchro-cyclotron are 6.34 yards (5.8 metres) thick and are fitted with two 200-ton concrete doors. Some 22,000 tons of concrete in all have been used in the protecting walls.

The better understanding of the sub-nuclear world to which research organizations such as CERN are leading may eventually give the world the virtually limitless power of nuclear fusion, the process which instead of splitting apart the components of heavy atoms, as in nuclear fission, unites the nuclei of light atoms. In both processes enormous amounts of energy are released. But fusion has a number of quite definite advantages over fission. For one thing its potentially best 'raw material' is not uranium which is costly to find and to process, but the hydrogen in ordinary water, available in virtually limitless supply. Yet the energy potential is enormous, a cubic foot of water containing as much energy as 10 tons of coal. And since none of the by-products need be radioactive, the unsolved problem of satisfactory waste disposal would disappear overnight.

For the last twenty-five years physicists and nuclear engineers in America, Britain, France and the USSR have been working on the problem of controlled nuclear fusion. There has been more than one rumour of coming success but none has been turned into fact. The heart of the problem lies in raising those atoms which are to be fused together to a temperature of about 100 million°C and holding them together in a sufficiently dense mass for a few seconds. At such temperatures the materials to be fused are turned into plasma, an ionized gas which has been called the fourth state of matter. There is no chance of physically confining the plasma at these temperatures but it can be influenced magnetically, and research has concentrated on heating it electrically and confining it by one of various magnetic methods.

Fusion reactions have been created in the laboratory for many years using the deuterium-tritium (D-T) reaction, but the amount of energy 'put in' has always exceeded the amount of energy 'got out'. The huge efforts now being put into fusion research are aimed at producing equipment which will yield a net energy gain, and only when this has been done will it be possible to tackle the problems of turning this fundamental source of energy into useable electricity. But that, it is already known, will involve engineering problems even more formidable than those which once barred the way towards power production by nuclear fission.

Inside the transformer holding the Joint European Torus at Culham where research into nuclear fusion is being carried out.

INTO SPACE

The most extraordinary of all man's engineering marvels are the two American Voyager probes travelling in the early 1980s through the immensities of outer space. Blasted off from Cape Canaveral, known as Cape Kennedy from 1963 to 1973, in August and September 1977, they have already sent back to earth spectacular photographs and scientific information from the approaches to such distant planets as Jupiter and Saturn, and are scheduled to continue their mission for a total of twelve years with an investigation of the even more distant planets Uranus and possibly Neptune.

These sophisticated vehicles, carrying cameras, infra-red and ultra-violet spectrometers and low-field magnetometers as well as a mass of other equipment are the latest results of man's ambitious plans to explore space, plans which even more directly than most engineering achievements, have come to fruition following the demands of war.

Ever since man realized that the twinkling points of stars seen in the night sky were physical entities, he has sought to stretch out into space, first of all towards the moon whose mountains and craters were revealed by Galileo and his telescope in 1609. Galileo's discoveries, and those of his near-contemporaries, Tycho Brahe and Johann Kepler, were quickly followed by fantastic speculations on the future. Francis Godwin, bishop of Llandaff and Hereford, wrote shortly before his death in 1633 *The Man in the Moon*, published in 1638 in which the moon was visited in a chariot drawn by swans. Shortly afterwards, and possibly as a result of Godwin's book, John Wilkins, bishop of Chester, published also in 1638 *The Discovery of a World in the Moone; or a Discourse Tending to Prove, that 'tis probable there may be another habitable World in that Planet*. A few years later Cyrano de Bergerac published his fantasy in which the moon could be visited in a vehicle carrying bottles of dew which were drawn upwards by the sun's rays. Significantly enough, the vehicle in a later version is propelled by rockets.

However, it was only with the publication of Isaac Newton's *Principia Mathematica* in 1687 that the real problems of space travel began to be spelled out. Newton's law of gravity gave men for the first time an idea of the force needed to carry them beyond the pull of the earth's gravity. At the same time, his statement that every action produces an equal reaction suggested the utility of the rocket, the method of propulsion that was centuries later to become the key to space travel.

Although Edgar Allan Poe sent Hans Pfall to the moon in a balloon and Jules Verne popularized the idea of space travel with his two-part novel, *From the Earth to the Moon* in 1865 and with *A Trip Round the Moon* also published as *All Around the Moon*, five years later, it was left to H.G. Wells with his *First Men in the Moon*, published in 1901, to present at least a pseudo-scientific plan for space travellers. Cavorite, a material which destroyed gravity, solved what potential moon travellers by now appreciated would be their first problem.

The Wellsian Cavorite was described shortly before the first serious approach to space flight was made. It came from Konstantin Tsiolkovsky, a Russian born in 1857 who during the last years of the nineteenth century wrote *Exploration of Cosmic Space by Reaction Apparatuses*, published in 1903. Tsiolkovsky not only decided that rocket propulsion offered the only chance of escaping from the earth's gravity but also maintained that liquid fuel would be far more efficient than solid. This was significant since the only rockets in common use were versions of the Congreve solid fuel rocket, developed from existing rockets as a weapon by Sir William Congreve almost a century before – and the source of 'the rockets' red glare' in Francis Scott Key's *The Star-Spangled Banner*. Tsiolkovsky also proposed the use of movable vanes in a rocket's exhaust to give directional stability and the principle of using one rocket engine to boost another engine in flight, both ideas to be used half a century and more later.

The Russian pioneer received little official support, and the same is true of the men and groups of men who during the first three decades of the twentieth century believed that exploration of space would one day be possible. In Britain, France, Germany and the United States, Societies for Space Exploration were founded, and in Germany Hermann Oberth studied physiological aspects of space flight and drew up plans for a long-range rocket which he sent to the Army authorities. They dismissed the plans as worthless.

It was in America that Robert Hutchings Goddard eventually gained support for his experiments in rocketry, although after some difficulty and delay. In 1919 Goddard submitted to the Smithsonian Institution *A Method of Reaching Extreme Altitudes*, outlining a plan for sending up rockets for meteorological research but at the same time prophesying that man might be able to escape from the earth's gravity. Money for practical work was hard to obtain but support finally came from the Smithsonian Institution and on 16 March 1926 Goddard was able to make the first successful firing of a liquid fuel rocket. This was at Auburn, Massachusetts, where he turned on the liquid oxygen and petrol valves and lit the mixture with a blowlamp. The fuel lasted for 2½ seconds and moved the 10 foot (3 metres) rocket for about 184 feet (56 metres) at a height of 41 feet (12.5 metres) at a speed of 60 mph (97 kph). Goddard was next allowed to use an Army range for further tests – but only after a fall of rain or snow when the dangers of starting a fire were considered to be less.

During the next decade Goddard began to solve the problem of stabilizing and then controlling the rocket's course by the use of gyroscopes operating on the directing vanes. He developed fuel pumps to ensure an even supply of fuel to the engine, and he experimented with different methods of cooling the engine itself. But this work, useful as it was, concentrated more on the use of rockets for research such as meteorological investigation than on the ambitious project of sending a man to the moon.

It was in Germany that the next decisive step was taken, a step which was a direct outcome of the Treaty of Versailles, signed by the combatants at the end of World War I. The Treaty forbad Germany to make heavy artillery but it made no mention of rockets, and in 1929 Professor Doctor Karl Becker, Chief of the Ballistics and Munitions Branch of the German Army, decided that rockets might fill the gap in its armoury.

Becker chose for the task Walter R. Dornberger, an officer-engineer who has said that his orders were as follows: 'You have to make of solid rockets a kind of weapon system which will fire an avalanche of missiles over a distance of 5 to 6 miles, so as to get an area effect out of it. Next, you have to develop a liquid rocket which can carry more payload than any shell we have presently in our artillery, over a distance which is farther than the maximum range of a gun.'

The end-product of this historic order was eventually the V-2, the rocket missile which, ironically, was a starting-point for the rockets which were to put America's first astronauts into orbit. The V-2 was planned and built at Peenemunde on the Baltic coast of Germany, largely the brainchild of Wernher von Braun, much helped by Hermann Oberth who in 1938 started rocket work for the Luftwaffe and in 1941 joined the German teams working on a wartime rocket. In October 1942 a test rocket fuelled with liquid oxygen and alcohol, rose at Peenemunde to a height of over 50 miles (80 kilometres), reached a speed of nearly 3,500 mph (5,600 kph) in the process and landed only 2½ miles (4 kilometres) from its target, 120 miles (193 kilometres) from the launching point. 'The technical feasibility of a big guided rocket', Dornberger wrote, 'had been proven for the first time in history.'

In 1945 Peenemunde was overrun by the advancing Russian armies, and although about a thousand Germans moved south before the Russians arrived, and more than a hundred eventually settled in the United States to work for the American space programme, more than 4,000 went to the USSR. Both countries made use of the Germans, but the Russians were able to do so more thoroughly, experimentally firing hundreds of captured V-2 rockets and building successors that increased the range from 200 miles to 700 (320 to 1120 kilometres).

The exploitation of the scientists at Peenemunde – and of the captured archives – marked the real watershed dividing the speculative investigation of

Robert H. Goddard, the US rocket
pioneer at a blackboard in Clark University, Worcester, Mass., in 1924.

A German V-2 rocket, being prepared for test-firing
by the British in 'Operation Backfire' in north Germany in September 1945.

space to the potentially practical. Development of the V-2 carried with it the
promise that rockets would soon be able to be pushed up to the point where they
would escape the earth's gravity and could be put into orbit round the earth.

There were, however, other contributory reasons which were almost as
important in making the exploration of space a practical proposition. The
electronic computer, now on the verge of development, was an essential tool for
working out the courses that rockets should be made to follow in space. The
development of radar which could keep track of rockets in flight was another
near-essential. So was the mass of electronic equipment which has grown out of
the war, and especially the art of miniaturization which in such equipment as the
proximity fuse had allowed the most sensitive devices to be incorporated in
individual shells.

These war-time advances enabled both Americans and Russians to embark
on space programmes within less than a decade after the war ended in 1945. Little
is known of the first decisions which put Russia on the path to space. More, as one
might expect, is known of the developing American story.

Within a few weeks of the war ending, a hundred captured V-2s had arrived
in the United States. Von Braun arrived a few months later and was soon
cataloguing the thousands of documents which the Americans had captured, and
helping to test-fire a number of V-2s. In 1950 most of the remaining rockets, and
most of the German scientists, who had now been joined by Dornberger, were
moved to the Redstone Arsenal in Huntsville, Alabama, where work was started
on improving the rockets. The outbreak of the Korean War in the summer of
1950 speeded up the work and led to an Army Ordnance demand for a mobile

rocket with a range of 500 miles (800 kilometres).

None of this was directly linked with attempts to investigate space, and it was only in July 1952 that the US National Advisory Committee for Aeronautics passed a resolution urging that 'NACA devote modest efforts to problems of unmanned and manned flights at altitudes from 50 miles to infinity and at speeds from Mach 10 to escape from the earth's gravity.'

Now, for the first time, serious efforts were to be made to thrust satellites, and eventually man himself, outside his earthly environment. And now, for the first time, there began to be appreciated the multiplicity of problems that would have to be solved.

Where space begins is a matter of definition. The density of the atmosphere and the pull of gravity decrease as height is gained and above 18½ miles (30 kilometres) the earth's enclosing envelope is usually called the upper atmosphere. While the density of the atmosphere falls off comparatively quickly, gravity decreases only slowly.

If it is to escape entirely from the earth's gravitational pull and sail out into space, a rocket on the earth must be launched with a velocity increment – the escape velocity – of about 25,000 miles (about 40,000 kilometres) an hour. If it is merely to go into orbit a lesser velocity increment is sufficient, that needed to put a rocket into synchronous orbit – one in which such a satellite makes one complete revolution in the twenty-four hours in which the earth rotates once on its axis – being about 23,500 miles (about 37,800 kilometres) an hour. [The momentum change to put a mass in orbit has two components of velocity increment. The first, to raise the mass to the required altitude is vertical to the earth; the second component, to give it adequate velocity to stay in orbit balancing the gravity force, is parallel to the earth's surface. A figure of 7,200 mph (11,500 kph) is the parallel component, the vertical component being 16,300 mph (26,200 kph) for a total velocity increment of 23,500 mph (37,800 kph)]. At the pre-determined height the main rocket will be automatically switched off, the rocket will be tilted by auxiliary motors into a flight parallel to the earth's surface and, the centrifugal pull on the rocket being equal to the earth's gravitational pull, the rocket will continue to circle the earth.

The higher above the earth that the rocket – or artificial satellite as it will have become – is put into orbit, the longer the time taken for it to make one circuit round the earth. At 125 miles (200 kilometres) above the earth the time for one circuit is about ninety minutes. At about 22,000 miles (about 35,400 kilometres) the orbiting time is twenty-four hours; and, since this is the time during which the earth itself makes one revolution, a satellite put into orbit at 22,000 miles (35,400 kilometres) above the equator will remain above the same spot on the earth, an important fact which has always made this orbit of particular use for communications satellites.

The computations needed to ensure that a rocket is correctly placed in orbit would be complex enough even if the earth were a true sphere. But, as long suspected, and as was shown by the early space probes, the earth is slightly 'pear-shaped', its equatorial radius being about 13 miles (roughly 20 kilometres) longer than its polar radius. There is thus a greater gravitational pull above the equator, a fact which further complicates the calculations to be made before a satellite is put into orbit, and which make more necessary the electronic computers that were becoming available in the 1950s.

The problems of correctly putting a satellite into orbit were compounded by ignorance of what would happen when it was circling the earth in conditions about which very little was known. At least a million meteoroids from outer space enter the earth's atmosphere every day, and while most burn themselves away through the heat caused by friction, some reach the earth as meteorites. What, it was asked, would be the danger to satellites of contact with meteoroids? What, it was asked, would be the dangers from cosmic rays, the very energetic radiation reaching the earth from outer space? What, in addition, would be the effect on a satellite of the solar wind, that fluctuating stream of electrically charged particles emitted by the sun which produces the spectacular displays of the *aurora borealis*?

These were all questions which arose as unmanned satellites were considered. Putting a human being in orbit raised innumerable other queries concerning the effect of weightlessness on the human body, and the psychological effects of movement in outer space.

Some of these questions could be resolved with the use of equipment simulating conditions that astronauts would face and the Lockheed-California Company devised a number of ZEROG (Zero Gravity) simulators. These were tested by future astronauts in a Lunar Topographical Simulation Area in Arizona where the physical conditions of the landscape were nearest to those on the moon, after 1961 the main target of space flight.

A further encouragement for those seeking exploration of space rather than more effective military rockets came with the inauguration of the International Geophysical Year. For this event both the United States and Russia agreed to develop satellites which would make scientific observations. In America, three different groups competed for the chance of putting into orbit the world's first artificial satellite. One was led by Von Braun, another by a US Navy group, a third by a US Air Force group. All relied on different rockets that had evolved in the United States from the V-2. The Navy team with its Vanguard, developed from the Viking research rocket which was itself a by-product of the V-2, was finally given the task. But before it could be carried out America was shaken by an announcement from Moscow: that on 4 October 1957 the Russians had put into orbit the world's first artificial satellite, Sputnik I.

At first most Americans outside the space programme found it hard to believe the Russian announcement although there was, in fact, little justification for their scepticism. Since 1945, when the Russians had captured Peenemunde, they had set up at least three cosmodromes from which rocket missiles could be launched, one at Baykonur, about 190 miles (300 kilometres) north of the Aral Sea in Asiatic Russia, one at Tyuratam in Kazakhstan and another at Kapustin Yar about 45 miles (about 72 kilometres) east of what had been Stalingrad and had later become Volgograd. They had followed up their test-firings of V-2s by a major programme of rocketry and space research that included parachuting from great altitudes first scientific instruments and then experimental animals. Little was known about this work outside Russia. However, more than one public announcement foreshadowed in general terms what was to come.

Alexander Nesmayanov, President of the Academy of Sciences, had told the World Peace Council as far back as 1953: 'Science has reached a state when it is feasible to send a stratoplane to the Moon, to create an artificial satellite of the Earth.' Moreover in August 1957, only a few weeks before the launching of the first Sputnik the Russians announced that they had successfully tested a multi-stage Inter-Continental Ballistic Missile capable of carrying a nuclear weapon to any point on the globe. But just as wishful thinking had prevented most Americans from believing that the Russians could make their first nuclear weapons as quickly as they did, so did this claim for an inter-continental ballistic missile tend to be written off as little more than an unsubstantiated boast.

Scepticism of the Russian achievement might have lasted longer than it did but for the marvels of radio-astronomy, a new science whose foundations had been laid in the United States before the outbreak of war and developed in Britain after it. Until the 1930s man's only source of information about the solar system which space flight was soon to be investigating, was the light emitted by the sun and picked up by optical telescopes. In 1931 and 1932, however, Karl Jansky, a radio engineer working for the Bell Telephone Laboratories recorded the reception of microwaves – the shortest radio waves – from outer space and located them as coming from the direction of the constellation of Sagittarius.

Jansky's work was followed up before the outbreak of World War II by another American, Grote Reber, but it was only with the development of radar and the vastly increased sensitivity of receiving equipment that it became possible to exploit the new method of locating distant objects in the universe. In 1946 Alfred Charles Bernard (later Sir Bernard) Lovell, who had helped create the airborne radar device which could paint for a pilot a map of the ground over which he was flying, found that radar echoes could be obtained from showers of meteors that were invisible in daytime. From this, and the desire to learn more about the universe, there grew plans for the Jodrell Bank radio telescope outside Manchester. The engineering problems were formidable since the receiver was to be a 250 foot (76 metre) wide paraboloidal bowl made from 80 tons of sheet steel mounted on an 800 ton metal cradle. From the centre of the bowl there rose a 62½ foot (19 metre) aerial tower. The entire structure had to be fully steerable so that it could be pointed towards any part of the sky. The bowl and cradle, therefore, had to be mounted on trunnions 185 feet (56 metres) above the ground and the whole 2,500-ton superstructure had to be horizontally movable on a double railway track 350 feet (106 metres) in diameter and accurate to ¹⁄₁₆ of an inch (1.6 millimetres). At the top of the 185 feet towers, 25 foot (7.6 metre) diameter racks from the 15 inch (38 centimetre) gun turrets of the battleship *Royal Sovereign*, dismantled after the war, were used to control the bowl's elevation.

Problems, both engineering and financial, were even more considerable than had been expected, and the radio telescope was not fully completed when, in October 1957, Lovell was asked if he could track the launching rocket that had put Sputnik 1 into orbit. After numerous initial difficulties Lovell and his team tried again on October 11. 'This time', he has written, 'the equipment was clearly functioning. The cathode ray tube was full of the radar echoes from meteor trails and although there was absolutely nothing to guide us as to what an echo from a rocket would look like we were reasonably satisfied that one of the responses was at such a range and of such a character that it was a response from the rocket. The next evening (Saturday 12 October) there was no doubt at all. Just before midnight there was suddenly an unforgettable sight on the cathode ray tube as a large fluctuating echo, moving in range, revealed to us what no man had yet seen – the radar track of the launching of an earth satellite, entering our telescope beam as it swept across England a hundred miles high over the Lake District, moving out over the North Sea at a speed of 5 miles per second. We were transfixed with excitement. A reporter who claimed to have had a view of the inside of the laboratory where we were, wrote that I had leapt into the air with joy.'

Throughout the next few days, as the Russian satellite circled the world with regular precision, Lovell and his team were able to confirm the Russians' achievement beyond any shadow of doubt. They had initially planned to launch Sputnik 1 on the centenary of Konstantin Tsiolkovsky's birth, 17 September 1857, but there were a number of hitches and it was not until 4 October that the satellite, shaped like a football and weighing about 184 pounds (83.5 kilograms) soared up into space on top of its three-stage rocket from the cosmodrome.

Sputnik 1, the first of nine spacecraft to be launched by the Russians within the next forty months, circled the earth in 96.2 minutes and continued doing so for ninety-two days until on 4 January 1958, it re-entered the atmosphere as planned and burned itself up. During that period its four long antennae had been transmitting scientific information which was automatically radioed back to the Russian Space Control Centre outside Moscow.

If this record-breaking Russian achievement astonished the United States – the only other nation which had at that time been in the space race – the next two satellites hammered home the lesson that Russian technology in space was more than equal to American. A month after the launch of Sputnik 1 a satellite nearly six times as heavy – 508.3 kilograms (1,120 pounds) compared with 83.5 (184 pounds) – was put into orbit. It contained the mongrel dog 'Laika', the subject of an elaborate biological experiment whose results were relayed back to earth before the satellite burned up after being in orbit for 162 days.

'On board the second artificial satellite' reported the official Russian news agency Tass, 'there are: apparatus for studying solar radiation in the short-wave ultra-violet and Röntgen region of the spectrum; apparatus for studying cosmic rays; apparatus for studying temperatures and pressure; hermetically sealed container with an experimental animal – a dog inside (it being air-conditioned and having a supply of food and instruments for studying living functions under cosmic space conditions); and measuring apparatus for transmitting various scientific observations to the Earth.

'Two radio transmitters are working on frequencies 40.002 and 20.005 mc/s (wavelengths about 7.5 and 15 metres). The total weight of all the above apparatus, the experimental animal, and the power supply is 508.3 kilograms. According to observation data the Sputnik was given an orbital speed of about 8,000 metres per second. According to calculations which are being made more exact by direct observations, the maximum distance of the Sputnik from the earth's surface is over 1,500 kilometres. The time of one complete circuit is about 1 hour 42 minutes. The angle of inclination to equator is approximately of 65 degrees. . .'

The Russian achievement of putting into orbit a satellite weighing half a ton was very great. So much so that the London *Daily Telegraph* went so far as to suggest that the Russian announcement might be a hoax. Once again Jodrell Bank dispersed doubts. A radar echo was quickly obtained from the rocket when it was above the Arctic Circle, 931 miles (1,500 kilometres) away, and the same equipment recorded the burning up of the carrier rocket on 1 December. Six months passed and then on 15 May 1958, the Russians launched into orbit what they called a geophysical laboratory, the 2,866 pound (1,300 kilogram) Sputnik 3 which contained instruments for recording the composition and pressure of the atmosphere, primary cosmic ray data, and information on the concentration of ions at the heights at which the satellite was orbiting.

By this time the United States was already responding to the Russian challenge. As long ago as 20 August 1953, they had fired the first of their Redstone missiles, developed by Von Braun and his team at the ordnance Guided Missile Center at Redstone Arsenal and based on a mixture of V-2 and American plans. The Redstone was a 62,000 pounds (28,123 kilograms) rocket 69 feet 6 inches (21 metres) long, of 70 inch (178 centimetre) diameter, producing for 121 seconds 75,000 pounds (34,019 kilograms) of thrust from liquid oxygen and an alcohol-water mixture.

The first major development of the Redstone came after a proposal that it might be adapted for launching the American satellite in the International Geophysical Year which was to begin in July 1957. It was then modified to burn more fuel and to utilize the more powerful Hydyne which increased its thrust to 83,000 pounds (38,000 kilograms). In addition, it was given two more stages; one was a cluster of 11 solid-fuel rockets each of which gave an additional thrust of about 1,500 pounds (680 kilograms) when fired into space. The other consisted of three similar rockets. The result of these modifications was Jupiter C, a rocket which in August 1957 tested for re-entry survival a rocket nose-cone which was recovered undamaged after being jettisoned more than 285 miles (459 kilometres) above the earth – sometimes claimed as the first man-made object to be recovered from space.

It was an adaptation of Jupiter C, christened Juno I and made up of the Jupiter rocket plus a fourth-stage single rocket, that was launched by the Americans on 31 January 1958. On top of the fourth stage there rested Explorer I, a 30.8 pound (14 kilogram) satellite. Despite its small weight, which suggested that the Russians possibly had a sizeable lead in rocket techniques, Explorer's information, automatically relayed back to earth, produced the first important discovery from space research: the existence of the Van Allen radiation belts, two belts or zones of charged particles trapped within the earth's magnetic field, the lower at heights of between 621 to 3,726 miles (1,000 to 6,000 kilometres) and the upper layer at between 9,315 and 15,525 miles (15,000 and 25,000 kilometres).

The second American satellite, put into orbit on 17 March 1958, also achieved important results. This was the 6 inch (15 centimetre) diameter Vanguard I satellite containing two radio transmitters, one operated by batteries inside the satellite, the other by solar cells on its outside. The information sent back to earth soon revealed the true, pear-shaped form of the earth with more accuracy than ever before.

From these early weeks of 1958 it became increasingly obvious that the Soviet Union and the United States were taking part in a space-race which had the aim of keeping ahead of their rivals in both military rocketry and in the exploration of space. The rivalry has continued ever since with the Americans concentrating on putting men on the moon, achieved in 1969, and the Russians dedicating most of their energies to a more detailed exploration of inner space. The record of the last quarter-century and more has therefore been one of leap-frogging achievements, with one country at first appearing to take the lead before being overtaken by the other. At times there has been co-operation while the development of earth-orbiting satellites has itself made it more and more difficult for either side to conceal its achievements.

On 7 October 1958 the Americans established the Mercury Project aimed at putting a manned spacecraft in orbit round the earth. This was an ambitious project, as is shown by two figures. The Mercury-Redstone to be used for sub-orbital manned missions had a thrust of 78,000 pounds (35,980 kilograms); the Mercury Atlas, to be used for putting a vehicle in earth-orbit, required 367,000 pounds (166,468 kilograms) of thrust.

Early in January 1959 the Russians opened a major campaign to investigate the moon, sending up Lunik I packed with equipment to measure radiation and ionization and with a device to release a cloud of sodium vapour in space. Lunik I passed within 2,187 miles (3,519 kilometres) of the moon and then went into solar orbit. Some nine months later Lunik II was launched. It was intended to hit the moon, which it succeeded in doing on 12 September, landing near the Sea of Serenity. Jodrell Bank was able to confirm that it reached the moon on target and virtually on time. 'The predicted impact time was 1 minute after 10 pm BST', Lovell has written. 'The excitement remains vivid in the memory as the time approached. At 10.01 the bleeps were still loud and clear, at 10.02 we began to think that it may have missed, but 23 seconds later the bleep ceased – the first man-made object had reached the moon!' On the journey of 236,875 miles (381,203 kilometres) the craft had taken just 83 seconds longer than predicted.

In America scepticism remained. Vice-President Nixon was dismissive in his: 'None of us know that it is really on the moon', while the former President Harry Truman, commenting that the achievement was 'a wonderful thing', added 'if they did it.' Luckily J.G. Davies at Jodrell Bank had measured the Doppler shift – the apparent change in the frequency of sound due to relative motion between source and observer – during the rocket's last hours of moon-approach and the results ruled out any trickery. It had sent out information continuously after its launch and had automatically fired into space the sodium cloud that was photographed from a number of Soviet observatories.

These two Russian probes were followed in October 1959 by an enterprise considerably more ambitious. This was the launching of a probe which would photograph the 'dark side' of the moon which, due to its equal periods of rotation and revolution, is never seen from the earth. The photographs, taken from a distance of about 40,000 miles (64,000 kilometres), were automatically processed in Lunik III and then transmitted to earth by television. From thirty of them, later published in the Russians' *Atlas of the Other Side of the Moon* it was possible to produce for the first time a map of that far side of the moon never seen by man. *Soviet News* described the successful operation in some detail. 'First', it said in explaining how the photographs were taken from Lunik III, 'its lower endplate was trained on the Sun with the aid of solar pick-ups; in this way the optical axes of the cameras were trained in the opposite direction, on the Moon. Then the appropriate optical device, in whose line of vision the Earth and the Sun could no longer appear, switched off the orientation on the Sun and ensured accurate orientation on the Moon. A signal received from the optical device, showing that the moon was in focus, permitted automatic photographing. During the entire photographing time the orientation system ensured that the station was constantly trained on the Moon.'

The next advance came from the Americans who in March 1960, under the auspices of the National Aeronautics and Space Administration which had been formed two years previously, made the first successful probe deep into space with their Pioneer V, a craft originally designed to investigate Venus. By the time that Pioneer V was ready, the ideal conditions for a Venus probe had passed, and the probe was launched to carry out a series of more general experiments.

This time Jodrell Bank was used not only in the passive role of a receiver for transmissions from space, but as an active transmitter of instructions, since it was still the only radio-telescope capable, after installation of a new transmitter, of sending powerful enough signals across the millions of miles which Pioneer V would travel. Shortly after 1.00 pm BST Pioneer V was blasted off from Cape Canaveral. 'Twelve minutes later', Lovell has written, 'the probe came over the Jodrell horizon and its signals were immediately acquired by the telescope. At 1.25 pm, when Pioneer V was 5,000 miles from the earth, a touch on a button in the trailer at Jodrell transmitted a signal to the probe which fused the explosive bolts holding the payload to the carrier rocket. Immediately the nature of the received signals changed and we knew that Pioneer V was free, on course and transmitting as planned. For the rest of the day Pioneer responded to the commands of the telescope and when it sank below our horizon on that evening it was already 70,000 miles from earth. The next evening it was beyond the moon.' From now until the end of June, when a leak in Pioneer V's solar energy batteries ended its transmissions, the batteries were switched on and off every day by a signal from Jodrell Bank.

While the Americans now put into orbit a number of small satellites which measured the size of the solar system, studied the solar wind, and in August 1960 showed how a space capsule could safely be recovered from orbit, it was the Russians who with the successors to their first three sputniks, made two big steps forward with the five sputniks they successfully launched between May 1960 and May 1961. They were very different from the packages previously put into orbit. Each weighed more than 9,900 pounds (4,500 kilograms), all were described as spaceship satellites, and all were obviously designed and launched to discover, and help resolve, the difficulties of putting a human being into orbit.

The first, launched on 15 May 1960, had a dummy man strapped into a chair in its pressurized cabin, and instruments powered by chemical and solar batteries which continuously sent back to earth information on the physical conditions in the cabin. The second, on 19 August 1960, carried two dogs 'Strelka' and 'Belka' who were brought safely back – the first orbital recovery made. The experimental sputnik launched on 9 March 1961, not only carried a dog, guinea pigs, mice, insects and plant seeds, but brought them all safely back to earth. This performance was repeated after the launching which took place sixteen days later. After orbiting the earth, the space capsule carrying the dog 'Zvedzochka' (Little Star) was safely brought down with its passenger on to a pre-determined target in Asiatic Russia.

The safe return of the two Soviet space vehicles showed that the Russians had solved a problem of space research second only to that of providing enough thrust to push the vehicle away from the earth's gravity – the problem of re-entry. As the satellite loses speed and begins to drop down into the earth's atmosphere, the friction between the satellite and the atmosphere creates an enormous temperature which would burn up the satellite and its contents unless preventive measures were taken. The solution, it was eventually found, lay in coating the satellite with ablative materials which would dissipate the heat by, in effect, burning it away. The final descent to earth was slowed by parachutes – which were ineffective in the higher ranges of the atmosphere.

Once it was found possible to bring living specimens back from orbit without injury, the way was open for sending men into space, an operation accomplished by both the Russians and the Americans within a few weeks of each other during the spring of 1961. The Russians succeeded first, putting Yuri Gagarin into orbit in a space-ship, the first of a series christened Vostok, on 12 April 1961. The 10,395 pound (4,725 kilogram) vehicle, orbited the earth in 89 minutes 6 seconds, and made slightly more than one orbit before it was brought back to earth.

Vostok-1 consisted of a pressurized cabin and an instrument section. The first included air conditioning and pressure control systems, food and water, physiological recording equipment, a television system which allowed the pilot to be watched from earth, and a manual landing control system. The instrument section of the craft contained telemetry equipment and sensors, electrical supplies, the craft's retro-rocket control for slowing down its speed, and a variety of other equipment. The complexity of the space ships involved is indicated by figures issued by the Russians. Vostok-1 contained 241 electronic valves, 6,000 transistors, 56 electric motors, some 800 electric relays and switches, the systems being connected by 33 pounds (15 kilograms) of wire and 880 plug-type connectors.

The American response to Russia's success in putting the first man into space came less than a month later when Alan B. Shepard Junior on 5 May went off in a Mercury Redstone launch vehicle from the base on the Florida coast that was to become the Kennedy Space Center. But Shepard, and Virgil I. Grissom, who was launched in a similar vehicle in July, did not go into orbit. Instead, they made ballistic flights of more than 300 miles (480 kilometres), reaching heights of more than 100 miles (160 kilometres) before being safely recovered from the Atlantic Ocean. These were considerable achievements, but they lagged behind the Russians, a fact that was underlined when on 6 August 1961 Gherman Stepanovich Titov was launched in Vostok-2. He made seventeen orbits of the earth and spent more than twenty-five hours in space before being successfully landed in Asiatic Russia. Although it was stated immediately after the flight that Major Titov was in good physical condition the Russians later admitted that when in orbit he had suffered from nausea and other effects brought on by prolonged weightlessness. The effects had disappeared as Vostok-2 re-entered the earth's atmosphere but they pointed towards the need for more research on the phenomena of weightlessness.

During the next two years the Russians launched four more Vostoks. They put the first woman into space, Valentina Tereshkova. They put space ships into orbit from which television programmes were transmitted to earth. They increased the time that their cosmonauts spent in space until it reached more than seventy hours – a feat accomplished by Valentina Tereshkova – and in the middle of March 1965 they enabled Colonel Aleksey Leonov to move into a spacesuit while within the ship, move into the ship's airlock and then take a ten minute walk in space, linked to the ship only by a fragile lifeline.

By the end of 1964, it was clear that while the Russians and the Americans had both embarked on space exploration with roughly similar aims, their programmes had developed along different lines with the result that each was ahead in certain fields and behind in others. The Russian cosmonauts had spent far more time in space than the Americans and they had carried out more investigations into the problems of space medicine. In addition, their unmanned rockets are believed to have made more detailed investigations of the moon. On the other hand, the Americans had collected far more scientific information from space; in particular, their satellites had given them a great advantage over the Russians in the use of space for communications and weather forecasting.

The Americans had, moreover, one clear objective. This had been laid down in May 1961 when President Kennedy announced to a joint session of the US Congress that it would be an American ambition to land men on the moon and return them safely to earth before the end of the decade. Many of the space probes in the 1960s had been designed to gather information needed for the enterprise, christened the Apollo Project.

These included development of the Ranger series of spacecraft which transmitted close-up photographs of the moon before crashing on to its surface; and the use of Lunar Orbiters to map large areas of the moon, and of Surveyor spacecraft which soft-landed on the moon and whose television transmitters then began sending thousands of photographs back to earth.

By the mid-1960s it had been decided that the Apollo spacecraft to be sent to the moon by a three-stage Saturn rocket would consist of three components: a command module, a service module, and a lunar-excursion module comprising one section in which the astronauts would descend to the surface of the moon and a second section, attached to the first, in which they would ascend from the moon to the spacecraft which would have been kept in lunar orbit above them.

On 30 July 1965 the tenth Saturn rocket was tested, and the following February an uprated liquid-fuelled Saturn blasted an unmanned Apollo space-ship into a sub-orbital flight that ended some 4,500 miles (7,200 kilometres) away near Ascension Island in the South Atlantic. A second, similar, test successfully took an unmanned Apollo craft for a sub-orbital 18,000 mile (29,000 kilometre) flight to the Central Pacific. Then, early in 1967, three lives were lost before the start of what was to have been a fourteen-day orbital mission when, on 27 January, fire broke out inside the command module, apparently due to faulty electric wiring. The three men inside died before they could be rescued.

Despite the tragedy, further tests were successfully carried out, and in October 1968 a further step forward was taken when three men were put into orbit in an Apollo spacecraft for an eleven-day mission. After travelling the equivalent of 163 orbits, the Apollo command module safely landed by parachute some 300 miles (480 kilometres) from Bermuda.

By this time the tempo of the American space effort can be judged from the fact that in 1968 no less than sixty-four craft were launched into space. Their tasks ranged from those of the Radio Astronomy Explorer whose antennae, more than 1,500 feet (457 metres) across, monitored low frequency emissions from the Milky Way and the sun, to others monitoring auroral phenomena, taking lunar photographs, reporting on solar radiation, gathering charged particle data and studying interplanetary physics.

These successes all helped pave the way for the first flight of men round the moon which began on 21 December 1968 when three American astronauts were

An unmanned Saturn rocket
seven seconds after ignition during lift-off from Cape Kennedy early in 1968.

launched into earth orbit, directed towards the moon by the firing of subsidiary rockets, and then put into lunar orbit. They made ten orbits of the moon before firing the command module's engine to bring themselves back into earth orbit. After their six-day mission they splashed safely down into the Pacific.

The orbiting of the moon had been made with the command and service modules only but early in March 1969 three astronauts took off for critical tests of the lunar module. In earth orbit this module, which would eventually have to take men from lunar orbit to the moon's surface and bring them back, was successfully separated from, and then rejoined to, the rest of the space-ship. After ten days, and 152 orbits of the earth, the three astronauts made a safe splash-down in the Atlantic.

There was to be one final test before the vital moon-landing flight was begun. This came in May 1969 when the manned lunar-excursion module was put into a series of lunar orbits which brought it to within less than 10 miles (16 kilometres) of the moon's surface.

Everything was now ready for the great attempt.

By mid-morning on Wednesday 16 July 1969, more than a million visitors arrived at Cape Kennedy, as Cape Canaveral had been re-named, to see the start of man's first voyage to the moon. Another 520 million were to watch it on television. What they saw was the 281 feet (86 metre) high, 33 feet (10 metre) diameter Saturn on top of which sat the space vehicle consisting of the command and service module named Columbia and the lunar module Eagle, a vehicle consisting of descent and ascent stages, weighing about 35,200 pounds (16,000 kilograms), 22 feet 11 inches (7 metres) in height and 31 feet (9 metres) in diameter, and a total height of 363 feet (111 metres). Inside were Neil A. Armstrong, a civilian in command of the operation, Colonel Edwin E. Aldrin Junior and Colonel Michael Collins.

At 9.32 am EDT (14.32 BST) the Saturn rocket lifted off as its five engines threw out brilliant sheaths of red flame amid a roar of sound. At first the huge rocket appeared to move slowly. But it quickly gathered speed and within less than twelve minutes was 118 miles from the earth. Here it was put into earth orbit in which it stayed for 2½ hours before being put on course for the moon.

By 10.30 am on Thursday 17 July Apollo had reached a point mid-way between the earth and the moon and was travelling, on schedule, at 3,600 mph (5,790 kph). The launch had been so good that the first planned course-correction did not have to be made, and at 1.22 pm on Saturday the 19th Apollo entered lunar orbit. By this time television transmissions had begun from the craft and the world saw for the first time how it looked from the moon.

On the morning of Sunday 20 July Armstrong and Aldrin entered the lunar module and at 1.47 pm, while behind the moon and out of contact with the earth, detached the unit from the command and service module. They were now on their own, at more than 40,000 feet (12,200 metres) above the moon, but fully under their own control, and as they emerged from the far side of the moon Armstrong radioed to earth: 'The Eagle has wings.'

The lunar module was now put into its own lunar orbit and then, by the firing of its descent rockets, slowed and brought lower down towards the moon's surface. As it neared its target the rocket thrust was decreased. When 500 feet (152 metres) up the vertical velocity was 27 feet (8 metres)/sec. At an altitude of 75 feet (23 metres), it was down to 3 feet (0.914 metres)/sec.

During the final moments of descent Armstrong had to manoeuvre the craft to avoid landing in a big boulder-strewn crater. As he did so listeners back on earth heard his words. 'Lights on. Down 2½. Forward. Forward. Good. 40 feet down 2½. Kicking up some dust. 30 feet 2½ down. Faint shadow; 4 forward, 4 forward, drifting to the right a little, 6 down a half. Contact light on. Engine stopped.' Then came the long-awaited words: 'Houston, Tranquillity Base here. The Eagle has landed.'

The next six and a half hours were spent checking the control systems in the module, radioing information back to earth, and getting into the Portable Life Support System and other specialized clothing. Then, at 3.56, 20 secs. BST, the door of the craft was opened and Armstrong stepped down a ladder and on to the moon's surface with the words: 'That's one small step for a man, one giant leap for mankind.'

Aldrin, who now also descended from the craft, and Armstrong were to spend slightly more than 21½ hours on the moon, gathering samples of soil and rocks, carrying out scientific experiments and setting up equipment which would continue to send information back to earth after they had returned. They also left the American flag, medals commemorating American and Russian astronauts who had died, and a plaque signed by themselves and President Nixon saying: 'Here men from the planet Earth first set foot upon the moon, July 1969 AD. We came in peace for all mankind.'

All these activities were televised and most photographed. Then, at 18.54 BST, after the two men had returned to the craft and sealed themselves in, the rocket engines for the ascent were fired. Eagle began by rising vertically, leaving as a take-off pad the descent portion of the craft which had been designed to act as a launch platform.

Eight seconds after take-off Eagle was 165 feet (50 metres) up from the moon and here the controls were set to bring the craft into lunar orbit. When this was obtained, further guidance corrections with the rocket motors took the capsule into the same orbit as the Columbia in which Collins had been left circling the moon while his two colleagues were at work. The two craft were delicately brought closer to one another by a number of small course corrections and braking manoeuvres. Then, at 22.35 BST on the 21st, they met.

After the docking of the two craft with each other, the pressure between them was equalized and Aldrin and Armstrong's suits, as well as everything they had brought from the moon, were vacuumed to remove any contamination by lunar dust. Only then did they move into the Columbia, before the Eagle was jettisoned into lunar orbit.

At 5.57 BST on Tuesday the 22nd the engines of Columbia were switched on to take the craft out of lunar orbit and set it on its path to earth. Wednesday was a comparatively quiet day as the Columbia sped on its predetermined route, although the members of the crew televised a programme illustrating life in a condition of weightlessness, including a demonstration of how to drink water from a spoon when the force of gravity was absent.

On the 24th the service module of the Columbia was separated from the command module in which the three members of the crew were now living and the latter was oriented so that its blunt heat shield faced forward in the direction of flight. There now came another crucial stage in the operation. This was the guidance of the spacecraft into the narrow 're-entry corridor' which spread out between the earth's atmosphere and outer space. For the astronauts' practical purposes the corridor was only some 25 miles (40 kilometres) wide. Below its lower level, the atmosphere was so thick that it would slow down the spacecraft to an unacceptable degree, while above it the atmosphere was too thin to slow it down sufficiently to bring it back to earth.

Re-entry into the earth's atmosphere began at 400,000 feet (122,000 metres). Later, as the craft passed through the sound barrier with its decrease of speed, there came the sonic boom which was heard by the recovery vessels. At 100,000 feet (30,500 metres) the command module was sighted by a waiting aircraft. At 24,000 feet (7,310 metres) the pilot and drogue parachutes were opened and at 10,000 feet (3,050 metres) the three main parachutes also opened.

At 17.50 BST, 195 hours 17 minutes 35 seconds after it had been blasted off from Cape Kennedy, the command module landed in the Pacific, some 13 miles (20 kilometres) from USS *Hornet*, the main recovery vessel, and 960 miles (1,536 kilometres) south-west of Hawaii.

For the three astronauts, it was only the main stage of the enterprise which was finished. As they waited in their craft, a swimmer carrying three Biological Isolation Garments reached it and handed the sterilized garments up to them. As they moved first into a liferaft and then into a helicopter to be taken to the USS *Hornet*, their command module, and the sea around it, was sprayed with acid to ensure that no lunar particles could contaminate the earth. On USS *Hornet*, where not even President Nixon was allowed to make physical contact with them, they were immured in an isolation van. Here they were to stay for eight days to prevent contagion from any unknown micro-organisms to which they might have been exposed.

The Apollo Project had thus far been an unqualified success. But landing of men on the moon had been but the first of the activities planned, and the rest were to be considerably reduced for financial reasons. Nevertheless, more successful flights were to be made, by Apollo 12 in November 1969 and by Apollo 14 in January 1971. Between those two flights, in which the astronauts brought back further specimens and placed more automatic equipment on the moon's surface, there was the mission of Apollo 13, during which a service module oxygen tank ruptured causing loss of power and it was brought safely back to earth with its crew only with great difficulty.

These moon landings were now to be followed by three others during which the astronauts took with them in their lunar module a Lunar Rover, a vehicle which greatly increased the area of the moon they could explore and allowed them to collect a wider variety of materials for transport back to earth.

The successful conclusion of the Apollo programme ended one era of space exploration. Once men had got to the moon, and could apparently go there again whenever the necessary finance was available, effort in both the United States and the Soviet Union began to concentrate on the establishment of permanent space-stations. The project, which had been discussed since men first began to think seriously about space travel, would use earth-orbiting craft in which a team of men – numbers mentioned varied from seven to fifty – could live and work indefinitely. They would be supplied from smaller craft which would dock with the space station as both orbited the earth and would be able to carry out an almost unlimited number of experiments from beyond the earth's atmosphere.

Both the United States and the USSR had been considering the

Cernan, an American astronaut walking on the moon towards a Rover, a vehicle landed for exploration of the moon's surface – Apollo 17 mission.

establishment of space stations, and much of the necessary groundwork had been done by the time of the moon landings. The Russians' first unmanned research workshop as they called it, Salyut-1, was put into orbit on 19 April 1971 and was manned five days later after a craft carrying a crew of three docked with the workshop. Consisting of several cylindrical sections, joined together and pressurized, Salyut-1 weighed about 20 tons, only a fifth of the weight of the 100 ton Skylab which the Americans launched in May 1973. The following decade witnessed a succession of technological advances with which Russia and America leapfrogged each other with more efficient ways of manning a space station.

Prominent among US developments has been the Space Shuttle, a craft which is shot into earth orbit as a rocket, manoeuvres in orbit as a spaceship and comes back like an aircraft. After launching, an operation which submits the crew to only about three times the force of gravity – a third of that faced by earlier astronauts – the shuttle jettisons its external tank containing 1½ million gallons (7 million litres) of fuel at take-off just before entering earth orbit which may be 110 to 225 miles (177 to 362 kilometres) up. After completing the mission, which may have lasted up to a month, the engines are used to slow down its speed and bring it out of orbit into the earth's atmosphere from which it descends much like an ordinary aircraft.

While the establishment of permanent space stations has for more than a

The launch of an American Space Shuttle.

The flight deck of the Space Shuttle Orbiter Columbia. The commander sits on the left, the pilot on the right. Between the two seats are the flight computer and navigation aids console.

decade been one aim of space-exploration, another has been the, equally ambitious, investigation of the planets which the Americans began as far back as 1962 when they inaugurated the Mariner programme. The programme had started badly when, on 22 July 1962, their Mariner I had gone out of control soon after being launched and had had to be destroyed. The situation was counterbalanced the following month when Mariner II was successfully launched on a track which was to take it past Venus and enable it to transmit back to earth details of the planet's temperature and other scientific data. As late as January 1963 it was still sending back information across the 53.9 million miles (87 kilometres) that by then separated it from the earth.

Further Mariner probes launched in 1964, 1967, 1969, 1971 and 1973 examined Mars, Venus and Mercury, and took excellent pictures of the surface of Mars, while the 1964 probe set up a new record by transmitting information while 190 million miles (306 million kilometres) from earth.

Then, in the late summer of 1977, there began the most spectacular of all man's engineering ventures into space. In August and September two ¾-ton spacecraft were blasted off from Cape Canaveral on journeys to study Saturn, Jupiter and Neptune. Voyager II was, confusingly perhaps, launched before Voyager I but overtook Voyager I and reached the area of Jupiter first. Voyager I continued on its way out of the solar system. Voyager II meanwhile continued on

Satellite/Spacecraft receiving station,
Wallops Island, Virginia, photographed during the 1970 solar eclipse.

its way towards Uranus, the area of which it was scheduled to reach in 1986 before travelling on towards Neptune 2,000 million miles (3,200 million kilometres) away from earth.

By this time both American and Russian space probes had provided much information about Mars, Venus, Mercury and the Moon, all of which are made up of the heavier elements. The more distant planets such as Jupiter, Saturn, Uranus and Neptune, are composed mainly of hydrogen and helium and are believed to have retained most of the material from which they started to evolve five billion years ago. The Voyager probes will investigate them all, and had in fact begun doing so by 1979 when they photographed Jupiter and its satellite Io. Some of the information relayed back to earth was unexpected and dramatic. Thus the photographs transmitted from the Voyager probes showed a giant ring of particles circling Jupiter about 33,000 miles (53,100 kilometres) above the planet's atmosphere. But while the ring was some 4,000 miles (6,400 kilometres) wide it was only a single mile thick (1.6 kilometres). The cameras also photographed for the first time Amalthea, the innermost of Jupiter's satellites, and four of her bigger satellites, Io, Europa, Ganymede and Callisto.

The ambitious space probes which have produced these spectaculars from outer space are different in several ways from those with which the American National Aeronautics and Space Administration began its investigation of the

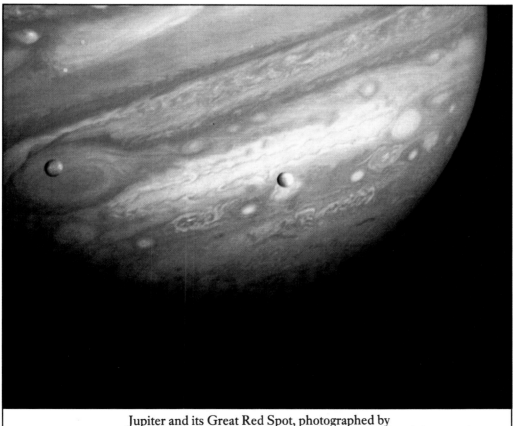

Jupiter and its Great Red Spot, photographed by
Voyager I in 1979 from a distance of 17.5 million miles (28.2 million kilometres).

solar system. Correction of their trajectory through space is made not by a single main rocket engine, as in the earlier probes, but by the use of hydrazine rocket propellant in sixteen small thrusters. By activating these as required the probes can be positioned either by reference to the Sun and the Star Canopus or through the use of the gyroscopes on board which create an inertial reference system.

The two Voyager probes will eventually pass beyond the nine known planets and at least thirty-nine moons which make up the solar system. But that will not be the end of their usefulness. If all goes as well as it has gone already, listeners on earth will be able to receive messages from them for another thirty years, at the end of which time they will be some 9,300 million miles (14,966 million kilometres) away.

Further investigation of space seems likely to be limited by lack of finance and material resources rather than by lack of technological expertise. In both the United States and the Soviet Union the demands on money, manpower and scarce materials are immense and the case for international co-operation in space research is overwhelming. For long the rivalry and distrust between America and Russia limited co-operation to the comparatively minor activities of the International Geophysical Year. But a change came in the autumn of 1970 when five American space experts visited Moscow to discuss possible collaboration. The following year there were talks in Moscow between experts from the two countries to discuss what form the collaboration might take.

The first tentative moves were firmed up in May 1972 when President Nixon and Premier Kosygin of Russia signed in Moscow an agreement for many kinds of space collaboration. The most spectacular of these included plans for a link-up in space of American and Russian spacecraft. The docking equipment of the Soyuz and the Apollo spacecraft that were to be used had to be adapted so that they were completely compatible and it was only in the summer of 1975 that arrangements were complete for what was to be a significant joint mission. By that time it had been arranged that the Russians would take into space a number of American biological experiments, and space collaboration between the two countries began to grow in a number of ways. Nevertheless, it was the planned link-up of the two countries' spacecraft, and the exchange of their crews in space which was rightly seen as a good omen for the future.

On 15 July, Soyuz XIX was launched at 8.20 (EDT) from the Tyuratam Space Center 1,400 miles (2,250 kilometres) southeast of Moscow and was soon in orbit more than 100 miles (160 kilometres) above the earth. Seven and a half hours later an Apollo spaceship was thrust into space on top of a Saturn 1B rocket from the Kennedy Space Center in Florida. As it came into orbit, Apollo was some 4,400 miles (7,080 kilometres) behind the Russian vehicle, a distance which had been decreased to 1,500 miles (2,400 kilometres) by the evening of the 16th.

On the morning of the 17th, as the two craft traversed Central Europe, Apollo moved up from below and behind Soyuz and slowly narrowed the gap until all was ready for docking. As the two craft passed over Western Germany Apollo eased its docking mechanism into contact with the Russian craft. The docking latches were operated and Soyuz and Apollo flew on as one craft.

> A manned manoeuvring unit, photographed February 1984,
> from the Space Shuttle Challenger, manned by astronaut Bruce McCandless.

The pressures between the two craft had to be equalized before the Russian and American crews were able to join each other. Then, or during the next day, each of the cosmonauts visited the cabin of the other country's craft. Russians and Americans shared their meals in space together. Then the two halves of a commemorative plaque which had been carried into space in the two craft were symbolically joined.

After the two crews had returned to their respective craft, Apollo and Soyuz separated, the Apollo moved about 200 yards (183 metres) away from the Russian craft and a second space-docking was successfully carried out.

After three hours the two craft were again unlocked from each other and Soyuz soon began her flight back to earth, landing 300 miles (400 kilometres) east of the Baykonur Space Centre on 21 July. Apollo remained in earth-orbit for another three days during which her crew carried out experiments and photographed much of the earth from their vantage point. They landed safely on 24 July – after a near-disaster when a mishap had allowed toxic gas to flow into the craft, and gasmasks were put on only just in time.

This first major example of space co-operation between the two superpowers has not been followed up with the speed that many hoped. In the United States itself the purely scientific investigation of space appears to have suffered from the economic recession, although military applications are being continued with even greater effort. Much the same is probably happening in Russia, although its closed society shrouds most details.

Even so, the 'spin-offs' from space investigation that have already taken place continue to be utilized and are more varied than is often appreciated. At one level there has been the ubiquitous effect on engineering techniques and methods. One obvious result has been the production of valves, pumps and switches which would operate with unprecedented reliability in the face of unprecedented conditions of acceleration and cold. Materials had to be developed which would withstand temperatures of 3,000°F (1,649°C) on the heatshield of spacecraft as they re-entered the earth's atmosphere at 25,000 mph (40,000 kph). An electromagnetic hammer had to be devised for smoothing the welding seams on the Saturn rocket without weakening the metal – and was soon being used in industry. New alloys, adhesives and lubricants had to be developed to do jobs in conditions that had never before been met.

Medical equipment specifically designed for the space programmes was quickly adapted for more everyday uses. Thus electronic sensors used to monitor the conditions of astronauts were soon being used to measure pulse and respiration rates, blood pressure and temperature of hospital patients. A device used to count meteorite hits on a spacecraft was adapted to record muscle tremours and may help to provide early detection of such illnesses as Parkinson's disease. And a meter used to measure stresses inside a solid-fuel rocket is already being used to measure the elasticity of bones in a living human.

Other medical results have been noted by Dr George E. Mueller, a former Associate Administrator, US National Aeronautics and Space Administration. 'For example,' he has written, 'medical doctors and engineers worked together to solve one basic problem. Ground monitors had to know on a "real-time" basis (instantaneously) while it was happening, how fast the hearts of the astronauts were beating.

'We had to know how much oxygen they were using; we had to know how their muscles were responding to their strange environment; so we invented a new system – biosensors reporting to computers and to data-gathering equipment, and, through the communications network, to the Manned Spacecraft Centre at Houston, Texas.

'The system works from 200 miles – or 800 miles or from a quarter of a million miles out in space. Half a dozen newly formed companies are now manufacturing adapted space-created instruments, for the use of doctors and hospitals on earth.'

In a totally different field, earth satellites have done more than augment world communications by their ability to transmit to all parts of the world the sights and sounds of events that take place anywhere on its surface. Pictures from Apollo 9 have revealed the progress of growing crops on a scale and with an accuracy never before possible and Landsat satellites increased the accuracy. The first signs of crop disease can be detected by satellite sensors and so can the varying state of the polar ice-caps. Water resources can be surveyed by satellite, and a watch kept for forest fires. The satellite prediction of hurricane Carla in 1968 allowed 500,000 people to be moved to safety in America, and as a long-term weather-predictor satellites provide a unique tool in the meteorologists' armoury.

✳ ✳ ✳

FUTURES

Forecasting the future – the science of futurology as it is sometimes over-grandly called – has become a growth occupation during the last quarter of a century. In World War II it was shown that ideas previously thought impossible of exploitation could under the spur of near-necessity be transformed into useful practice. The utilization of nuclear fission first into the bomb and then into power-plants, gave some idea of the great and unexpected possibilities which technology could achieve. So did the wartime development of radar and the post-war development of the computer. These events encouraged engineers, scientists and philosophers to prognosticate on what the world would be like towards the end of the twentieth century and during the first decades of the twenty-first.

Most of the forecasts – which assume that the world escapes a nuclear holocaust – rest heavily on the answers to two questions. Will new sources of energy, or new developments of old ones, be available? And will engineers have the use of new materials with characteristics radically different from those of today?

The first question is the more important. The great transformations of human life during the past few hundred years of industrial civilization have been due to the use of steam power from the eighteenth century, to the use of electrical energy from the late nineteenth century, and to the use of nuclear energy from the 1950s – although with the last it is still too early to be sure that the advantages genuinely counterbalance the disadvantages.

There is no difficulty in forecasting what could be the most important factor for change in the foreseeable future. Power from nuclear fusion, with its double bonus of providing virtually limitless energy without the problem of nuclear waste disposal, could change the world's prospects even more drastically than they were changed with the coming of steam power. But the necessary technological break-through, which could come in Russia, the United States or Britain, could well be delayed until the twenty-first century, and would almost certainly be followed by many years of research and development before the first man-controlled nuclear fusion was turned into useable electric power.

Three other alternative sources of energy have been investigated over the years – solar energy, tidal power and wind power. But conditions for the use of each vary from country to country, and there is still considerable dispute about the best technological methods of exploiting them, as well as much disagreement about the economics of their operation.

The earth intercepts a solar energy beam of about 17×10^{13} kw, 35 per cent of which is reflected back by clouds and 19 per cent absorbed by the atmosphere. However, the extent of the remaining 46 per cent can be judged by the fact that solar radiation received on the surface of the United States is estimated at 1,000 times the energy used by the nation as fuel and water power. If only 2 per cent of the US landmass were used for the recovery of solar energy, and if 10 per cent of this were converted into heat and power, then this would allow a five-fold increase in the country's use of energy.

It is not a new idea. Lenses found in the ruins of Nineveh, as well as Inca mirrors, are believed to have been used to concentrate the sun's rays to light sacred fires – the adult equivalent of the schoolboy setting leaves alight by focusing sunlight with a magnifying glass. Archimedes is claimed to have delayed the fall of Syracuse by setting the invading fleet alight with the help of mirrors which concentrated sunlight on to it. In the eighteenth century, Buffon used 140 mirrors in the garden of his Paris house to light a pile of wood 200 feet (61 metres) away, and almost thirty years later Lavoisier not only did much the same but devised a method of turning the mirrors so that they constantly pointed directly at the sun.

John Ericsson of *Monitor* fame built his first 'sun motor' in 1870 and within five years had built another six, managing to obtain one horsepower of energy for every 100 square feet (9 square metres) of mirror concentrating the sun's rays on a cylindrical boiler. But despite his development of a calorific engine, designed specifically to use solar energy, Ericsson never succeeded in putting his ideas on a commercial basis.

In the twentieth century various other schemes were proposed to tap energy from the sun, including one by the physicist Sir Charles Boys, who advised on the building of a solar plant in Cairo shortly before World War I. In America where the scientist J.B. Conant had considerable faith in solar energy, many solar heaters were developed in the southern states, but it was left to France seriously to investigate commercial possibilities after the end of World War II. Their most important site is at Odeillo, some 6,000 feet (1,830 metres) up in the French Pyrenees. Here there has been built a 130 foot (40 metre) high parabolic reflector comprising 9,500 mirrors. Sunlight is thrown on to this mirror-array by a group of sixty-three flat mirrors that can be automatically steered for eight hours a day to use the sun's rays most efficiently. From the parabolic reflector the light is concentrated on to a solar oven nearly 60 feet (18 metres) away and here temperatures of up to 6,872°F (3,800°C) have been obtained. The Odeillo solar oven has been employed mainly for research but the French have big plans for utilizing solar energy in agriculture and industry and hope that by the end of the century it will be providing 5 per cent of the country's energy requirements and saving 17 million tons of oil annually.

A start has already been made in practical applications at the Themis solar station at Targassone, also in the Pyrenees, where about 2,400 hours of sunshine a year is normal. Here sunlight falling on 52^2m. of mirrors is directed to the top of a 109 yard (100 metre) tower where it heats a mixture of melted salts to a temperature of 450° which in turn powers a turbo-generator.

In Britain, as in many other countries, the prospects for solar power are poorer, although it has been estimated that by AD 2025 houses and offices designed to make the best use of sunlight falling on them could be saving the equivalent of 3 million tons of coal annually.

There is no doubt about the potential in wind power. Electricity was generated by a windmill in Holland as long ago as 1890. Various attempts to utilize 'fuel-less power' were made during the following ninety-odd years, and a

report in 1983 from Britain's Energy Technology Support Unit estimated that wind could in theory produce about 20 per cent of the United Kingdom's electricity requirements. The problem is not only financial but also environmental, because of the large number of windmills – or aerogenerators as they are called – that would be needed.

The problems are illustrated by the ambitious project centred on the 100-metre high tower built for Western Germany's Federal Ministry for Research and Development near Brunsbüttel in Schleswig-Holstein. On top of the tower is the highest windmill in the world, with 44 yard (40 metre) long 18 ton sails which form part of a complex 310-ton structure including generator, transmission and sail-adjustment mechanism. The windmill feeds current into the local electricity grid sufficient to meet the requirements of 250 detached houses. But successful as this example of wind power appears to be, it reveals the limitation of the method. The generator is without doubt one of the most efficient of its kind; but to equal the output of a nuclear power station supplying 1.3 million kilowatts seven hundred similar windmills would be needed.

Tidal power, generated by allowing tidal waters to flow into a storage area unimpeded and, on the ebb tide, to flow out through sluices operating water-wheels, is limited by the comparatively small number of suitable estuaries. But it has been successfully operated in France where the Rance River estuary has been supplying up to 544 gigawatt-hours of electricity annually since the mid 1960s. Into the barrage across the estuary there have been built twenty-four 10,000 kilowatt turbo-generators which operate in both directions of tidal flow. In Britain, construction of a tidal barrage across the Severn Estuary has been discussed for many years, and a new government study of the possibilities was started in 1983. Five per cent of the country's electricity, it has been proposed, might be produced by it. A barrage across the Wash, which cuts into the eastern coast of England has also been suggested.

In addition to these major potential sources of energy, all of which appear to be technologically viable and whose success depends largely on economic considerations, other localized and more esoteric possibilities have been considered during the past decade. In Britain, experiments are already in progress to discover if it is practicable to bring into use geothermal aquifers, pockets of hot water lying deep in the earth and equivalent to the hot springs of New Zealand and parts of North America. Up to 1½ million tons of coal might be saved annually, it has been suggested, but the necessary technology is still at an early stage. Meanwhile, an imaginative hydro-solar project is being studied in Egypt. This involves nothing less than the creation of a $7,500^2$ mile (19,400 square kilometres) lake in the Quattara Depression west of Cairo, the lowest point on the African Continent. The depression would be filled by means of a channel cut to the Depression from the Mediterranean 100 miles (160 kilometres) to the north, the flow of water producing 3bKWh of electricity.

The prospects for new materials seem brighter than those of really significant new sources of energy. Here the growing knowledge about what really governs the strength of materials has led to a variety of possibilities, including those of

---

The Solar furnace at Odeillo, southern France. In front
of the parabolic glass reflector there rises the control laboratory and its tower.

'whisker' technology. In the early 1950s it was found that extremely fine and short crystals of both metals and ceramics – 'whiskers' – had a strength far greater than was foreseen. It was discovered, for instance, that a perfect crystal whisker of Epsom salts would be far stronger than piano wire. Difficulties in growing perfect crystals were known to exist but possibilities have been increased by the events of the space age for it has been established that in gravity-free space there is no limit to the growth of crystals. Eventually, it has been suggested, it may be possible to produce a material with the strength of steel but the weight of balsa wood.

These prospects for advance may very well be overshadowed by progress in other fields in which engineering plays a less important, and in some cases only a peripheral, part. Thus control of the weather – aided like so much else by the application of ever more powerful computer techniques – will probably move from the experimental to the practical. Here the problems are likely to be social and political rather than scientific: the potential benefits for crop production are certain to be countered by the complex implications of moving rain-or-shine from one part of the earth to another.

It is, however, in biology that work already in progress may provide the most potentially important – and most debatable – changes during the next few decades. Biochemical general immunization against bacterial and viral diseases seems almost certain to increase, as does the use of genetic engineering to control some hereditary defects. It has been speculated that early in the twenty-first century chemical control of aging may be able to extend human lifespan by fifty years, and the same is true of increasing human intelligence by the use of drugs.

The speed with which these developments take place and the extent to which they are carried out are likely to be governed not so much by improvements in science or technology as by the public view of their utility or morality. Just as engineering advance, typified by flight, must conform to economic criteria, so will future advance in the adjustment and control of human beings be governed not by what is possible but by what is considered acceptable.

BIBLIOGRAPHY

BELL, Sir Westcott Stile, *The Shipwright's Trade*, Cambridge, The University Press, 1948

PPLEYARD, Rollo, *Charles Parsons: His Life and Work*, London, Constable, 1933

SHTON, Thomas Southcliffe, *Iron and Steel in the Industrial Revolution*, Manchester, The University Press, 1924

SHTON, *The Industrial Revolution, 1760-1830*, London, Oxford University Press, 1948

AILEY, Cyril (ed.), *The Legacy of Rome*, Oxford, Clarendon Press, 1923

AINES, Sir Edward, *History of the Cotton Manufacture in Great Britain*, London, H. Fisher & Co., 1835

AYNES, Ken, and PUGH, Francis, *The Art of the Engineer*, Guildford, Lutterworth Press, 1981

EAMISH, Richard, *Memoir of the Life of Sir Marc Isambard Brunel*, London, Longman, Green, Longman and Roberts, 1862

ESSEMER, Sir Henry, *Sir Henry Bessemer, F.R.S.: an Autobiography with a concluding chapter by his son, Henry Bessemer*, London, Engineering, 1905

OURNE, John, *A Treatise on the Screw Propeller*, London, Longman & Co., 1852

RAUN, Wernher von, and ORDWAY, Frederick I., *History of Rocketry and Space Travel*, London, Nelson, 1967

RUNEL, Isambard, *The Life of Isambard Kingdom Brunel, Civil Engineer*, London, Longman Green, 1870

URSTALL, Aubrey Frederic, *A History of Mechanical Engineering*, London, Faber & Faber, 1963

ARDWELL, Donald Stephen Lowell, *Steam Power in the Eighteenth Century: A Case study in the application of Science*, London & New York, Sheed & Ward, 1963

AVALLO, Tiberias, *The History and Practice of Aerostation*, London, C. Dilly, 1785

HADWICK, George F., *The Works of Sir Joseph Paxton, 1803-1865*, London, Architectural Press, 1961

LARK, Edwin, *The Britannia and Conway Tubular Bridges*, 2 vols. London, Day & Son; John Weale, 1850

ONDIT, Carl Wilbur, *American Building*, Chicago & London, University of Chicago Press, 1968

ORLISS, Carlton Jonathan, *Main Line of Mid-America: The Story of the Illinois Central*, New York, Creative Age Press, 1950

RESY, Edward, *An Encyclopaedia of Civil Engineering, historical, theoretical and practical*, 2 vols. London, Longman, Brown, Green & Longmans, 1847

ICKINSON, Henry Winram, *Robert Fulton, Engineer and Artist: His Life and Works*, London, John Lane, 1913

ICKINSON, *A Short History of the Steam Engine*, Cambridge, The University Press, 1938

ICKINSON, and JENKINS, Rhys, *James Watt and the Steam Engine*, Oxford, Clarendon Press, 1927

ICKINSON, and TITLEY, Arthur, *Richard Trevithick: The Engineer and the Man*, Cambridge, The University Press, 1934

UNSHEATH, Percy, (ed.), *A Century of Technology, 1851-1951*, London, Hutchinson, 1951

UNSHEATH, *A History of Electrical Engineering*, London, Faber & Faber, 1962

MMERSON, George, *John Scott Russell: A Great Victorian engineer and naval architect*, London, John Murray, 1977

AIRBAIRN, William, *An Account of the Construction of the Britannia and Conway Tubular Bridges, with a complete history of their progress*, London, John Weale, and Longman, Brown, Green & Longmans, 1849

ORESTER, Tom, (ed.), *The Micro Electronics Revolution: The Complete Guide to the new Technology and its Impact on Society*, Oxford, Basil Blackwell, 1980

IBBS-SMITH, Charles H., *Sir George Cayley's Aeronautics, 1796-1855*, London, HMSO, 1962

ILLE, Bertrand, *The Renaissance Engineers*, London, Lund Humphries, 1966

LOAG, John, and BRIDGWATER, Derek, *A History of Cast Iron in Architecture*, London, George Allen & Unwin, 1948

HADFIELD, Charles, *The Canal Age*, Newton Abbot, David and Charles, 1968

HASTINGS, Paul, *Railroads: An International History*, London, Ernest Benn, 1972

HUGHES, James Quentin, *Military Architecture*, London, Hugh Evelyn, 1974

KAPLAN, Marshall H., *Space Shuttle*, USA, Aero Publishers, 1978

KIRBY, Richard Shelton, and LAURSON, Philip Gustave, *The Early Years of Modern Civil Engineering*, New Haven, Yale University Press, 1932

LESSEPS, Ferdinand Marie de, *Recollections of Forty Years*, Translated by C.B. Pitman, 2 vols., London, Chapman and Hall, 1887

MANTOUX, Paul, *The Industrial Revolution in the 18th Century*, London, Jonathan Cape, 1928

MCNEIL, Ian, *Joseph Bramah: A Century of Invention, 1749-1851*, Newton Abbot, David & Charles, 1968

NASA, *Aeronautics and Astronautics*, 1915-1960, Washington, DC, 1961, *et seq.*

NASA, *Apollo Expeditions to the Moon*, SP-350, Washington, DC, 1975

PANNELL, John Percival Masterman, *An Illustrated History of Civil Engineering*, London, Thames & Hudson, 1964

PARSONS, Robert Hodson, *The Development of the Parsons Steam Turbine*, London, Constable & Co., 1936

PAYNE, Robert, *The Canal Builders: The Story of Canal Engineers Through the Ages*, New York, The Macmillan Company, 1959

PETRIE, Sir William Flinders, *The Arts & Crafts of Ancient Egypt*, Edinburgh and London, T.N. Foulis, 1910

PLOMMER, W. Hugh, *Vitruvius and later Roman Building Manuals*, Cambridge, University Press, 1973

PLYMLEY, Joseph, *General View of the Agriculture of Shropshire*, London, Board of Agriculture, 1803

PRITCHARD, John Laurence, *Sir George Cayley: The Inventor of the Aeroplane*, London, Max Parrish, 1961

REES, Abraham, *The Cyclopaedia; or Universal Dictionary of Arts, Sciences, and Literature*, 39 vols., London, Longman, Hurst, Rees, Orme and Brown, 1819-20

RICKMAN, John (ed.), *Life of Thomas Telford, Civil Engineer*, London, Printed by James and Luke G. Hansard & Sons, 1938

ROBISON, John, *A System of Mechanical Philosophy*, Vol. II. 4 vols., Edinburgh, Printed for John Murray, 1822

ROLT, Lionel Thomas Caswall, *Isambard Kingdom Brunel*, London, Longman, Green & Co., 1957

ROYAL SOCIETY, THE, 'Obituary of William Henry Perkin, 1838-1907', signed R.M., *Proceedings of the Royal Society of London*, Series A. Vol. 80, (1907-08), pp. xxxviii-lix

RUSSELL, John Scott, *The Modern System of Naval Architecture*, 3 vols., London, Day & Son, 1864; 1865

SABIN, Edwin Legrand, *Building the Pacific Railway*, Philadelphia and London, J.R. Lippincott Company, 1919

SANDSTRÖM, Gösta E., *The History of Tunnelling*, London, Barrie & Rockliff, 1963

SANTOS-DUMONT, Alberto, *My Airships: the Story of My Life*, London, Grant Richards, 1904

SCOTT, J.D., *Siemens Brothers, 1858-1958: An Essay in the History of Industry*, London, Weidenfeld & Nicolson, 1958

SMILES, Samuel (ed.), *James Nasmyth: An Autobiography*, 3 vols., London, John Murray, 1883

SMILES, *Lives of the Engineers*, 5 vols., London, John Murray, 1874

STRAUB, Hans, *A History of Civil Engineering: An Outline from Ancient to Modern Times*, Translated by E. Rockwell, London, Leonard Hill, 1952

SWINTON, William, *The Twelve Decisive Battles of the War: a History of the Eastern & Western Campaigns, in relation to the actions that decided their issue*, New York, Dick & Fitzgerald, 1867

TOY, Sidney, *A History of Fortification from 3000 B.C. to A.D. 1700*, London, Heinemann, 1955

TREVITHICK, Francis, *Life of Richard Trevithick with an account of his inventions*, London, E. & F.N. Spon, 1872

VITRUVIUS POLLIO, Marcus, *Vitruvius: The Ten Books on Architecture*, Translated by Morris Hicky Morgan, New York, Dover Publications, 1960

INDEX

Page numbers in italic refer to illustrations

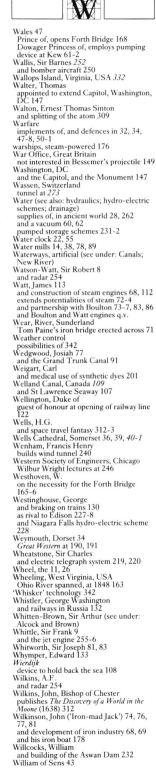

ACKNOWLEDGEMENTS

Project Editor Ian Jackson
Design Editor Nick Eddison
Design Assistant Amanda Barlow
Picture Researcher Faith Perkins
Proof Reader Jocelyn Selson
Production Bob Towell

The illustrations in this book have been selected by Eddison Sadd Editions who wish to acknowledge, with thanks, the following photographic sources:

Aerofilms 33; Aspect Picture Library 16, 101, 136, 225, 297, 301, 331; BBC Hulton Picture Library 187, 242 top; Birmingham Public Libraries 75 (Reproduced by permission of the Reference Library, Archives Department); Bridgeman Art Library 92 (The Guildhall, City of London) 117, 126-7 (Royal Holloway College), 145, 156-7; British Aerospace 257; Colorific! 164, 209 bottom, 325; James Davis Photographs 99; The Elton Collection, Coalbrookdale 268; Robert Estall 20, 27; Mary Evans Picture Library 237; Susan Griggs Agency 169, 259; Sonia Halliday 44; Michael Holford 24-5, 31, 37, 40-1, 43, 82, 121, 166, 172, 177, 185, 195, 216 (Royal Institution), 217, 224; Image Bank, London 212, 230, 233; Images Photos J. Winkley 135, 139, 40; Imperial War Museum 207, 315; Simon Lavington, Manchester University 286; Manchester Ship Canal Company 106; Mansell Collection 49, 66, 79, 188-9, 265; Marion and Tony Morrison 35; Peter Newark, Western Americana 192; Photo Source 141, 281; Royal Library Windsor Castle 53 (Reproduced by gracious permission of Her Majesty the Queen); Quadrant Picture Library 242-3; Heini Schneebeli 252; Science Museum, London 208, 242 bottom, 289, 292, 311, 314, 332; Ronald Sheridan 12, 29; Tony Stone 249, 279, 330; Daily Telegraph Colour Library 235, 329, 333, 335; John Topham 6-7; United Kingdom Atomic Energy Authority 304-5; University of Liverpool Archives 193; Vision International 46, 70, 109, 168, 209 top, 273, 277, 340; Walker Art Gallery, Liverpool 64-65; Derek Widdicombe 85.